PRODUCT SKETCHBOOK

国际产品设计师手绘集

创意—深化—表达

IDEA - PROCESS - REFINEMENT

度本图书（Dopress Books） 编著

赵侠 译

前言

产品设计手绘是设计师交流的语言，也是必备的职业技能之一。设计师通过快速手绘表现，捕捉设计灵感，推敲设计方案，与他人交流想法，并深化完成最终方案。对于学习产品设计的学生而言，不但要掌握精湛的手绘表现技巧，更重要的是，要将设计手绘作为自身拓展思维、与他人交流思想的工具。

本书收录了来自全球各地的产品设计师的七十余个案例的设计手稿。其中，既有个人手绘习作，也有在校课程设计、毕业设计以及竞赛作品，更有诸多真实商业项目。在案例分类上，包括了数码家电、生活家居、箱包鞋履以及交通工具四部分，基本涵盖了产品设计专业的学生通常接触的全部产品类型。

在每一个案例中，除了简明扼要的项目背景介绍或表现技法的文字提示以外，全书坚持“以图说话”的原则，通过大量绘制精妙、表意清晰的产品草图与效果图，让读者直观地体会设计师的设计思路。广大学生可以通过临摹、参考本书的众多案例，学习手绘表现技法以及设计的程序与方法。

目录

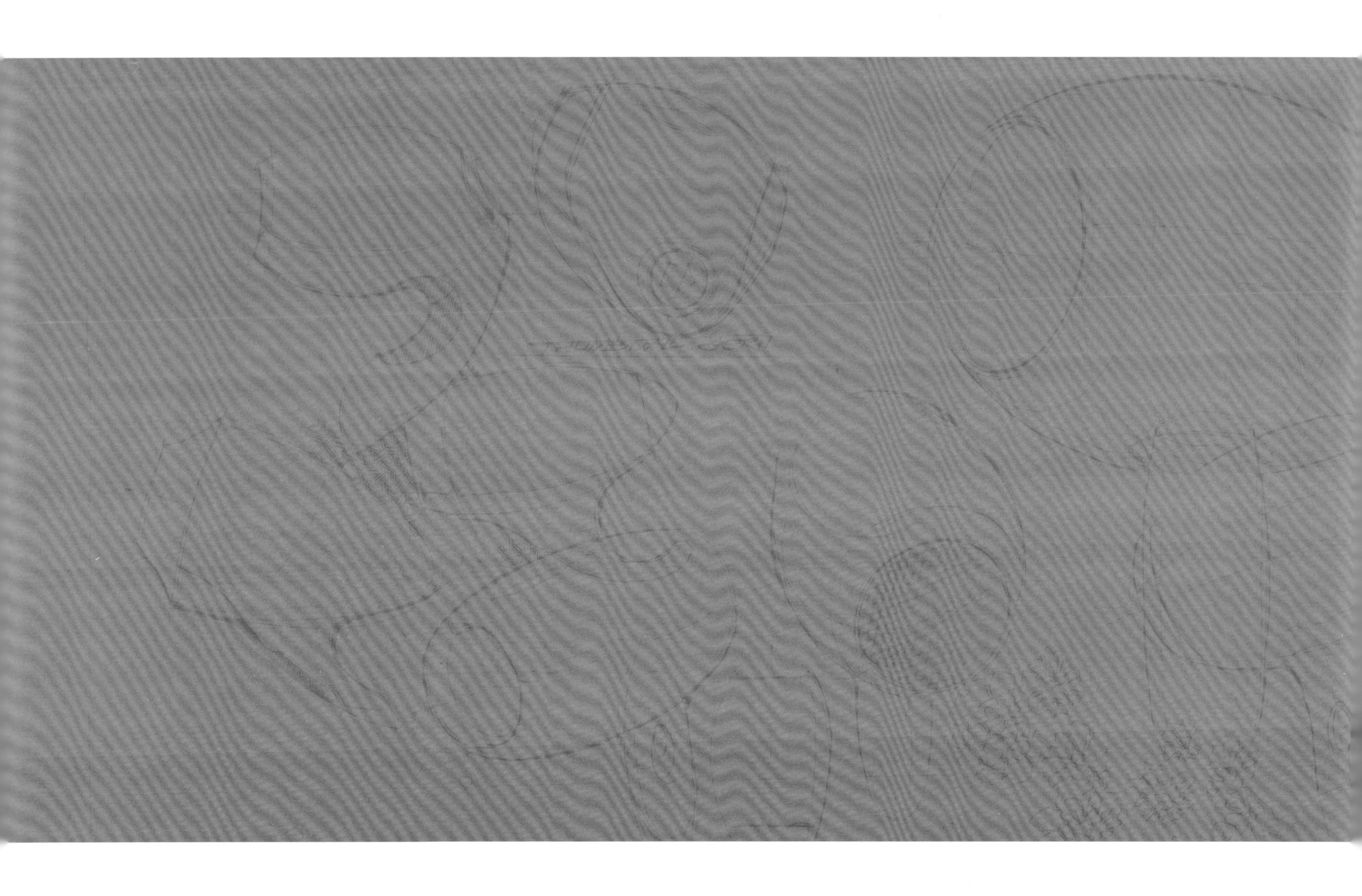

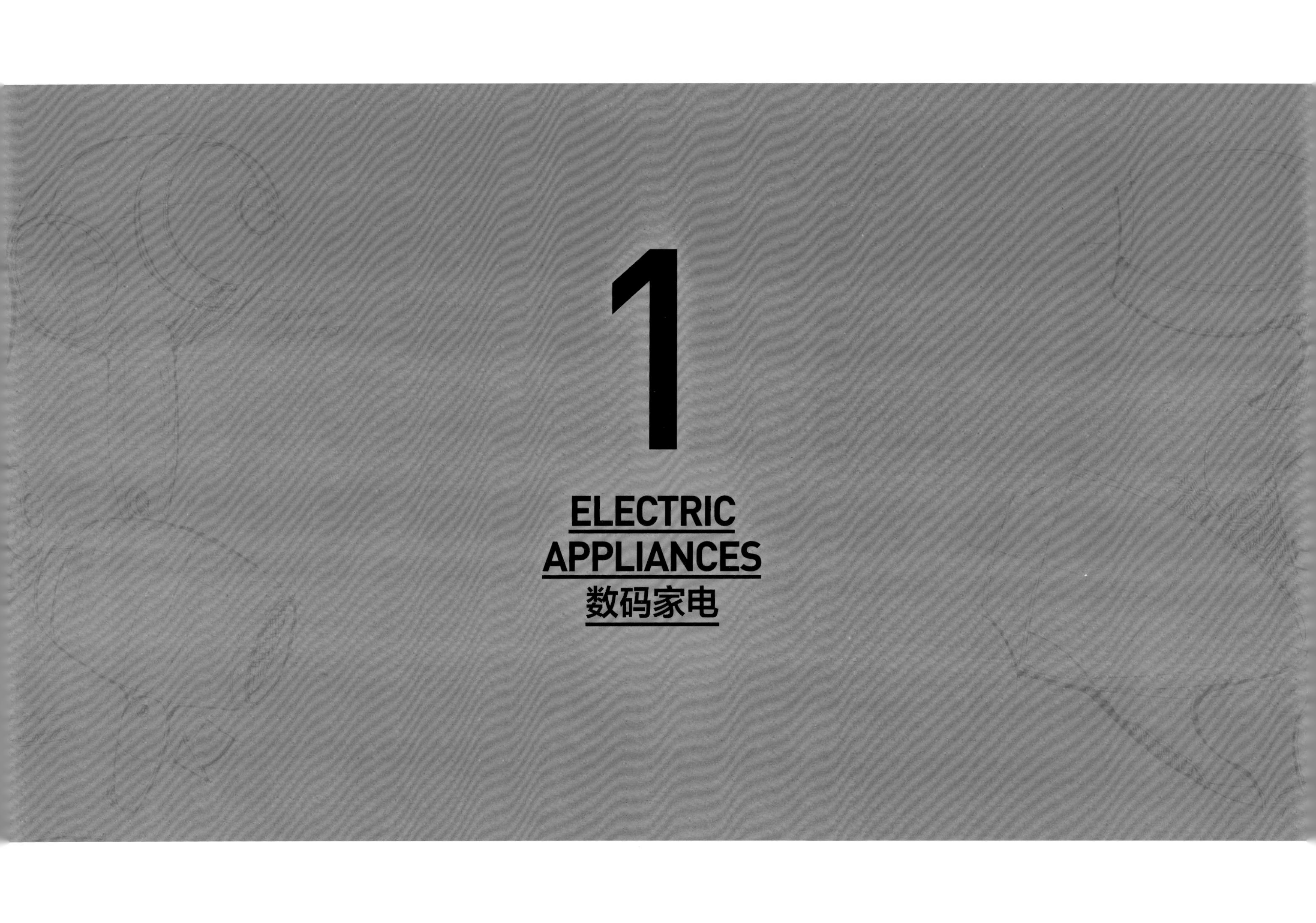

1

ELECTRIC APPLIANCES
数码家电

Polaroid Pocket

Polaroid Pocket相机

设计师：阿格尼·维斯尼奥斯凯特
（Agne Vysniauskaite）
客户：个人作品

Polaroid Pocket是一款便携易用的迷你相机，能够用一张小幅的即时成像照片捕捉生活中的美妙瞬间。设计从分析Polaroid品牌已有的老款相机及其部件开始，随后开始寻找新的样式和材料。在完成最终方案的草图之后，设计师还根据草图对该概念设计进行了计算机建模及渲染。

KASETĖ

FROG TONGUE

ĮORIS

15

85

58

DYDIS

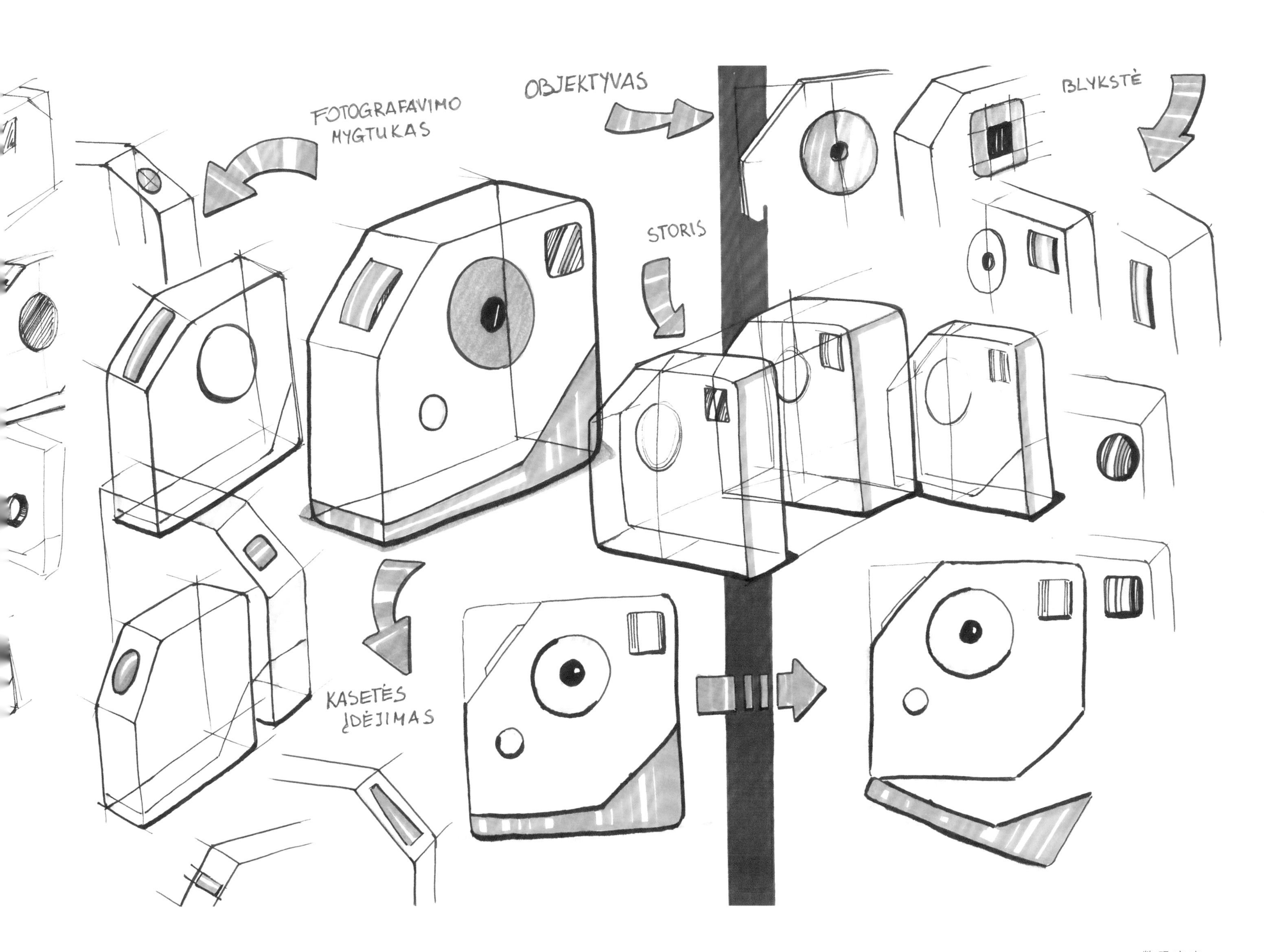
FOTOGRAFAVIMO
MYGTUKAS
OBJEKTYVAS
BLYKSTĖ
STORIS
KASETĖS
ĮDĖJIMAS

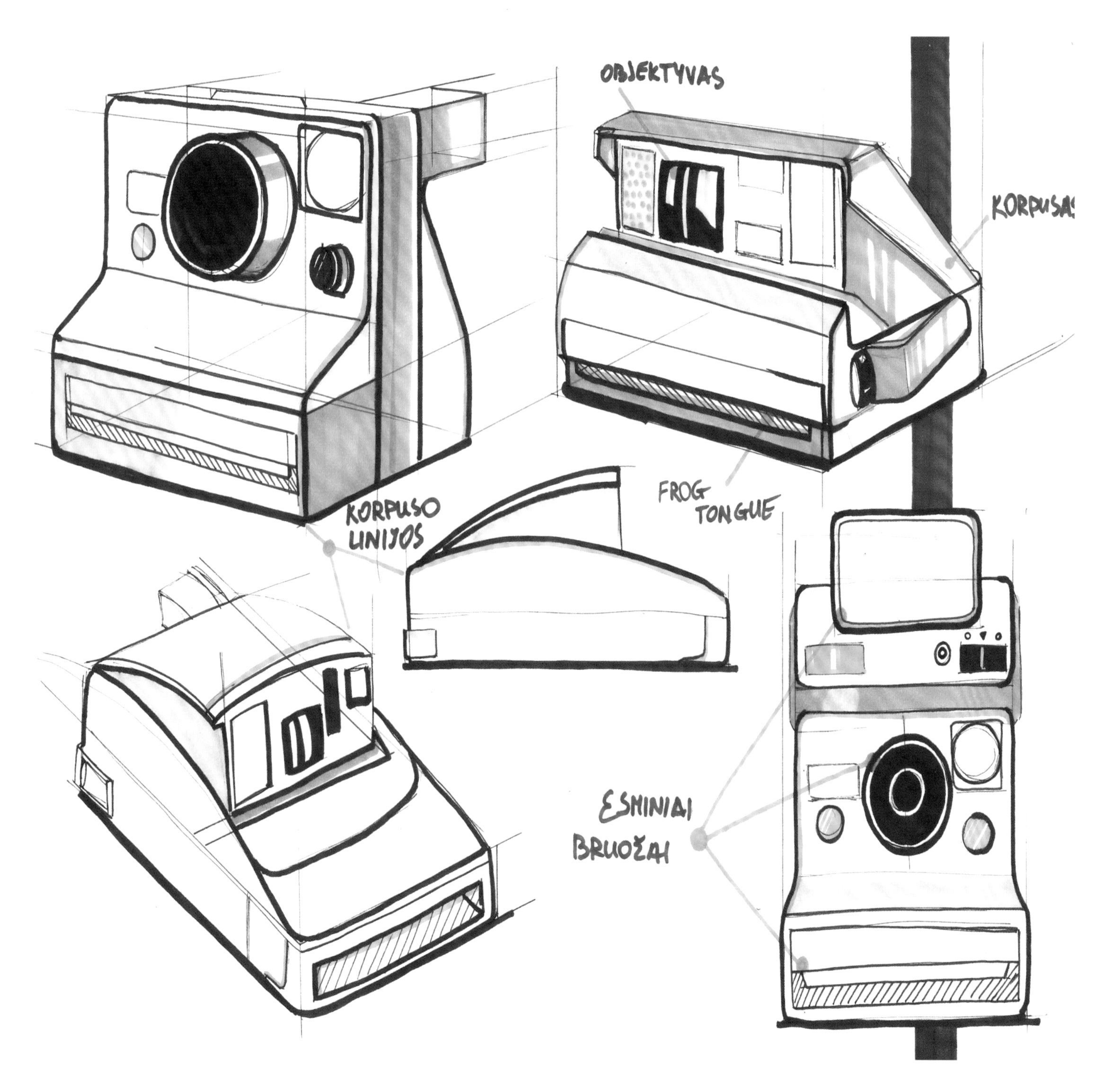
OBJEKTYVAS
KORPUSAS
KORPUSO
LINIJOS
FROG
TONGUE
ESMINIAI
BRUOŽAI

Polaroid
Polaroid

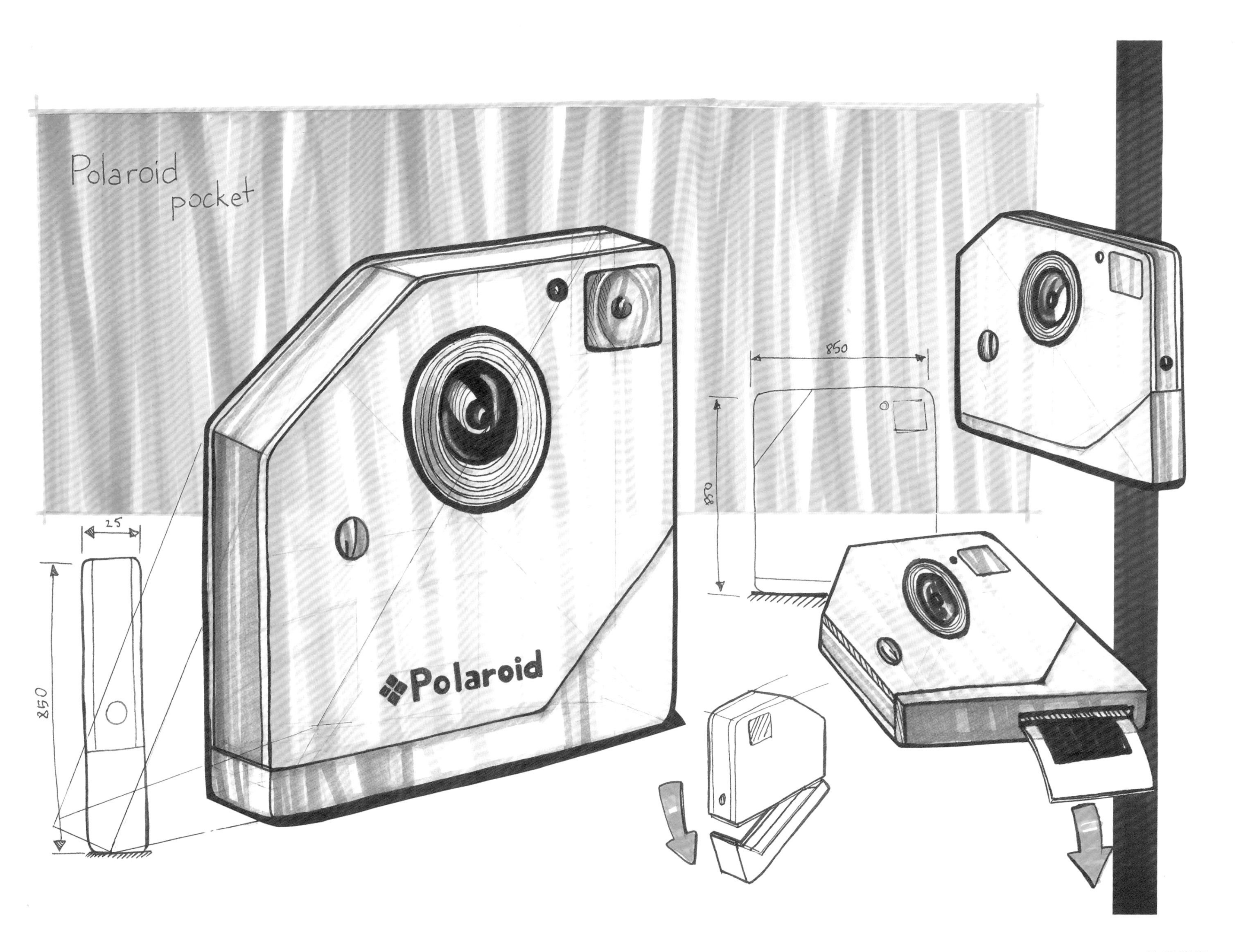
Polaroid
pocket
Polaroid
850
850
25
850

Potable Game Console
便携式游戏机

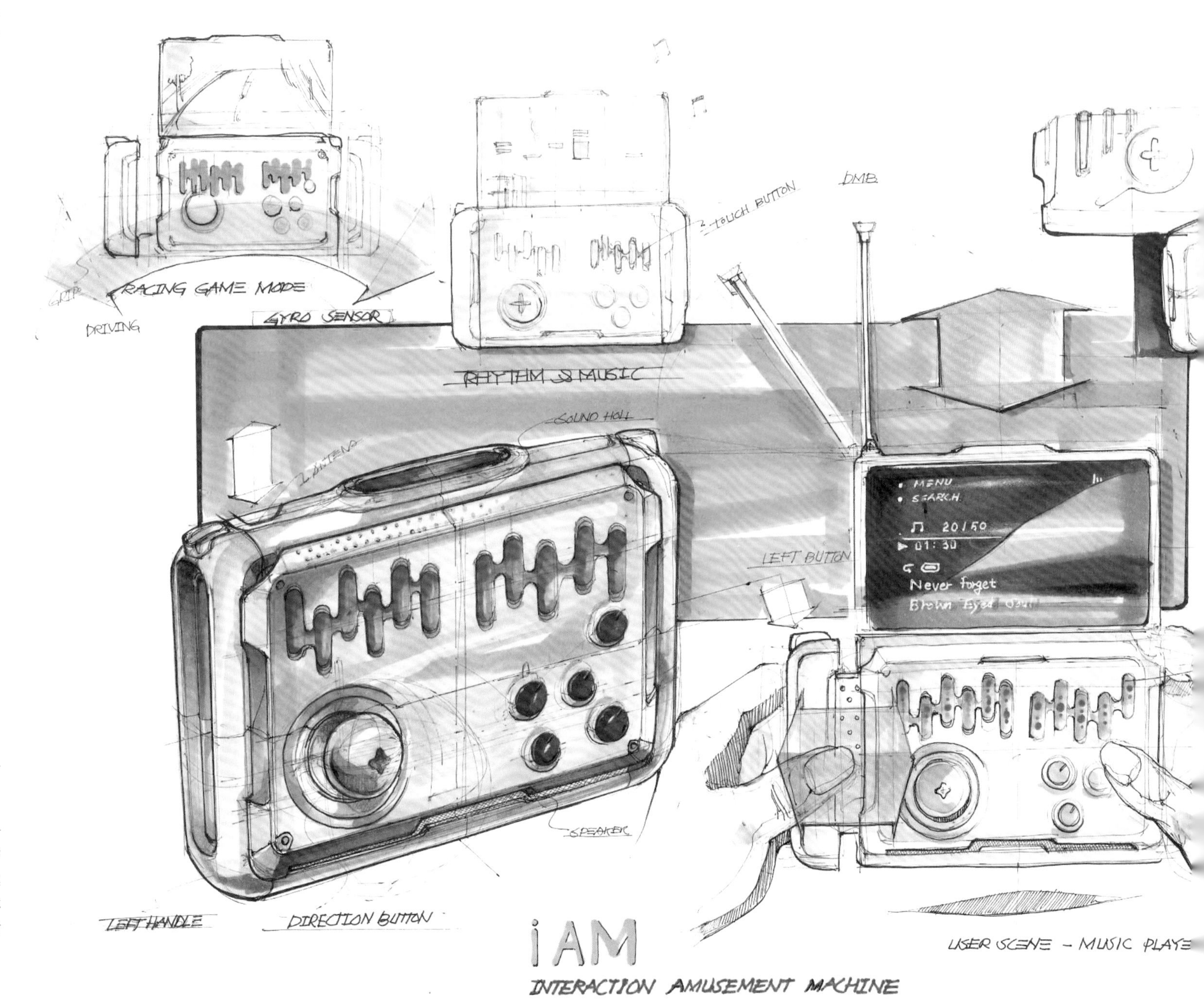

设计师：柳时形（Ryu Si Hyeong）
客户：个人作品

设计师主要使用两种设计方法设计了该产品，即了解（Understanding）和聚焦（Focusing）——了解用户，聚焦用户体验。其中大部分的手绘设计方案均来自对用户体验的研究。

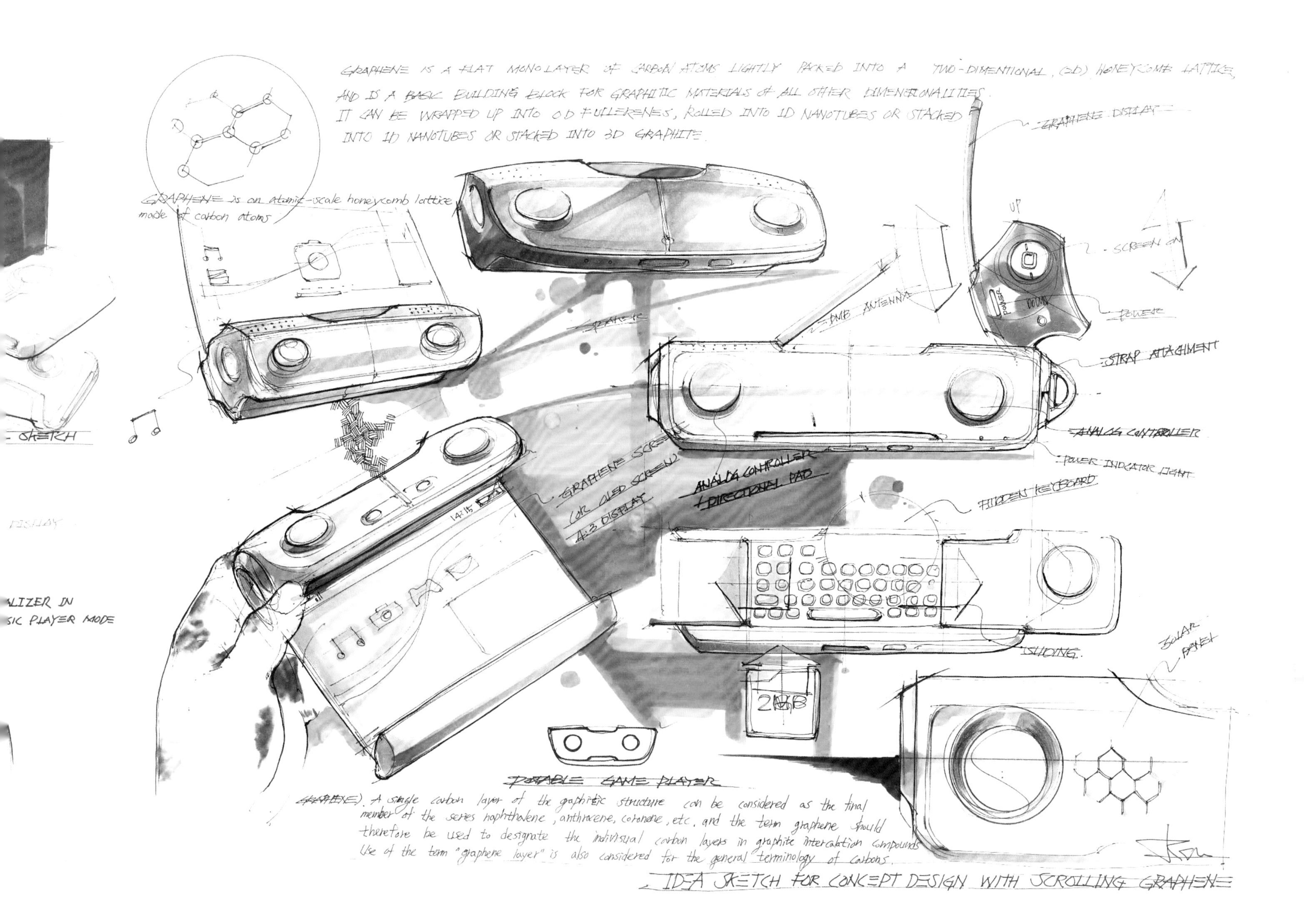
GRAPHENE IS A FLAT MONOLAYER OF CARBON ATOMS LIGHTLY PACKED INTO A TWO-DIMENTIONAL (2D) HONEYCOMB LATTICE, AND IS A BASIC BUILDING BLOCK FOR GRAPHITIC MATERIALS OF ALL OTHER DIMENSIONALITIES. IT CAN BE WRAPPED UP INTO 0D FULLERENES, ROLLED INTO 1D NANOTUBES OR STACKED INTO 1D NANOTUBES OR STACKED INTO 3D GRAPHITE.
GRAPHENE is an atomic-scale honeycomb lattice made of carbon atoms
GRAPHENE DISPLAY
UP
SCREEN ON
POWER
DOWN
DMB ANTENNA
STRAP ATTACHMENT
ANALOG CONTROLLER
POWER INDICATOR LIGHT
GRAPHENE SCREEN (OR OLED SCREEN) 4:3 DISPLAY
ANALOG CONTROLLER / DIRECTIONAL PAD
HIDDEN KEYBOARD
SLIDING
SOLAR PANEL
SKETCH
LIZER IN
SIC PLAYER MODE
PORTABLE GAME PLAYER
GRAPHENE) A single carbon layer of the graphitic structure can be considered as the final member of the series naphthalene, anthracene, coronene, etc. and the term graphene should therefore be used to designate the individual carbon layers in graphite intercalation compounds. Use of the term "graphene layer" is also considered for the general terminology of carbons.
IDEA SKETCH FOR CONCEPT DESIGN WITH SCROLLING GRAPHENE

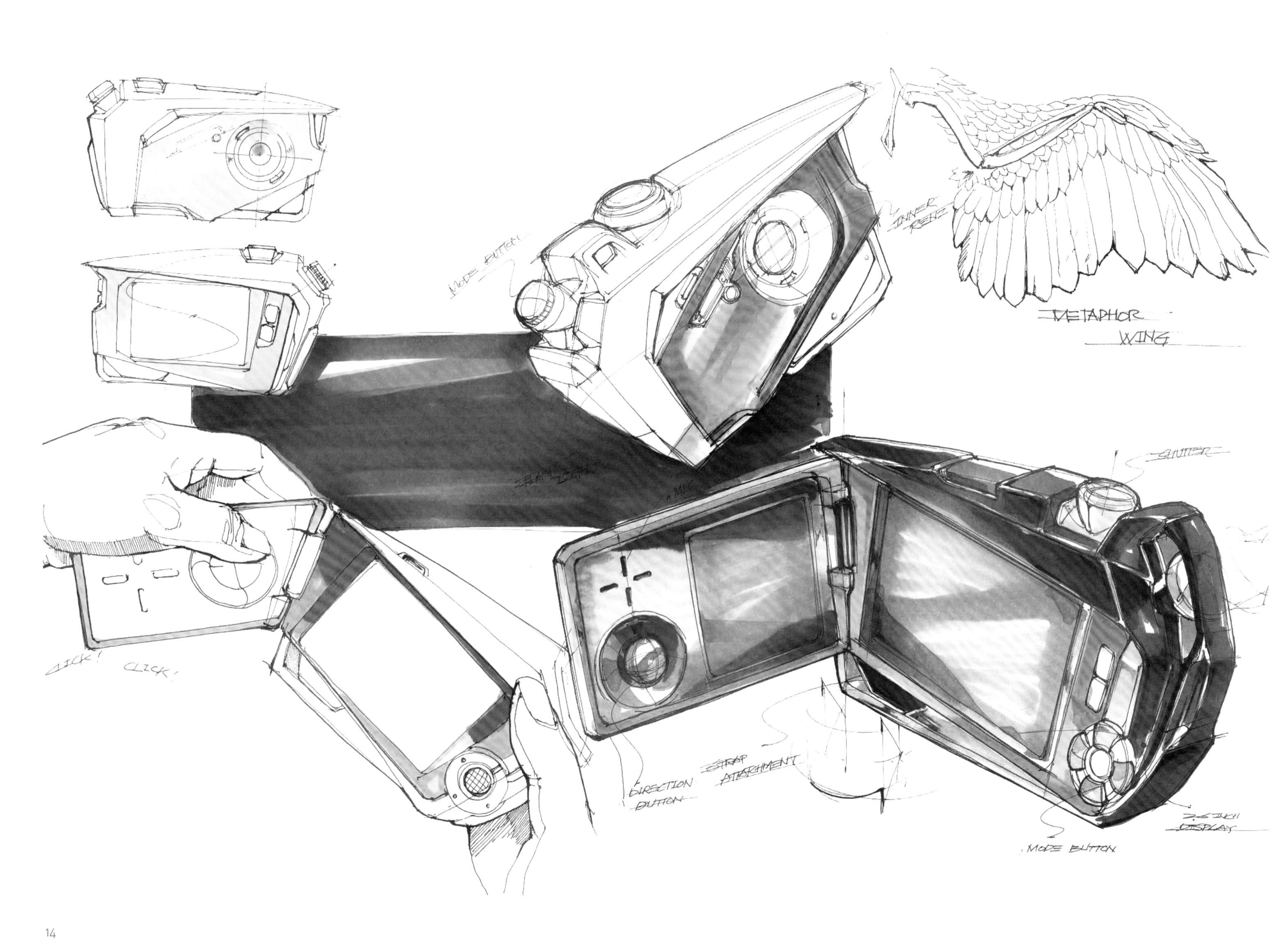
MODE BUTTON
METAPHOR
WING
SHUTTER
CLICK! CLICK!
DIRECTION BUTTON
STRAP ATTACHMENT
MODE BUTTON

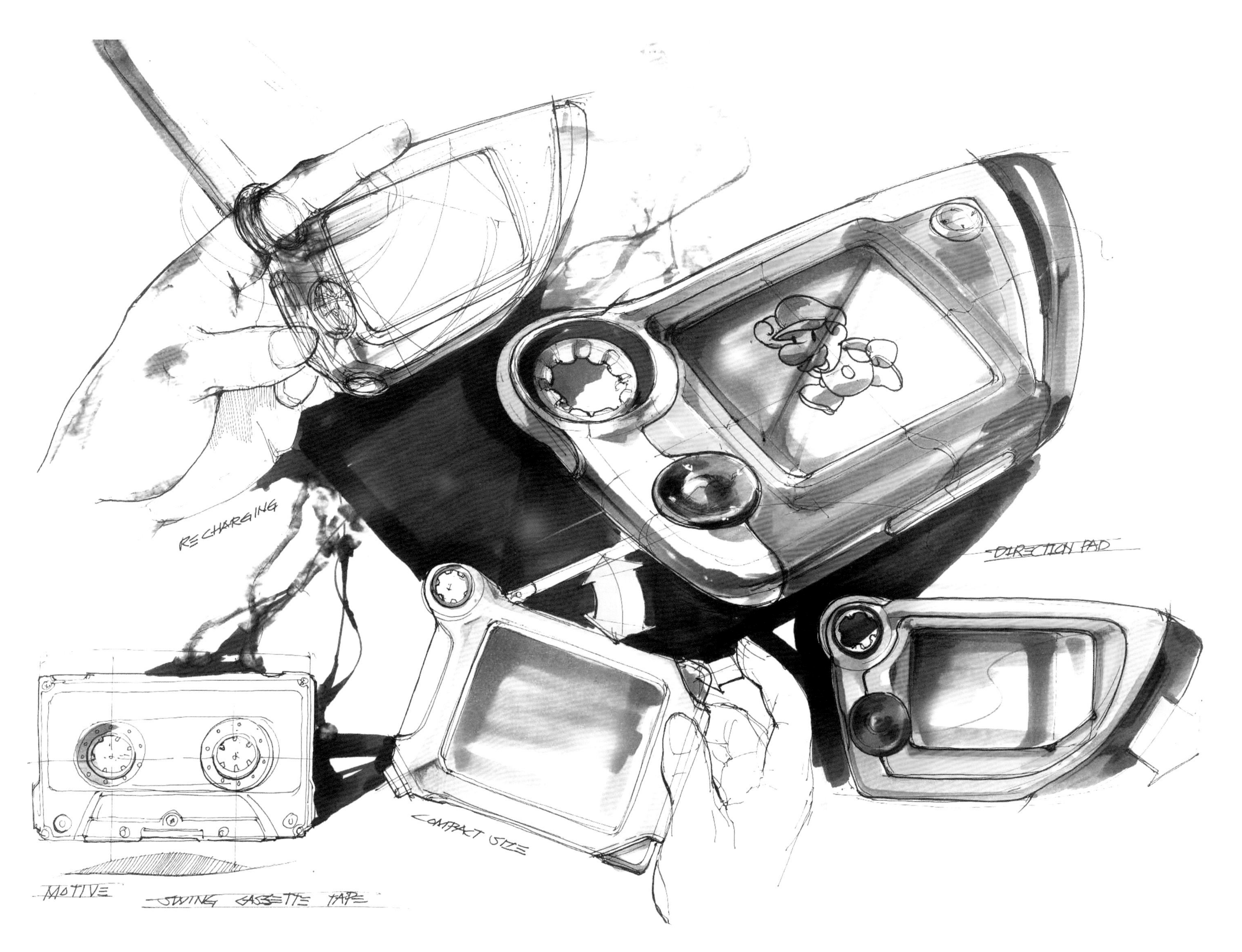
RECHARGING
DIRECTION PAD
COMPACT SIZE
MOTIVE
SWING CASSETTE TAPE

Kettle & Toaster

水壶和烤面包机

设计师：丹尼尔·法默（Daniel Farmer）
客户：个人作品

设计从探讨英国人品尝下午茶的习俗展开。在松散的线条上增加硬朗清晰的线条，可以增添产品手绘的空间立体感。

plastic

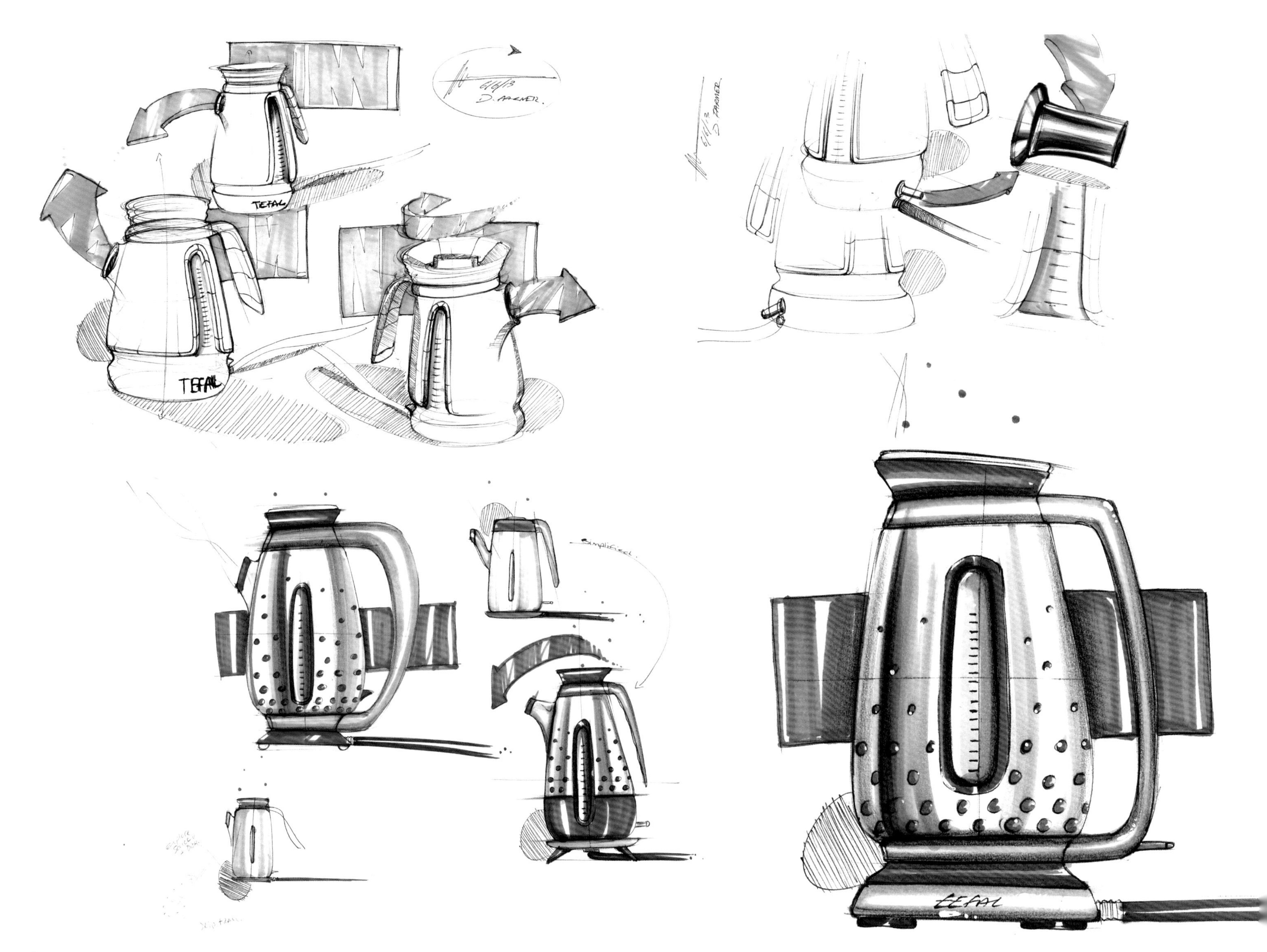
D. FARMER.
TEFAL
TEFAL
D. FARMER
Simplified.
TEFAL

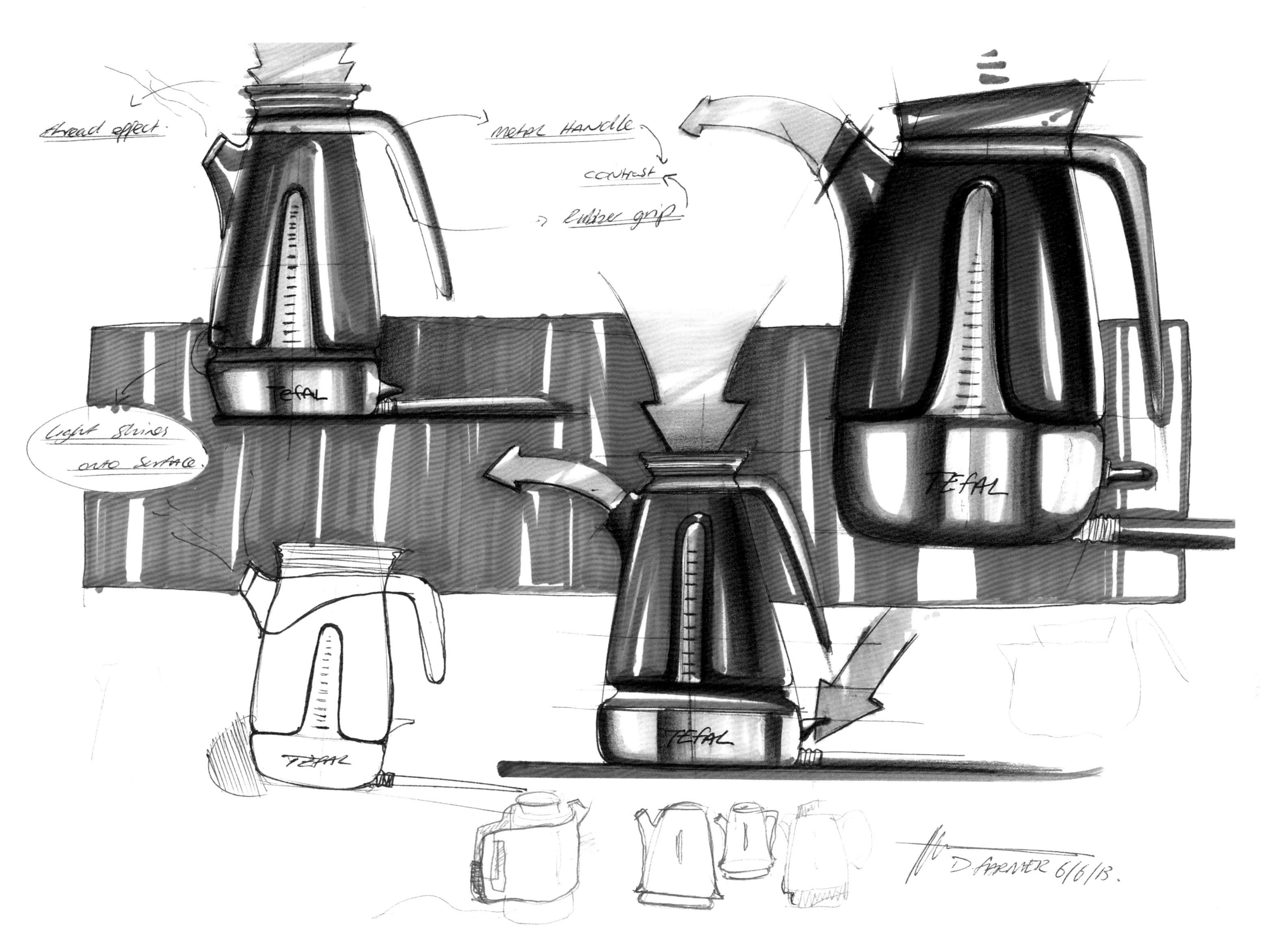
thread effect
metal handle
contrast
rubber grip
light shines onto surface
TEFAL
TEFAL
TEFAL
TEFAL
D.FARMER 6/6/13.

culinary
Tefal

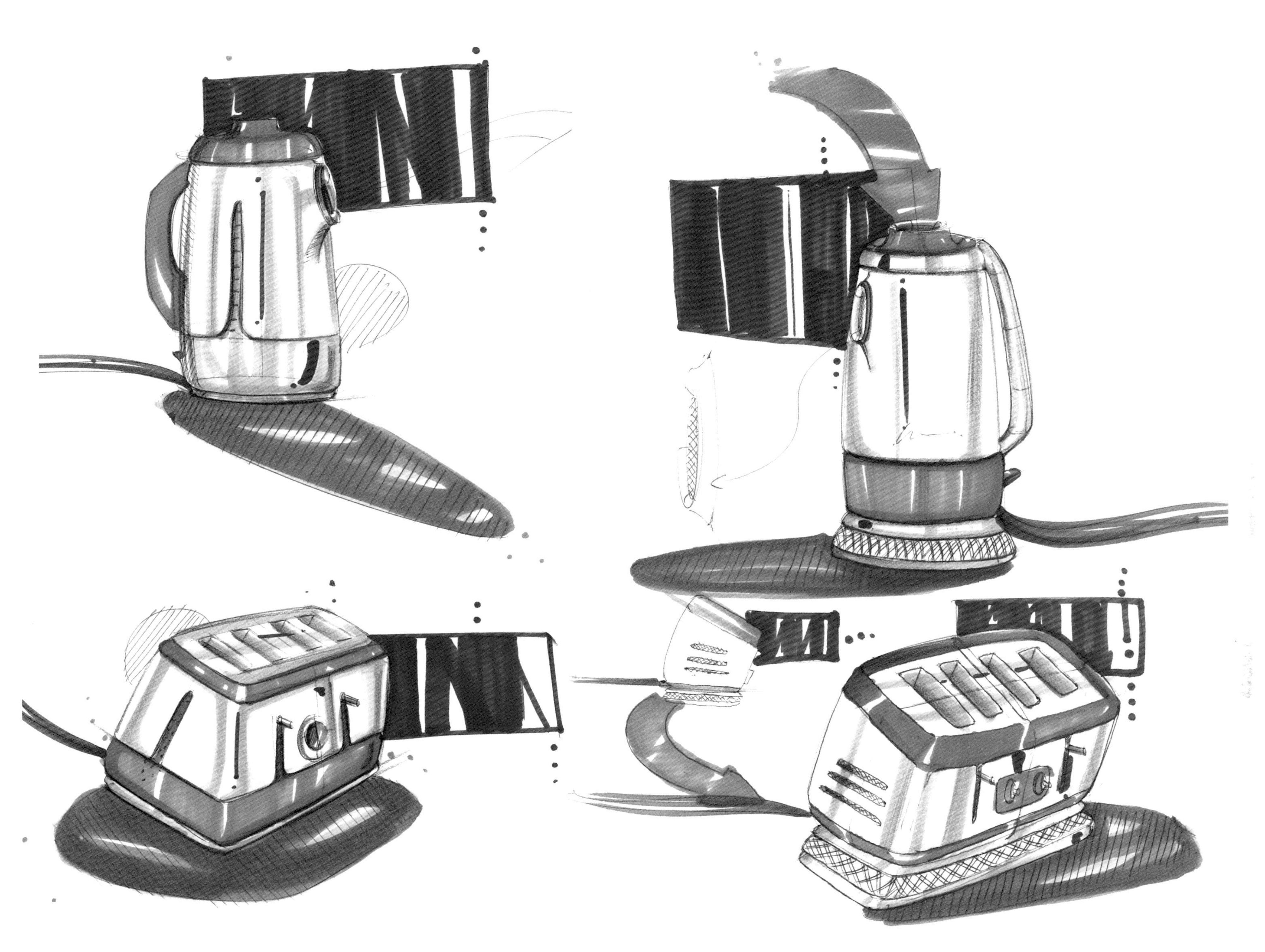

Cobb Cooker

Cobb炊具

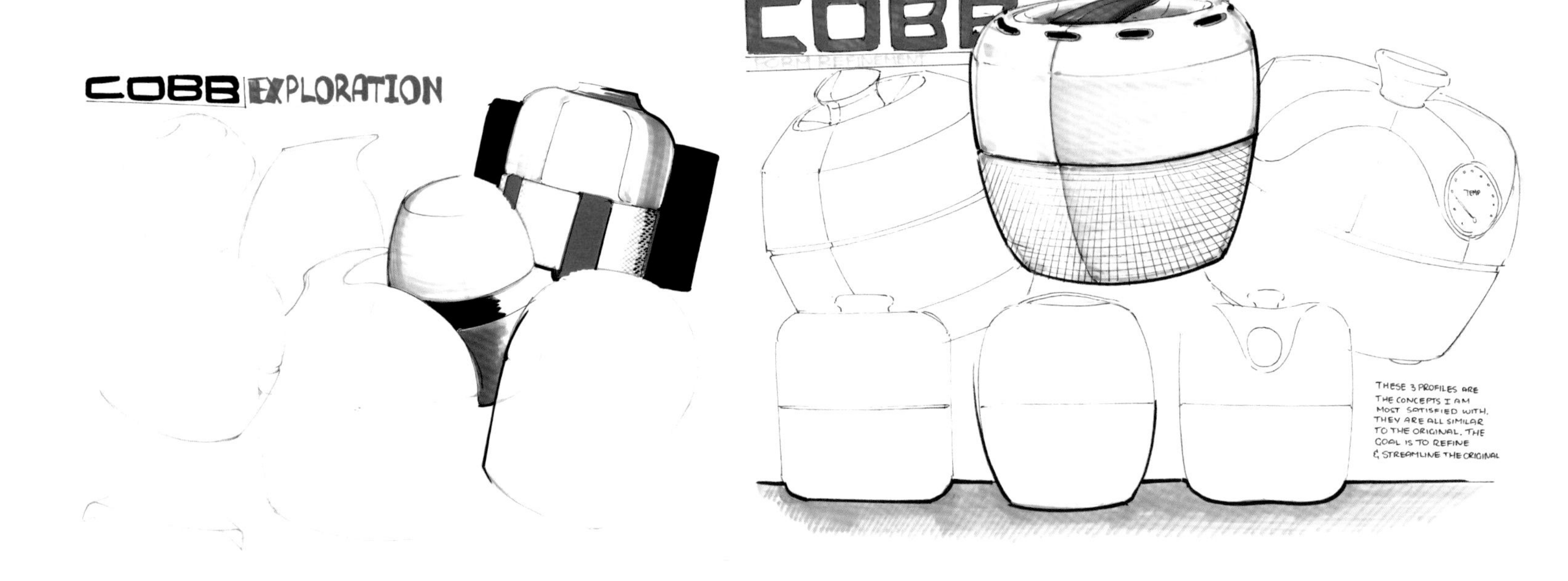

设计师：默里·夏普（Murray Sharp）

客户：Cobb公司

Cobb公司委托设计师为其打造新一代系列产品。该系列产品旨在为用户带来专属的“Cobb体验”，通过产品的使用方式体现一种全新的生活方式。设计简报要求在原有产品的视觉形象上寻求变化，同时使其更加实用。

该系列产品的设计围绕已有的内部组件展开，注重实用性。从美学角度来说，其造型更加精简、典雅，融合了斯堪的纳维亚风格。除简约的样式，这套炊具还设计有锁盖系统和通用的装配结构。同时，其最大限度减少了多余物料的使用，从而解决了包装运输等相关问题。

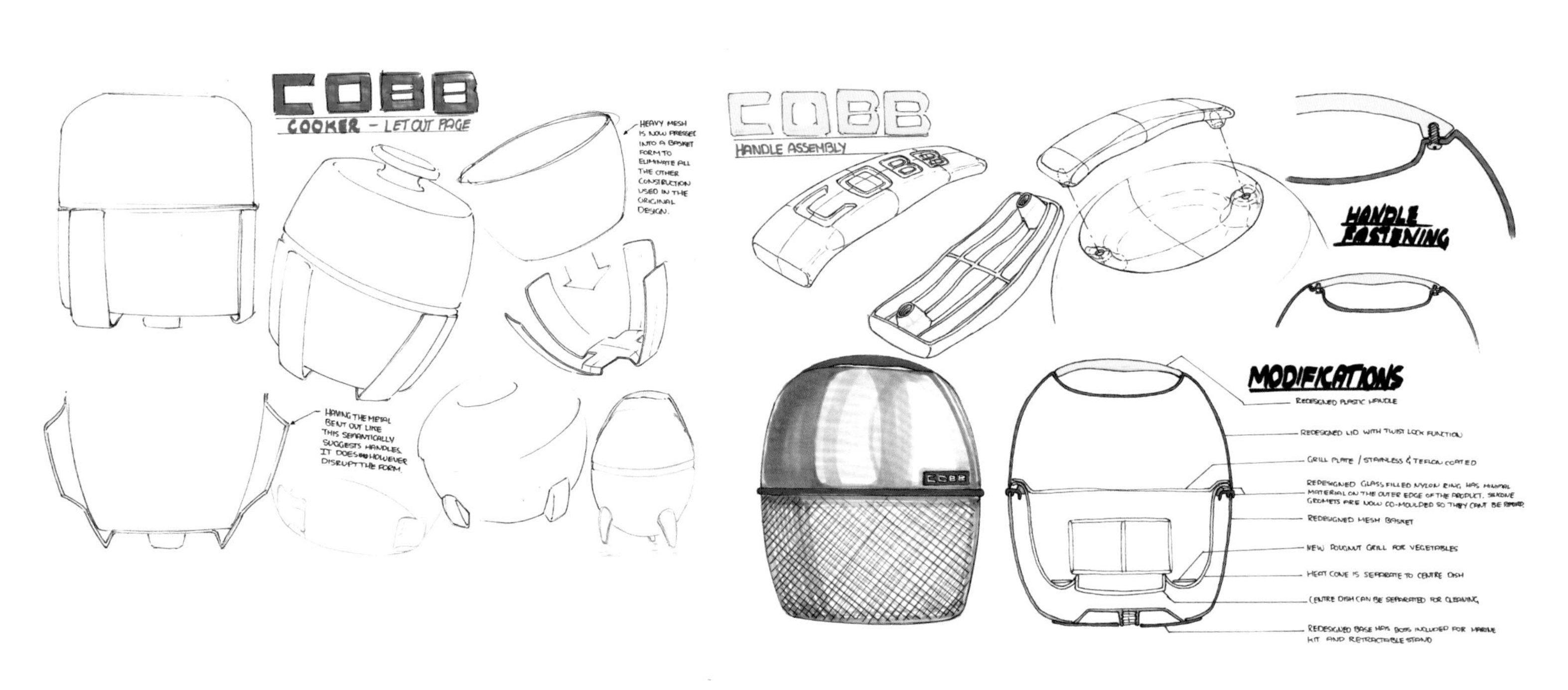

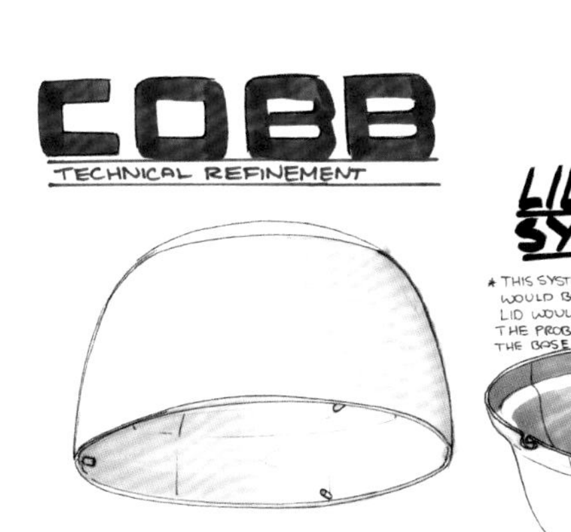

COBB
TECHNICAL REFINEMENT

LID CLIP SYSTEM

* THIS SYSTEM WOULD WORK WELL, CUT OUTS WOULD BE LAZER-CUT BEFORE PRESSING. LID WOULD SIMPLY TWIST LOCK INTO PLACE. THE PROBLEM IS THIS PART IS LOOSE FROM THE BASE.

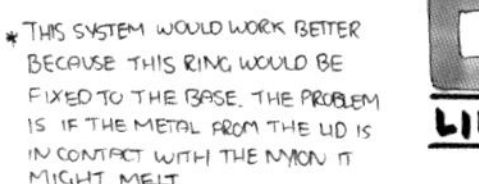

* THIS SYSTEM WOULD WORK BETTER BECAUSE THIS RING WOULD BE FIXED TO THE BASE. THE PROBLEM IS IF THE METAL FROM THE LID IS IN CONTACT WITH THE NYLON IT MIGHT MELT.

VEGI GRILL

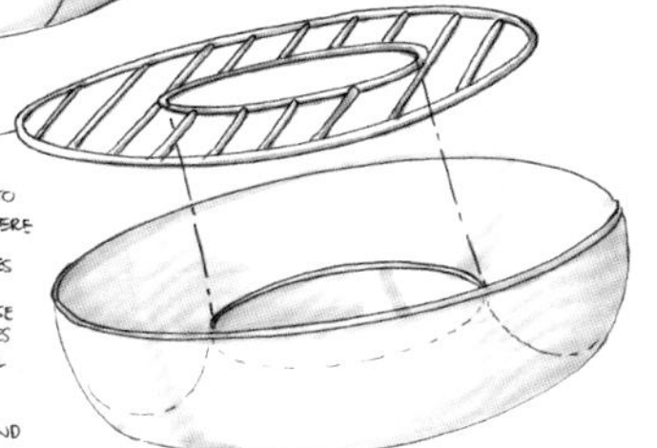

FROM TALKING TO A FEW USERS THERE ARE COMPLAINTS ABOUT VEGETABLES BURNING ON THE BASE DISH BECAUSE DRIPPING FAT MAKES IT TOO HOT. A SMALL STAINLESS GRILL WOULD LIFT THE VEG SLIGHTLY AND RESOLVE THIS.

UNIQUE MESH PATTERN

* UNIQUE MESH GRILL WOULD BE DIE CUT FROM ALUMINIUM PLATE AND THEN PRESSED INTO SHAPE.

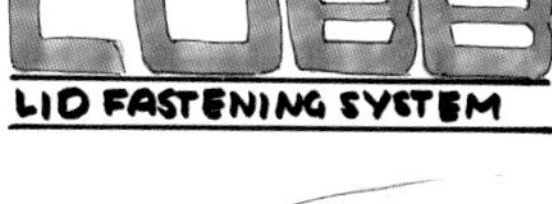

COBB
LID FASTENING SYSTEM

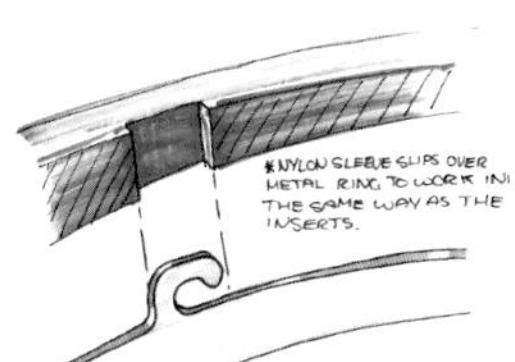

* NYLON SLEEVE SLIPS OVER METAL RING TO WORK IN THE SAME WAY AS THE INSERTS.

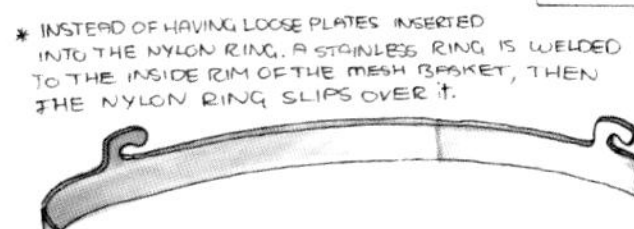

* INSTEAD OF HAVING LOOSE PLATES INSERTED INTO THE NYLON RING, A STAINLESS RING IS WELDED TO THE INSIDE RIM OF THE MESH BASKET, THEN THE NYLON RING SLIPS OVER IT.

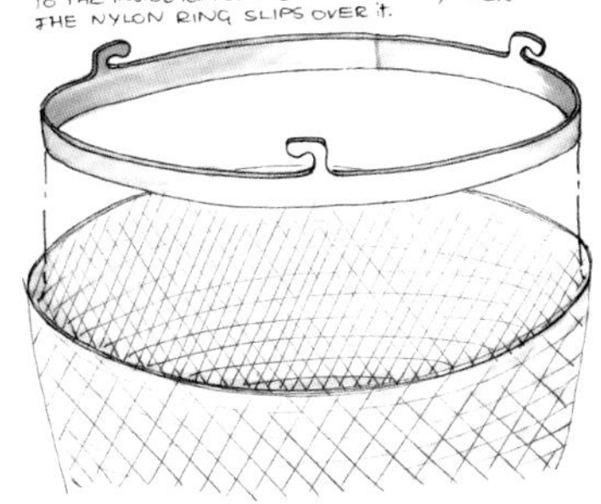

LID STUDS

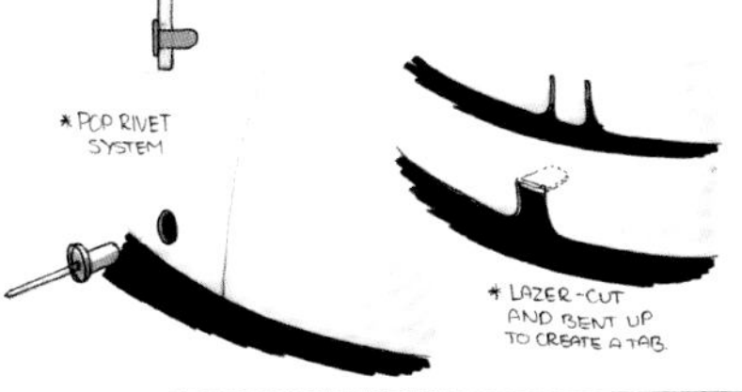

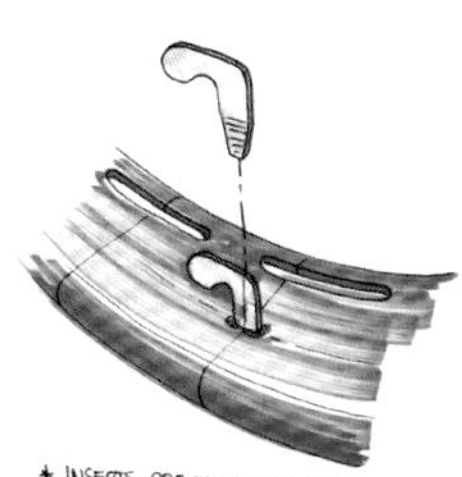

* INSERTS ARE EMBEDDED INTO THE NYLON RING IN THE SAME WAY METAL BOSSES ARE INSERTED INTO PLASTIC PARTS.

METAL INSERTS

* THIS SYSTEM MAY BE ABLE TO CARRY MORE WEIGHT THAN THE ABOVE INSERT. 'S' SHAPED PLATE IS FED THROUGH A HOLE IN THE NYLON RIM. THE BOTTOM HALF IS TAPERED AND KNURLED TO ENSURE A TIGHT FIT. A HAMMER IS THEN USED TO KNOCK THE PLATE UP INTO PLACE, SECURING IT.

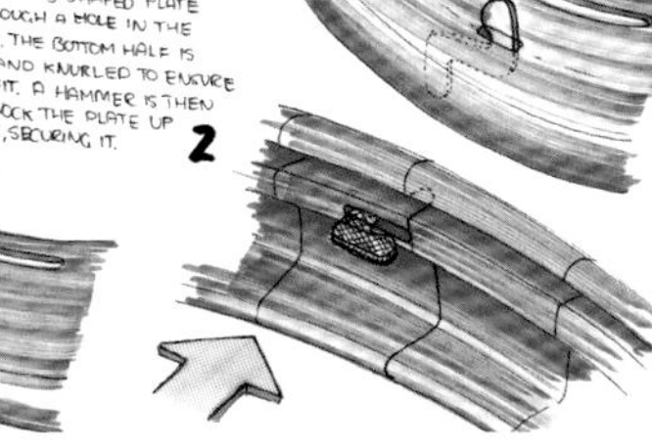

COBB
TECHNICAL REFINEMENT

* THE VEGETABLE DISH & COAL HOLDER MUST BE ABLE TO BE SEPARATED FOR CLEANING. GRIME & DIRT BUILDS UP IN THE PROBLEM AREA.

NYLON RING

RADIUSED EDGE BECOMES AN AESTHETIC ELEMENT IN THE OVER-ALL PRODUCT

SILICONE GROMMET WILL BE CO-MOULDED WITH THE NYLON RING SO THAT IT CANNOT BE REMOVED

THIS EDGE HAS BEEN DRASTICALLY REDUCED AS IT IS NO LONGER NEEDED FOR STRUCTURAL REASONS. I HAVE NOT ELIMINATED IT COMPLETLEY AS IT CAN BE USED AS A VISUAL ELEMENT & IT PREVENTS FINGERS FROM SLIPPING ONTO THE HOT LID WHILST CARRYING. IT ALSO SEMANTICALLY AIDS IN THIS SENCE.

I HAVE CHOSEN TO KEEP THE NYLON RING SIMILAR TO THE ORIGINAL FOR AIR FLOW & CONVECTION REASONS.

BASE ASSEMBLY

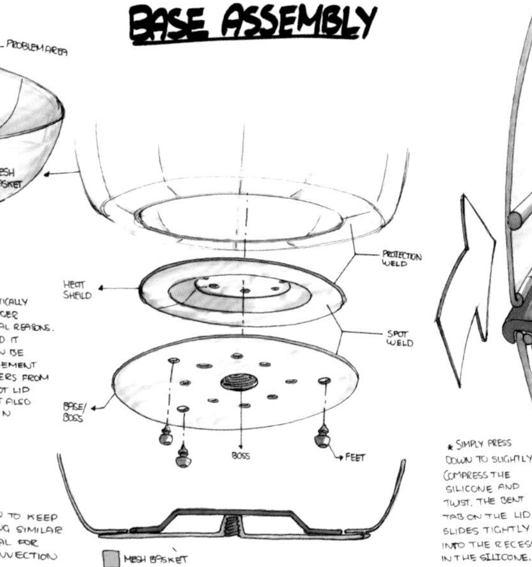

* RING OF WIRE IS WELDING TO THE LID AROUND 10mm OFF FROM THE BASE. THIS BECOMES THE NEW RESTING AREA, AND IT RESTS ON THE GRILL PLATE

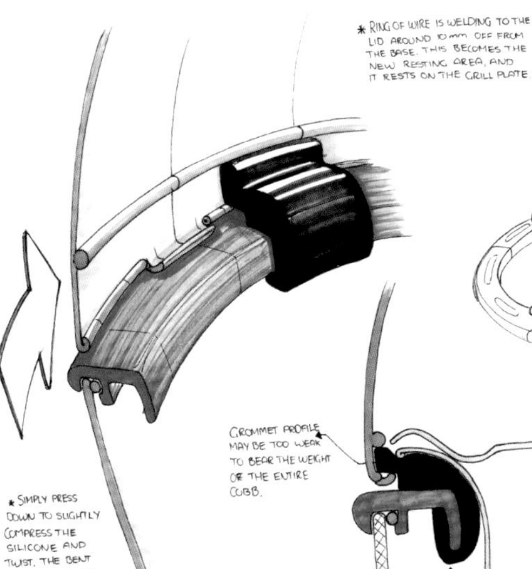

* SIMPLY PRESS DOWN TO SLIGHTLY COMPRESS THE SILICONE AND TWIST. THE BENT TAB ON THE LID SLIDES TIGHTLY INTO THE RECESS IN THE SILICONE.
GROMMET PROFILE MAY BE TOO WEAK TO BEAR THE WEIGHT OF THE ENTIRE COBB.
IF THE SILICONE IS CO-MOULDED ALL THIS EXCESS CAN BE REMOVED SAVING MATERIAL.

COBB
LOCK-ON LID SYSTEM

BECAUSE THE GROMMETS NEED TO BE ABLE TO LIFT THE ENTIRE COBB, AN EXTRA ONE WILL BE ATTACHED TO EVEN OUT THE LOAD. A DENSER SILICONE MAY ALSO BE REQUIRED.

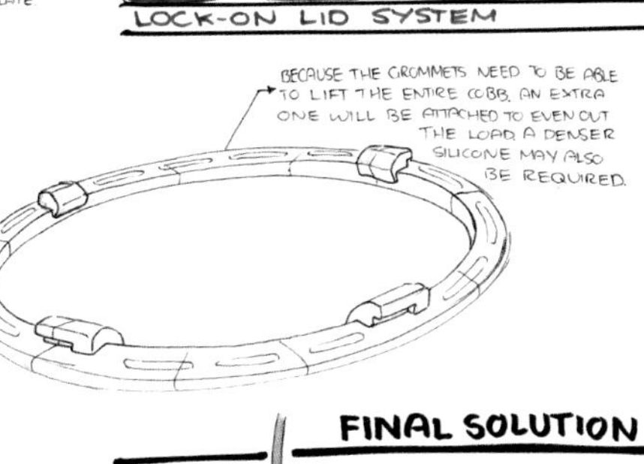

FINAL SOLUTION

RING OF WIRE RESTING ON GRILL PLATE. THIS ALLOWS EASIER TWIST THEN IF WIRE RESTED ON GROMMETS.

GROMMET DESIGN IS STRONGER TO BE ABLE TO CARRY MORE WEIGHT

GAP

ORIGINAL METHOD OF CONNECTING BASE TO NYLON RING.

CO-MOULDED SILICONE IS MOULDED WITH UNDER CUT, MAKING THE CONNECTION STRONGER

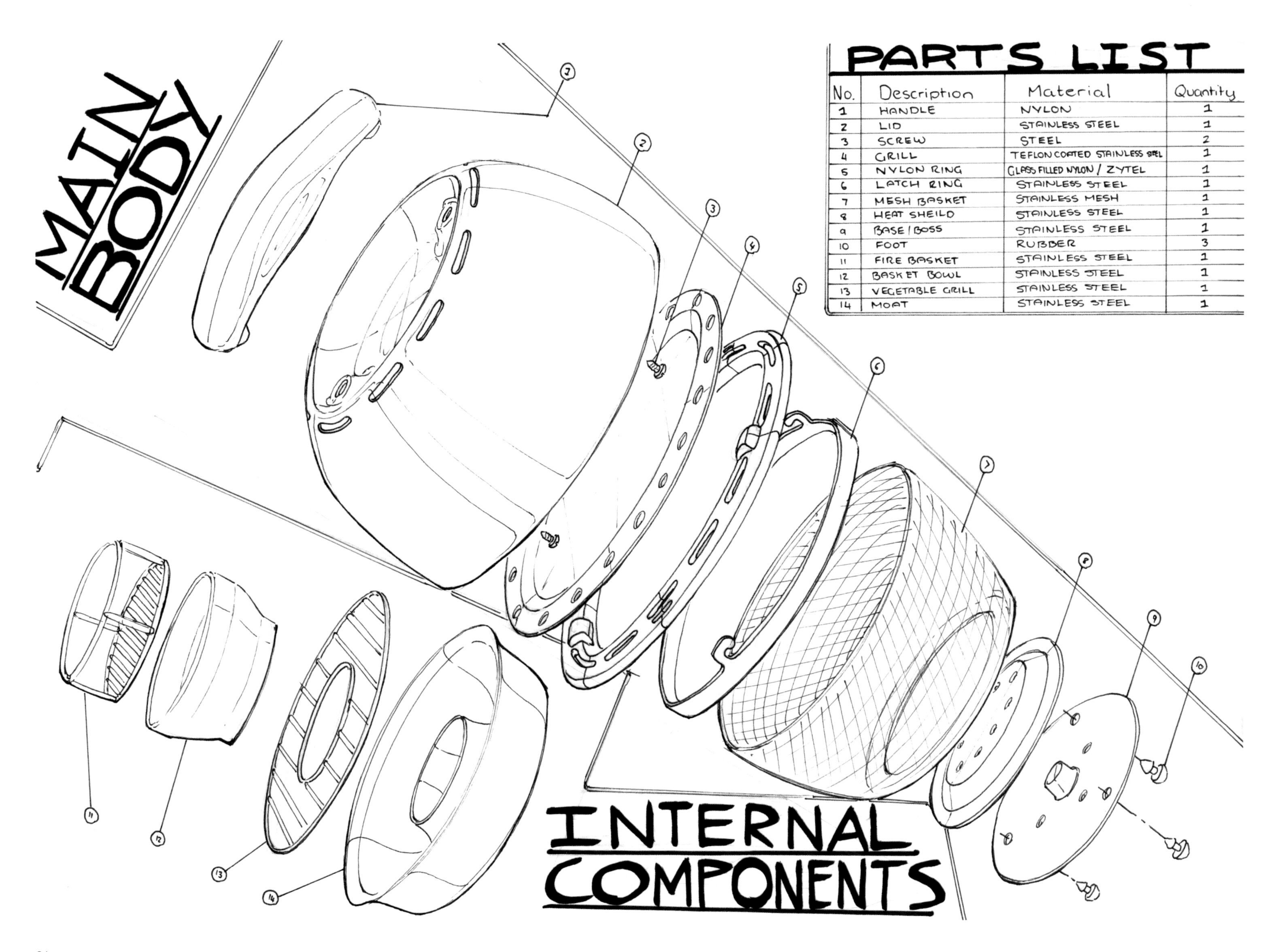

No.	Description	Material	Quantity
1	HANDLE	NYLON	1
2	LID	STAINLESS STEEL	1
3	SCREW	STEEL	2
4	GRILL	TEFLON COATED STAINLESS STEEL	1
5	NYLON RING	GLASS FILLED NYLON / ZYTEL	1
6	LATCH RING	STAINLESS STEEL	1
7	MESH BASKET	STAINLESS MESH	1
8	HEAT SHEILD	STAINLESS STEEL	1
9	BASE / BOSS	STAINLESS STEEL	1
10	FOOT	RUBBER	3
11	FIRE BASKET	STAINLESS STEEL	1
12	BASKET BOWL	STAINLESS STEEL	1
13	VEGETABLE GRILL	STAINLESS STEEL	1
14	MOAT	STAINLESS STEEL	1

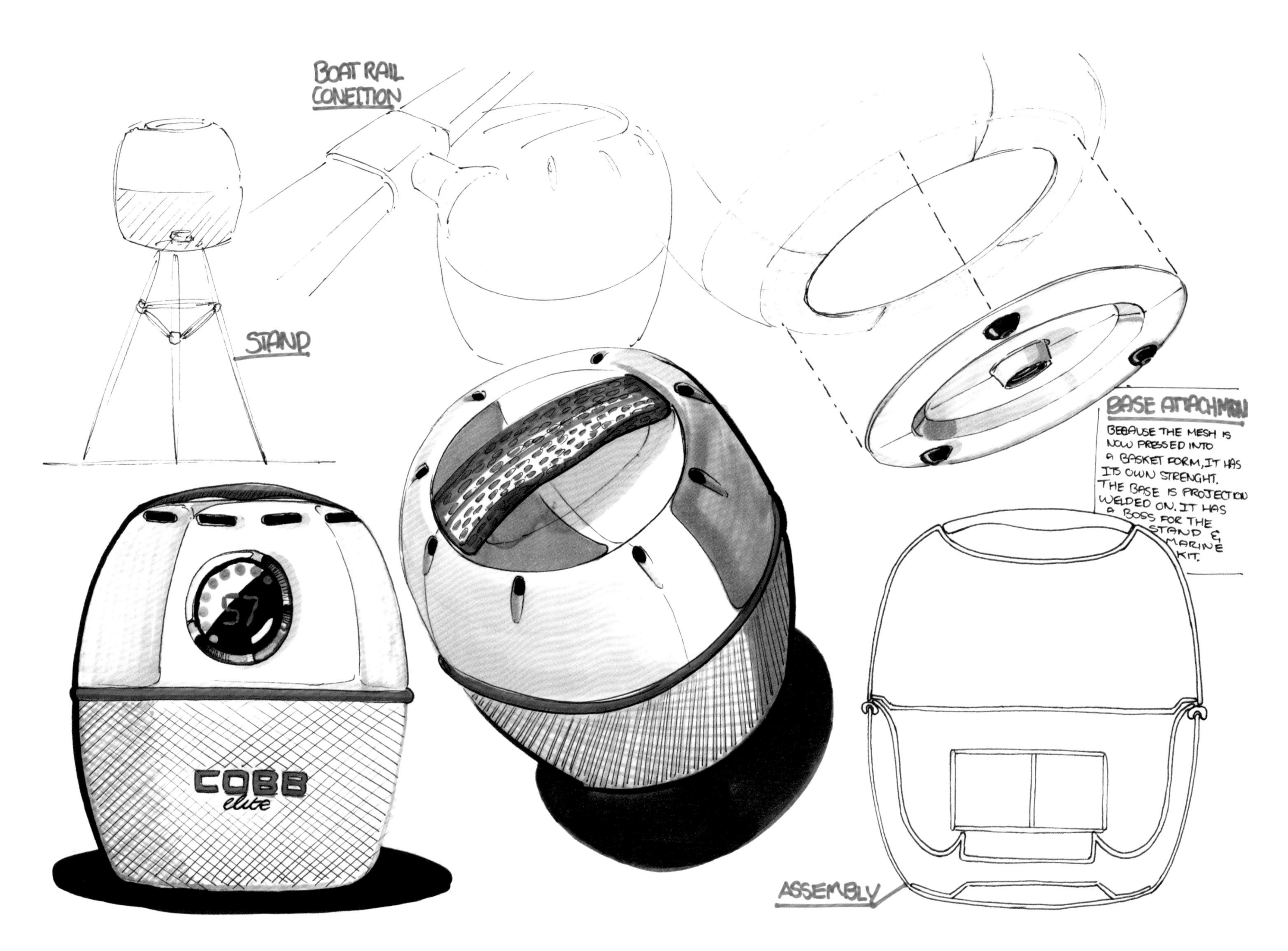

BOAT RAIL
CONEITION
STAND
BASE ATTACHMAN
BECAUSE THE MESH IS
NOW PRESSED INTO
A BASKET FORM, IT HAS
ITS OWN STRENGHT.
THE BASE IS PROJECTION
WELDED ON. IT HAS
A BOSS FOR THE
STAND &
MARINE
KIT.
57
COBB
elite
ASSEMBLY

Hawk Light

Hawk灯

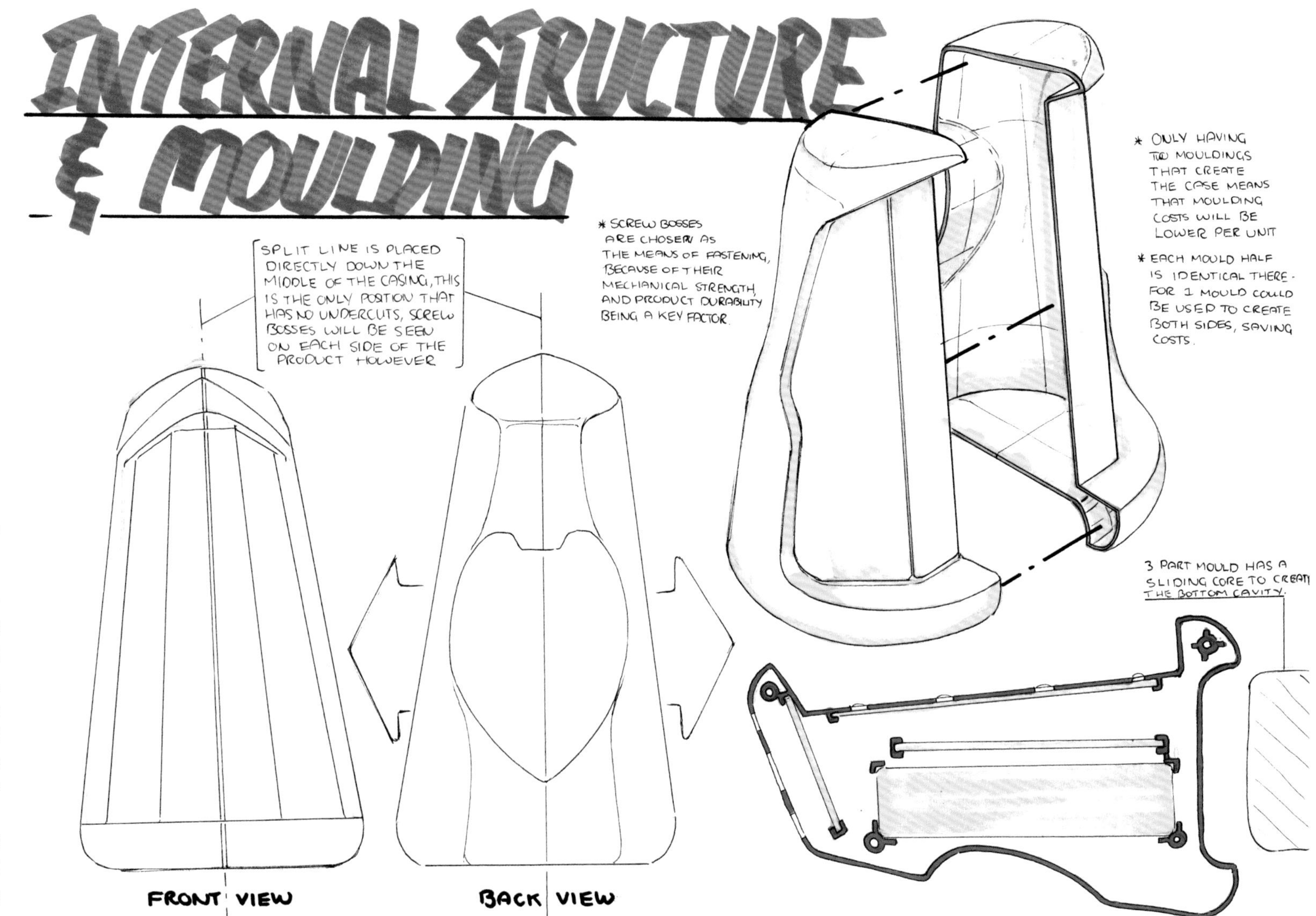

INTERNAL STRUCTURE
& MOULDING
SPLIT LINE IS PLACED DIRECTLY DOWN THE MIDDLE OF THE CASING, THIS IS THE ONLY POSITION THAT HAS NO UNDERCUTS, SCREW BOSSES WILL BE SEEN ON EACH SIDE OF THE PRODUCT HOWEVER
* SCREW BOSSES ARE CHOSEN AS THE MEANS OF FASTENING, BECAUSE OF THEIR MECHANICAL STRENGTH, AND PRODUCT DURABILITY BEING A KEY FACTOR.
* ONLY HAVING TWO MOULDINGS THAT CREATE THE CASE MEANS THAT MOULDING COSTS WILL BE LOWER PER UNIT
* EACH MOULD HALF IS IDENTICAL THEREFOR 1 MOULD COULD BE USED TO CREATE BOTH SIDES, SAVING COSTS.
3 PART MOULD HAS A SLIDING CORE TO CREATE THE BOTTOM CAVITY.
FRONT VIEW
BACK VIEW

设计师：默里·夏普

客户：约翰内斯堡大学，学校项目

南非约翰内斯堡东部地区的亚历山大小镇内的大部分区域没有通电，因此很多人使用危险且有害的石蜡灯，并且出现当地人非法偷电的现象。该设计图以图文结合的形式，从产品内部结构组件展开，并对太阳能板和插头进行了重新设计，使其与灯具主体的造型相呼应。Hawk灯由三个LED灯带组成，各自之间成60度角排列，从而扩大了光线的照射范围。为了节约成本，整个灯的设计只使用了三个模具。灯可以竖直放在桌子上，还可以被安装在墙上或者悬吊在顶棚上。

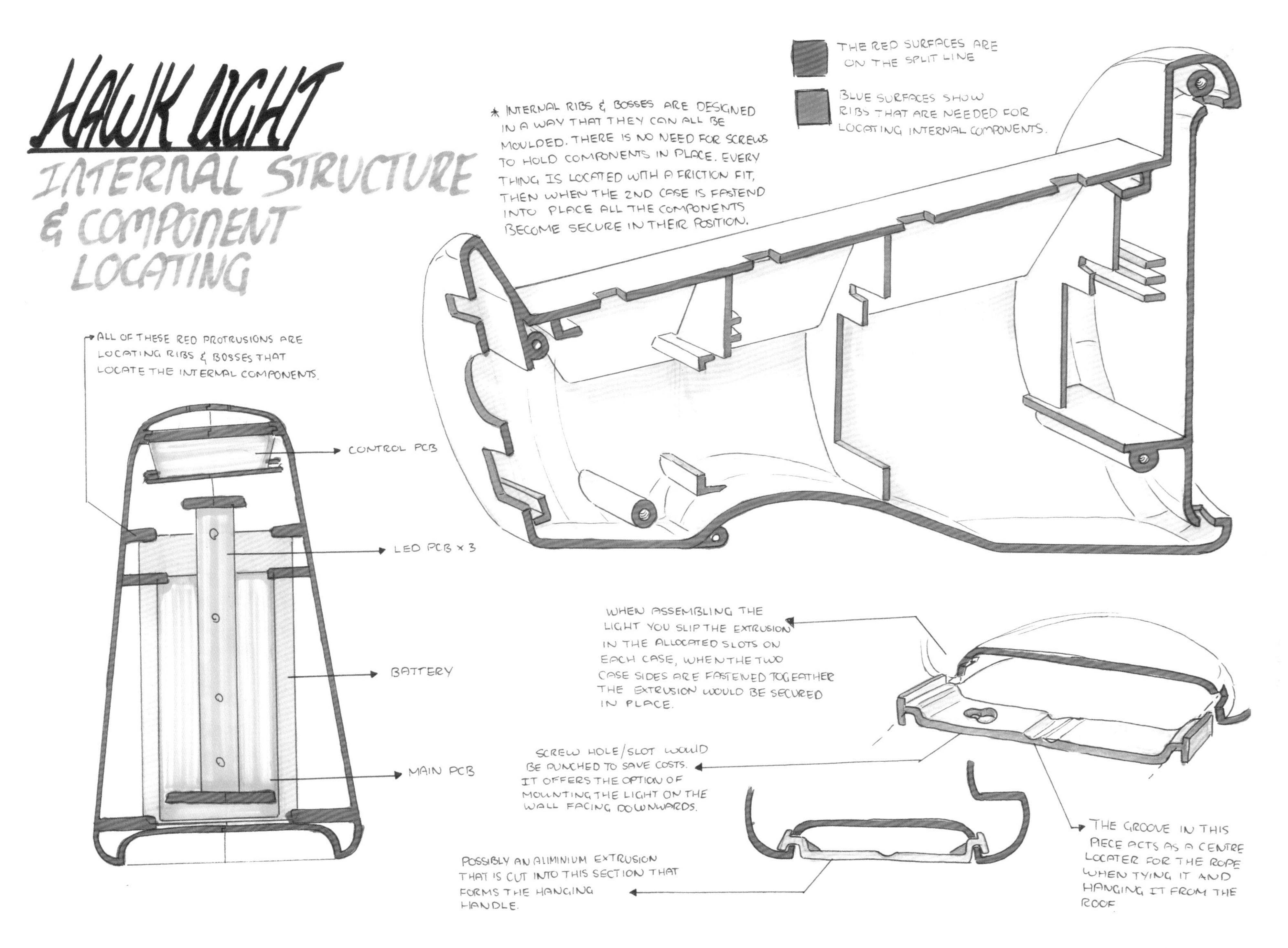
HAWK LIGHT
INTERNAL STRUCTURE
& COMPONENT
LOCATING
THE RED SURFACES ARE
ON THE SPLIT LINE
BLUE SURFACES SHOW
RIBS THAT ARE NEEDED FOR
LOCATING INTERNAL COMPONENTS.
* INTERNAL RIBS & BOSSES ARE DESIGNED
IN A WAY THAT THEY CAN ALL BE
MOULDED. THERE IS NO NEED FOR SCREWS
TO HOLD COMPONENTS IN PLACE. EVERY
THING IS LOCATED WITH A FRICTION FIT,
THEN WHEN THE 2ND CASE IS FASTEND
INTO PLACE ALL THE COMPONENTS
BECOME SECURE IN THEIR POSITION.
ALL OF THESE RED PROTRUSIONS ARE
LOCATING RIBS & BOSSES THAT
LOCATE THE INTERNAL COMPONENTS.
CONTROL PCB
LED PCB x 3
BATTERY
MAIN PCB
WHEN ASSEMBLING THE
LIGHT YOU SLIP THE EXTRUSION
IN THE ALLOCATED SLOTS ON
EACH CASE, WHEN THE TWO
CASE SIDES ARE FASTENED TOGEATHER
THE EXTRUSION WOULD BE SECURED
IN PLACE.
SCREW HOLE/SLOT WOULD
BE PUNCHED TO SAVE COSTS.
IT OFFERS THE OPTION OF
MOUNTING THE LIGHT ON THE
WALL FACING DOWNWARDS.
POSSIBLY AN ALIMINIUM EXTRUSION
THAT IS CUT INTO THIS SECTION THAT
FORMS THE HANGING
HANDLE.
THE GROOVE IN THIS
PIECE ACTS AS A CENTRE
LOCATER FOR THE ROPE
WHEN TYING IT AND
HANGING IT FROM THE
ROOF

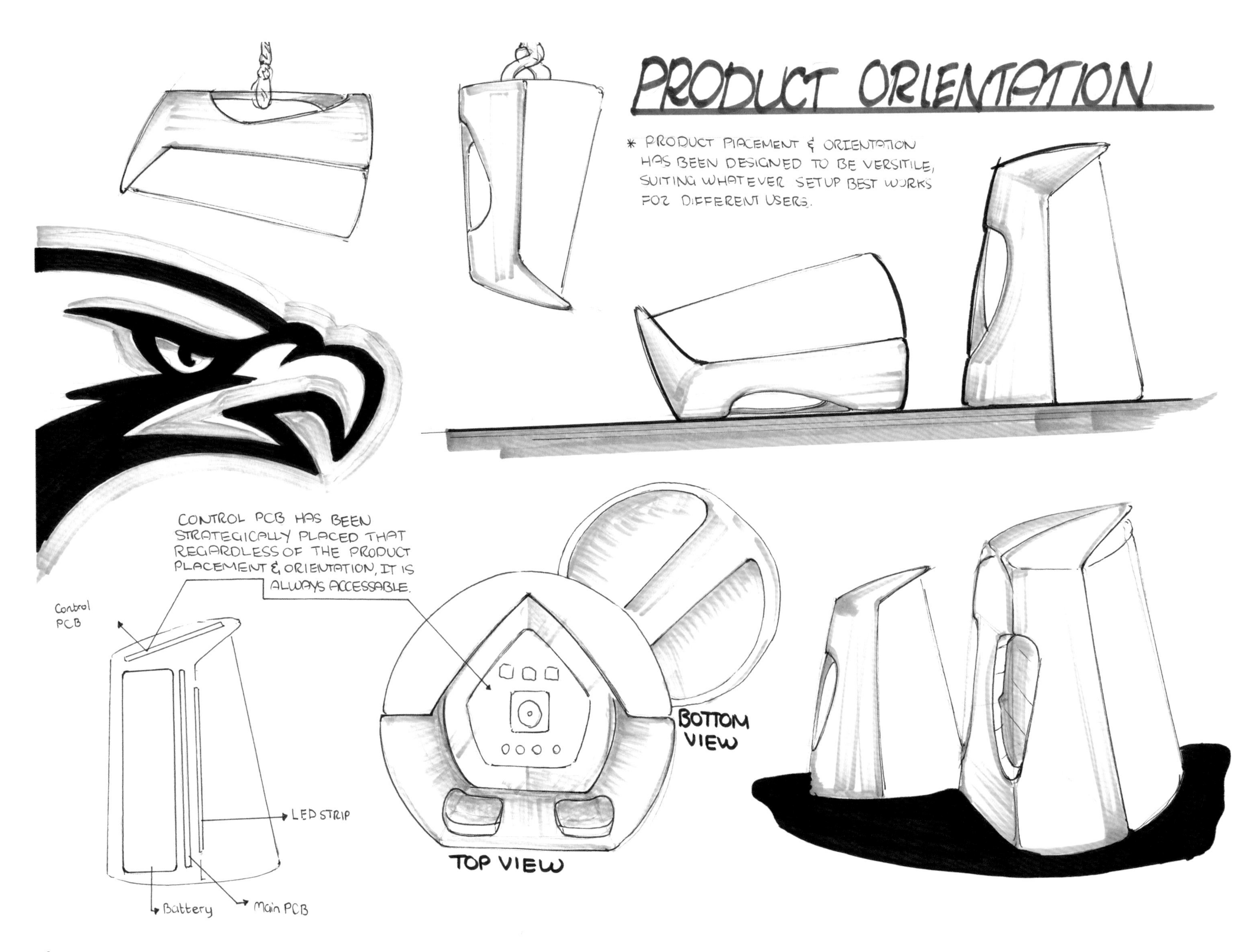
PRODUCT ORIENTATION
* PRODUCT PLACEMENT & ORIENTATION HAS BEEN DESIGNED TO BE VERSITILE, SUITING WHATEVER SETUP BEST WORKS FOR DIFFERENT USERS.
CONTROL PCB HAS BEEN STRATEGICALLY PLACED THAT REGARDLESS OF THE PRODUCT PLACEMENT & ORIENTATION, IT IS ALWAYS ACCESSABLE.
Control PCB
LED STRIP
Battery
Main PCB
TOP VIEW
BOTTOM VIEW

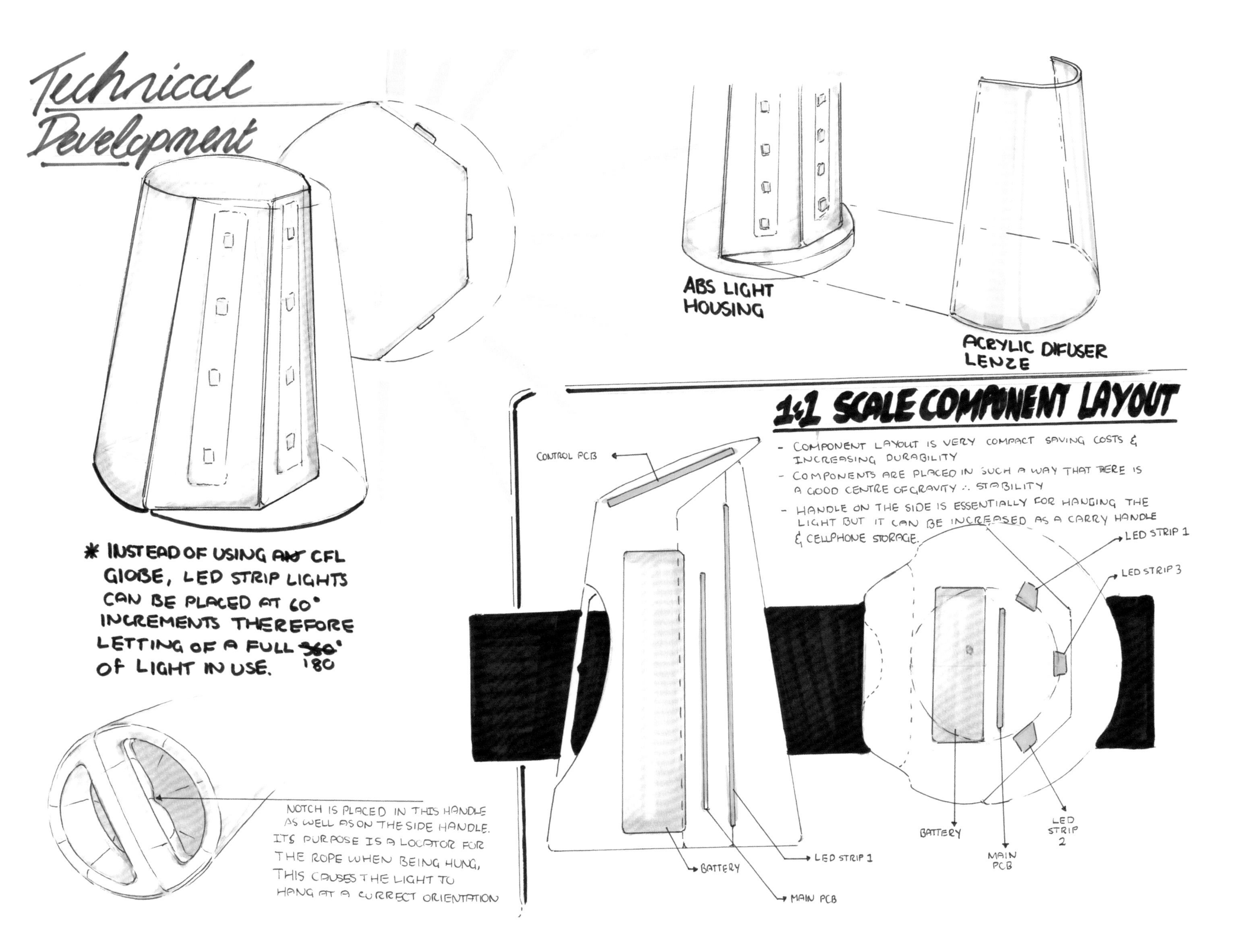
Technical Development
ABS LIGHT HOUSING
ACRYLIC DIFUSER LENZE
* INSTEAD OF USING AN CFL GLOBE, LED STRIP LIGHTS CAN BE PLACED AT 60° INCREMENTS THEREFORE LETTING OF A FULL 360° 180 OF LIGHT IN USE.
1:1 SCALE COMPONENT LAYOUT
- COMPONENT LAYOUT IS VERY COMPACT SAVING COSTS & INCREASING DURABILITY
- COMPONENTS ARE PLACED IN SUCH A WAY THAT THERE IS A GOOD CENTRE OF GRAVITY ∴ STABILITY
- HANDLE ON THE SIDE IS ESSENTIALLY FOR HANGING THE LIGHT BUT IT CAN BE INCREASED AS A CARRY HANDLE & CELLPHONE STORAGE.
CONTROL PCB
BATTERY
LED STRIP 1
MAIN PCB
LED STRIP 1
LED STRIP 3
BATTERY
MAIN PCB
LED STRIP 2
NOTCH IS PLACED IN THIS HANDLE AS WELL AS ON THE SIDE HANDLE. ITS PURPOSE IS A LOCATOR FOR THE ROPE WHEN BEING HUNG, THIS CAUSES THE LIGHT TO HANG AT A CURRECT ORIENTATION

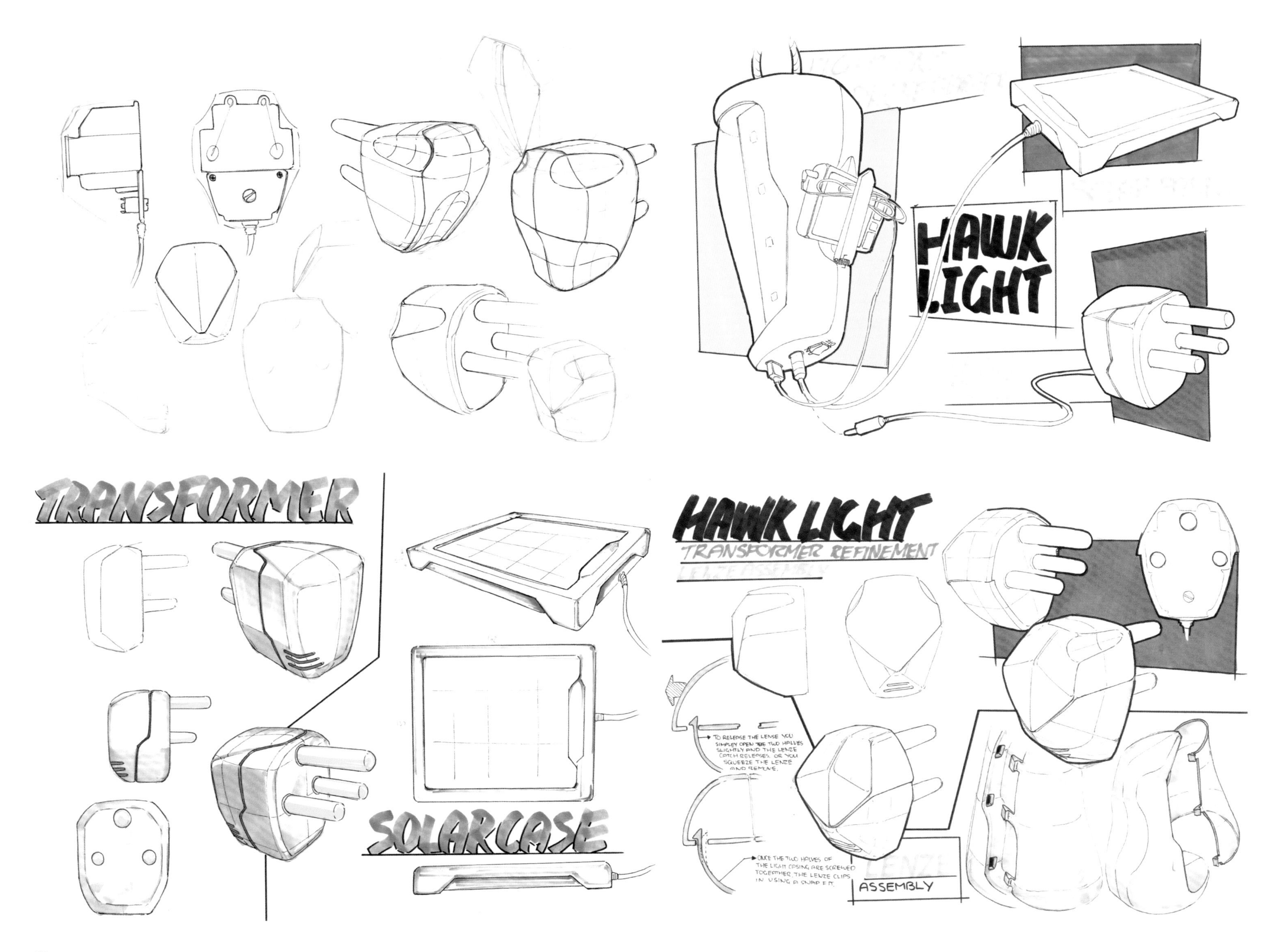

HAWK LIGHT
TRANSFORMER
SOLARCASE
HAWK LIGHT
TRANSFORMER REFINEMENT
TO RELEASE THE LENSE YOU SIMPLEY OPEN THE TWO HALVES SLIGHTLY AND THE LENZE CATCH RELEASES. OR YOU SQUEEZE THE LENZE AND REMOVE.
ONCE THE TWO HALVES OF THE LIGHT CASING ARE SCREWED TOGEATHER, THE LENZE CLIPS IN USING A SNAP FIT.
ASSEMBLY

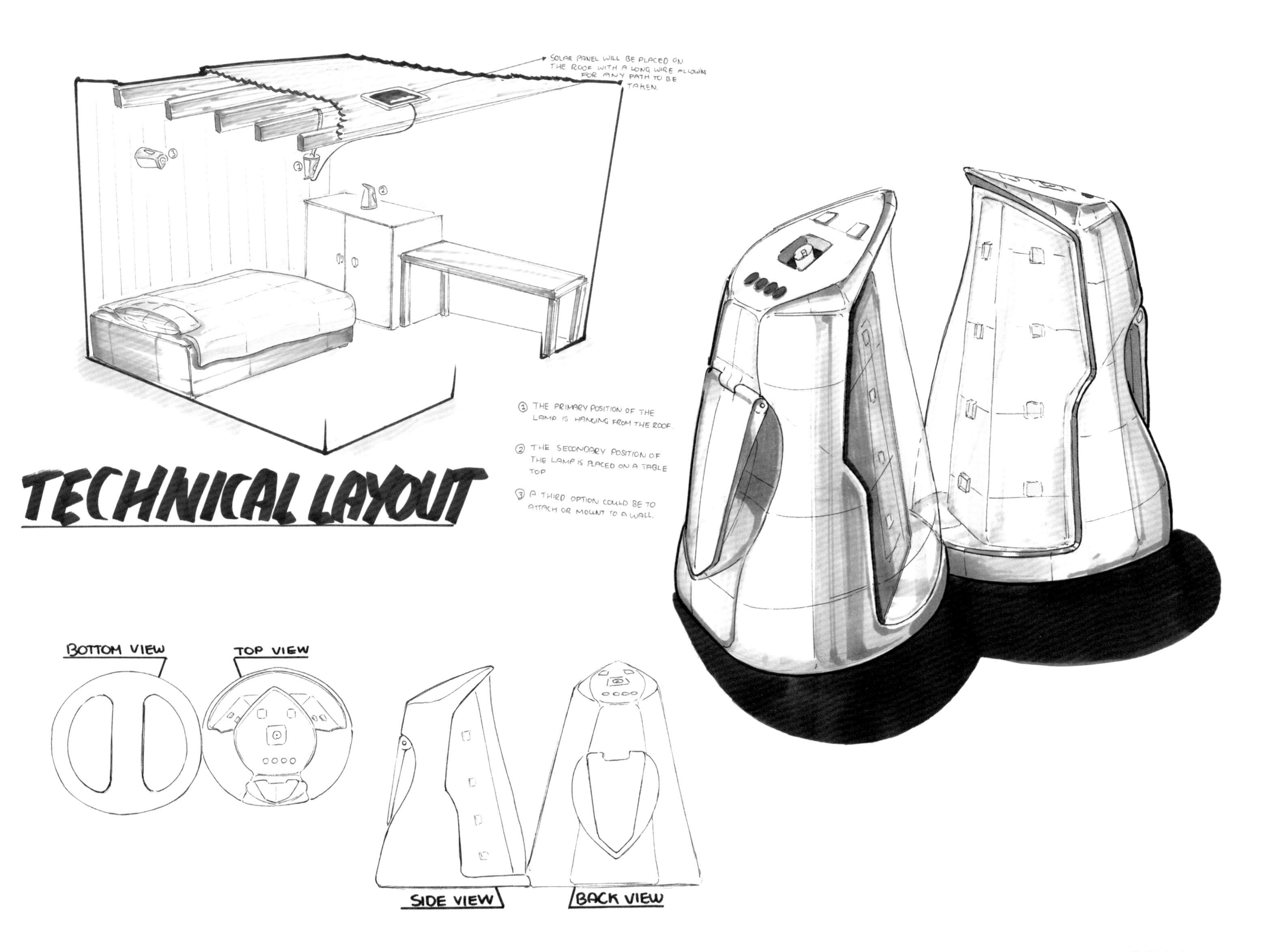
SOLAR PANEL WILL BE PLACED ON THE ROOF WITH A LONG WIRE ALLOWING FOR ANY PATH TO BE TAKEN.
TECHNICAL LAYOUT
① THE PRIMARY POSITION OF THE LAMP IS HANGING FROM THE ROOF.
② THE SECONDARY POSITION OF THE LAMP IS PLACED ON A TABLE TOP
③ A THIRD OPTION COULD BE TO ATTACH OR MOUNT TO A WALL.
BOTTOM VIEW
TOP VIEW
SIDE VIEW
BACK VIEW

Kumm

Kumm咖啡机

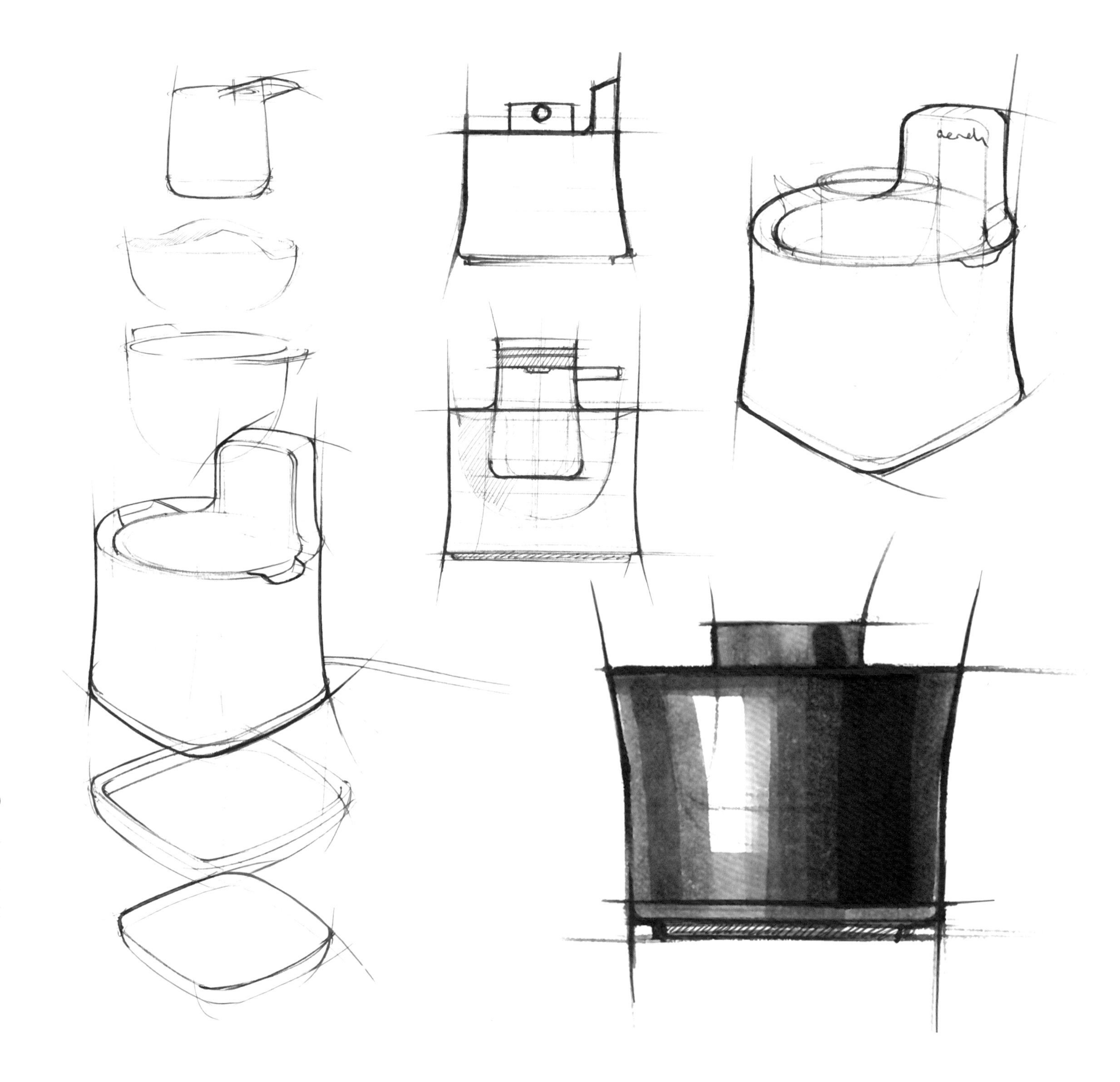

设计师：贝居姆·托姆鲁克（Begüm Tomruk）
客户：个人作品

这是使用传统咖啡烹煮技术的土耳其咖啡机的数码手绘设计图。设计师先用线条手绘了一些草图以推敲产品的造型，再用数码软件绘制出更为逼真的产品效果图。

kumm

Electric Razor

电动剃须刀

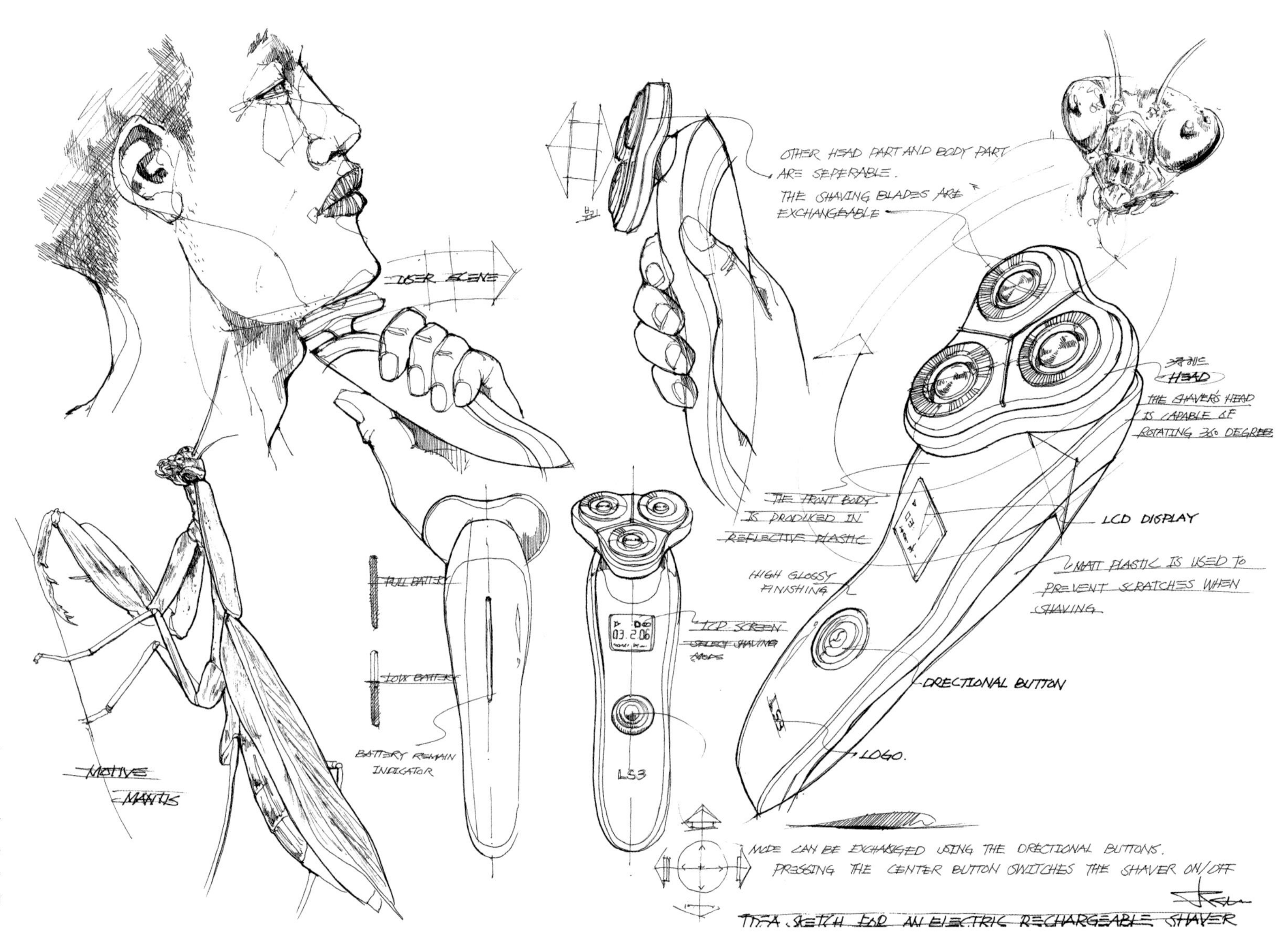
USER SCENE
OTHER HEAD PART AND BODY PART ARE SEPERABLE.
THE SHAVING BLADES ARE EXCHANGEABLE
HEAD
THE SHAVER'S HEAD IS CAPABLE OF ROTATING 360 DEGREE
THE FRONT BODY IS PRODUCED IN REFLECTIVE PLASTIC
LCD DISPLAY
HIGH GLOSSY FINISHING
MATT PLASTIC IS USED TO PREVENT SCRATCHES WHEN SHAVING
FULL BATTERY
LOW BATTERY
BATTERY REMAIN INDICATOR
LCD SCREEN
03.2.06
DRECTIONAL BUTTON
LOGO.
LS3
MOTIVE
MANTIS
MODE CAN BE EXCHANGED USING THE DRECTIONAL BUTTONS.
PRESSING THE CENTER BUTTON SWITCHES THE SHAVER ON/OFF
IDEA SKETCH FOR AN ELECTRIC RECHARGEABLE SHAVER

设计师：柳时形

客户：个人作品

这是一组电动剃须刀的概念设计草图。电动剃须刀使用起来简单便捷，是男人的必备物品。因此，设计师希望在这一产品的造型设计中体现出强烈的男人气概。

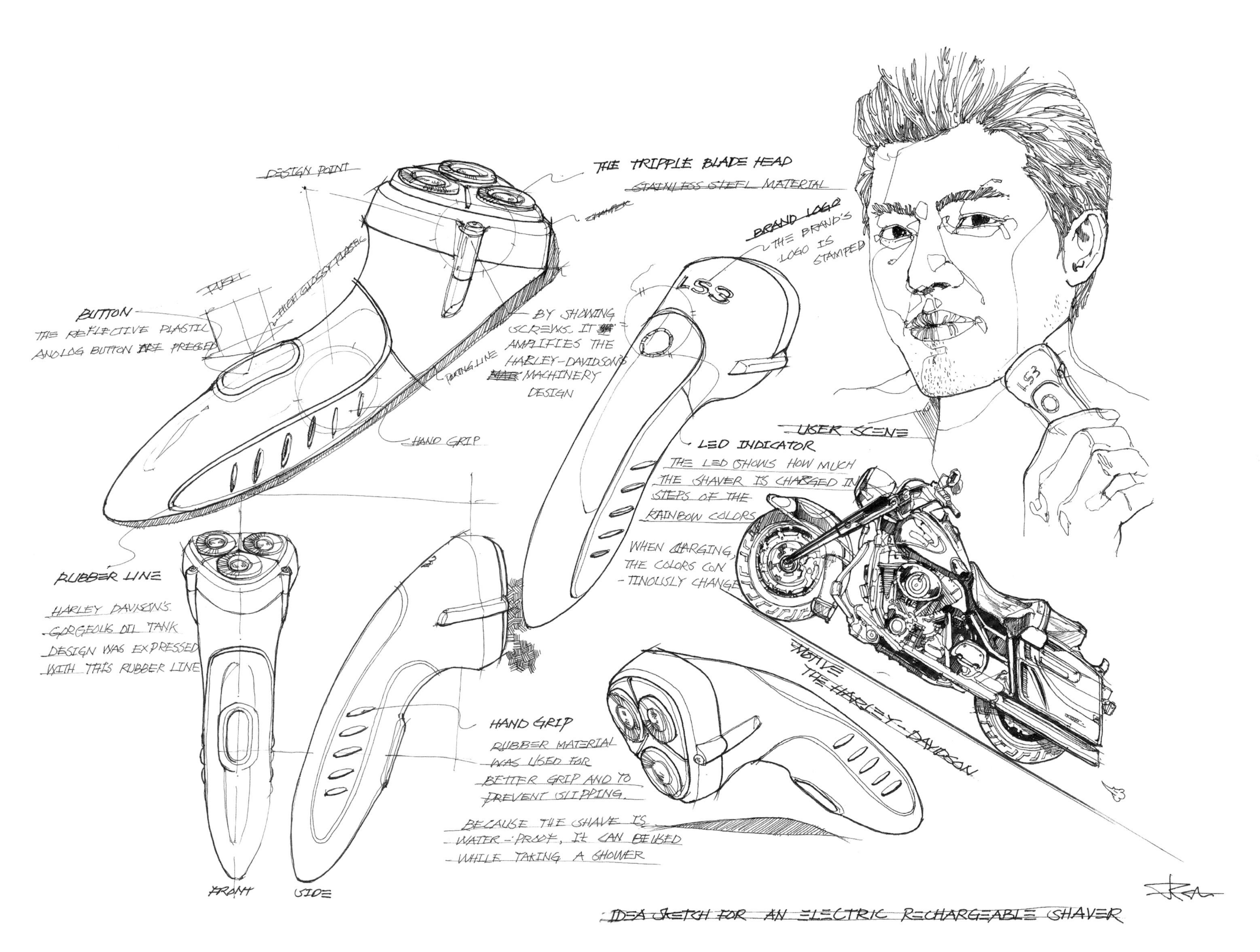
DESIGN POINT
THE TRIPPLE BLADE HEAD
STAINLESS STEEL MATERIAL
BRAND LOGO
THE BRAND'S LOGO IS STAMPED
PUSH
HIGH GLOSSY PLASTIC
BUTTON
THE REFLECTIVE PLASTIC ANOLOG BUTTON PRESSED
BY SHOWING SCREWS. IT AMPLIFIES THE HARLEY-DAVIDSON'S MACHINERY DESIGN
PARTING LINE
HAND GRIP
LS3
USER SCENE
LED INDICATOR
THE LED SHOWS HOW MUCH THE SHAVER IS CHARGED IN STEPS OF THE RAINBOW COLORS
WHEN CHARGING, THE COLORS CON -TINOUSLY CHANGE
RUBBER LINE
HARLEY DAVISON'S GORGEOUS OIL TANK DESIGN WAS EXPRESSED WITH THIS RUBBER LINE
HAND GRIP
RUBBER MATERIAL WAS USED FOR BETTER GRIP AND TO PREVENT SLIPPING.
BECAUSE THE SHAVE IS WATER-PROOF, IT CAN BE USED WHILE TAKING A SHOWER
MOTIVE THE HARLEY-DAVIDSON
FRONT
SIDE
IDEA SKETCH FOR AN ELECTRIC RECHARGEABLE SHAVER

Multi-Function Flashlight
多功能手电筒

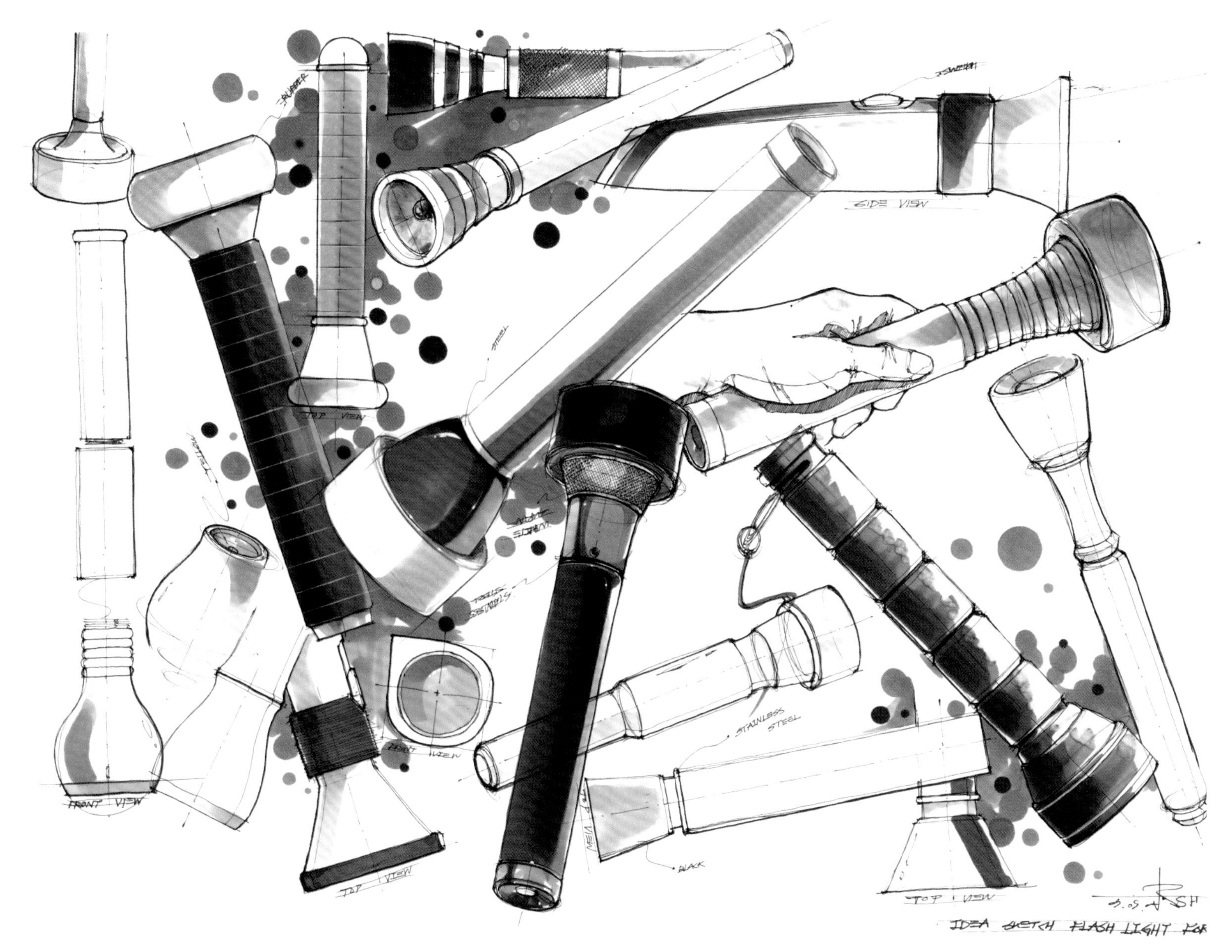

设计师：柳时彤
客户：个人作品

设计师试图设计一把多功能手电筒，以帮助人们在不同的环境中用其应对各种不同状况。设计师通过大量的草图绘制来不断调整设计的想法，重点对产品的功能性、人机尺度的把握以及用户的使用行为进行推敲。例如，其中的一个想法是设计一个可调节的灯头来调整光线。

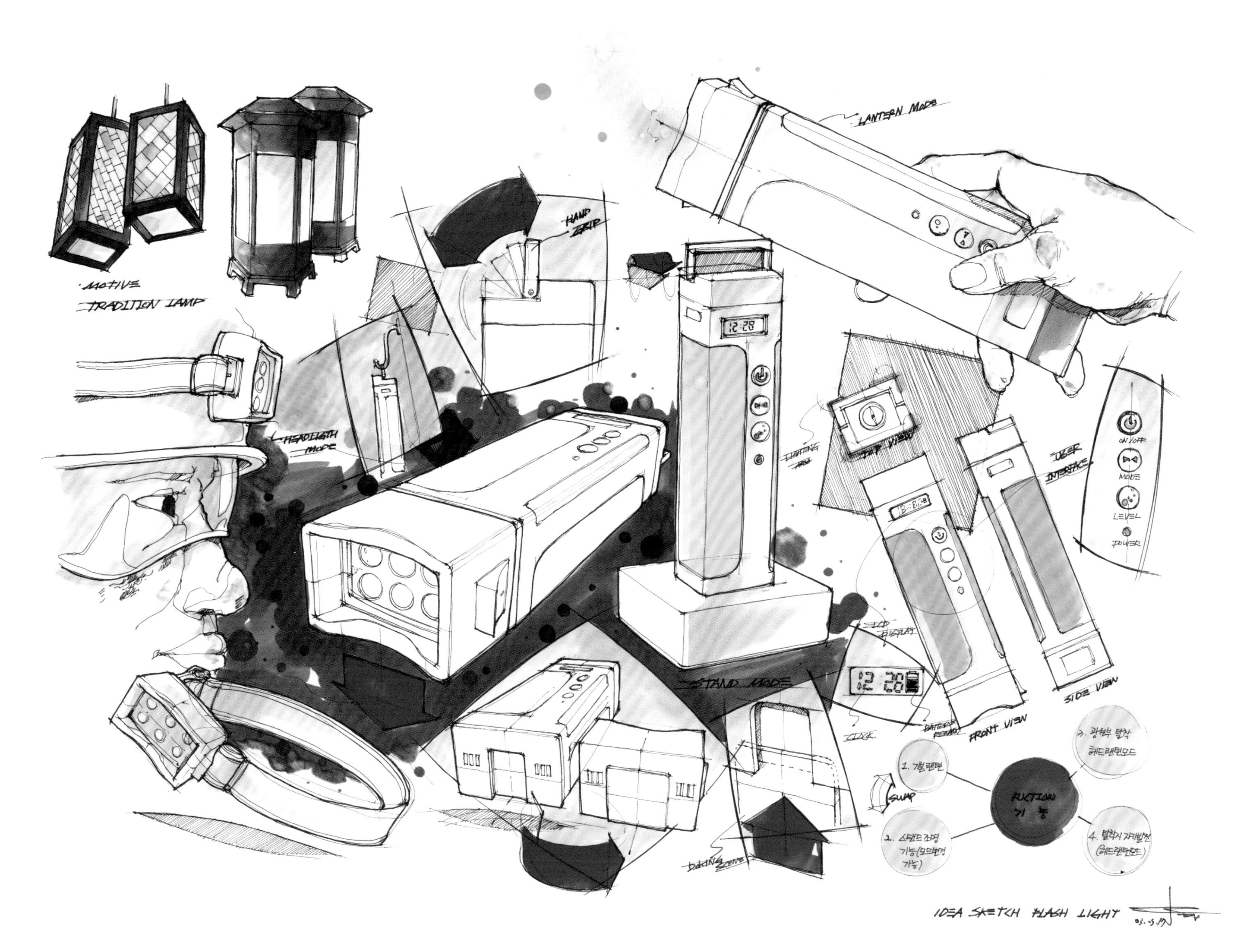
MOTIVE
TRADITION LAMP
LANTERN MODE
HAND GRIP
HEADLIGHT MODE
12:28
LIGHTING AREA
TOP VIEW
USER INTERFACE
ON/OFF
MODE
LEVEL
POWER
LCD DISPLAY
STAND MODE
CLOCK
BATTERY REMAIN
FRONT VIEW
SIDE VIEW
1. 기본랜턴
2. 스탠드조명 기능(모드변경 가능)
3. 광원부 탈착 헤드랜턴모드
4. 탈착시 자가발전 (헤드랜턴모드)
FUCTION 기능
SWAP
IDEA SKETCH FLASH LIGHT
05.5.17

Hair Dryer
吹风机

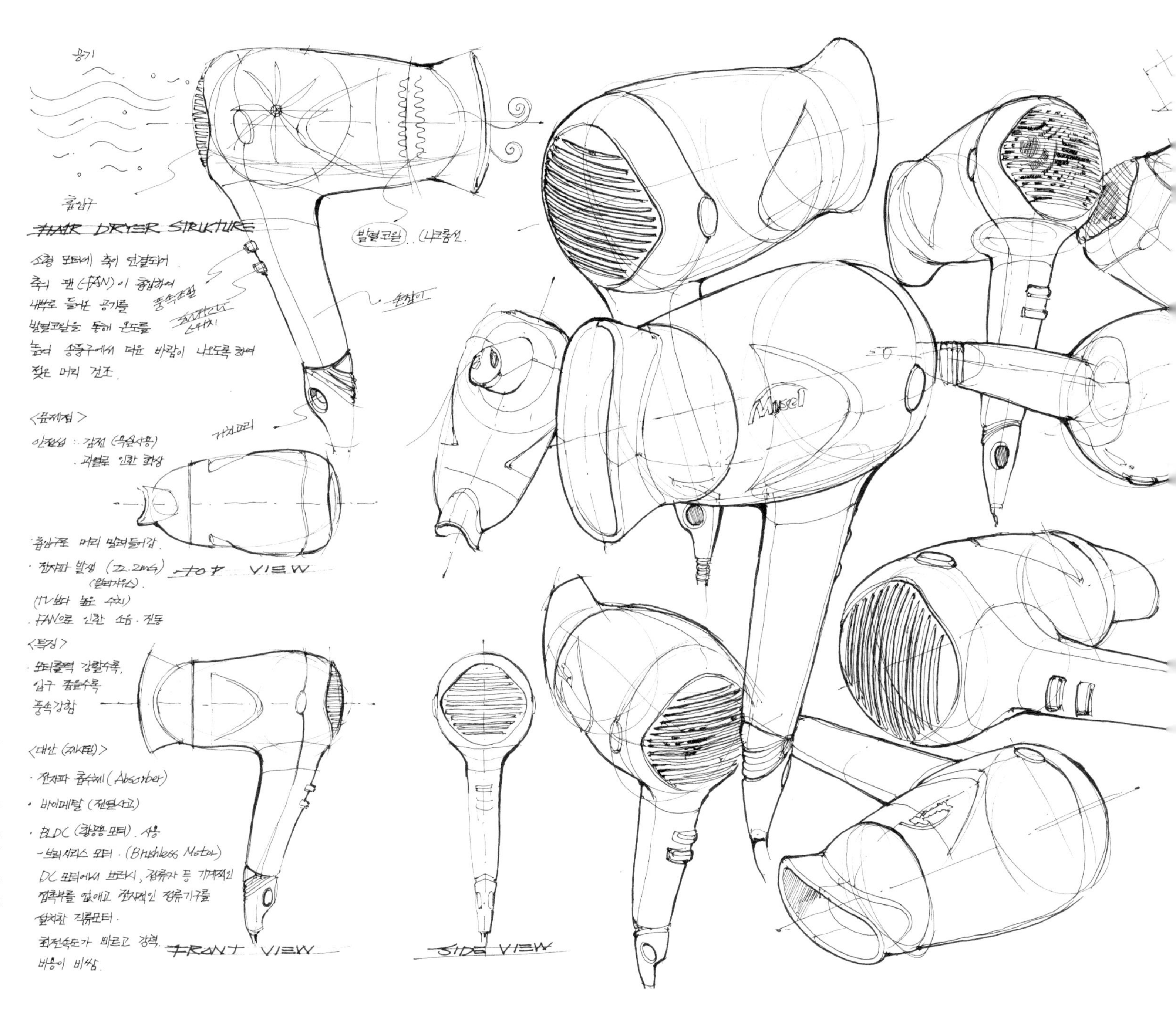

设计师：柳时彤
客户：个人作品

设计师研究自然形态，并将其运用到吹风机的概念设计中。设计师发现火山石拥有很多孔洞，从中获得启发，试图解决吹风机的散热问题。有些时候，自然可以给我们许多设计的灵感。

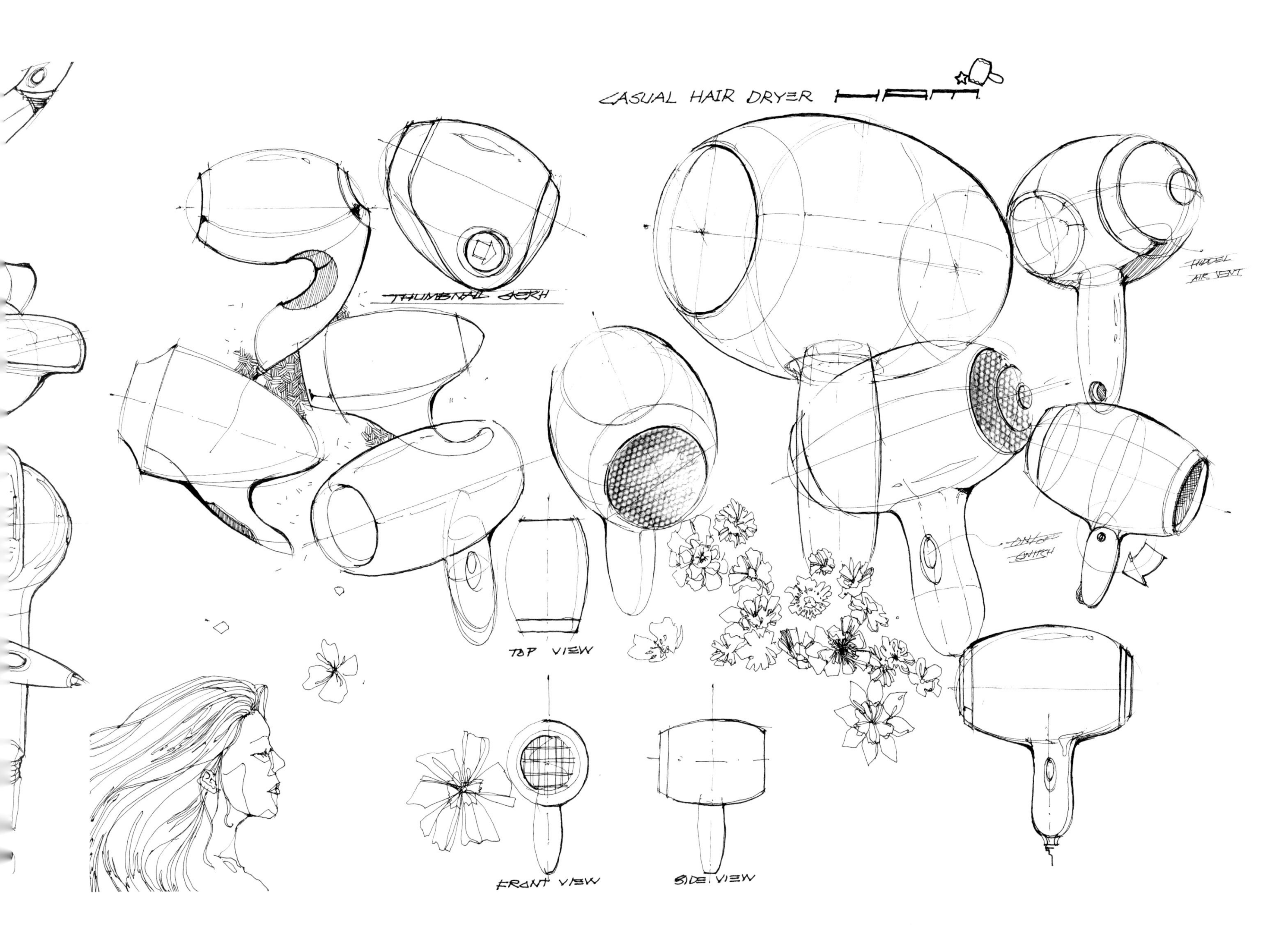
CASUAL HAIR DRYER
THUMBNAIL SKETCH
HIDDEL AIR VENT
ON/OFF SWITCH
TOP VIEW
FRONT VIEW
SIDE VIEW

Fastrack Watch: Press

Fastrack Press 手表

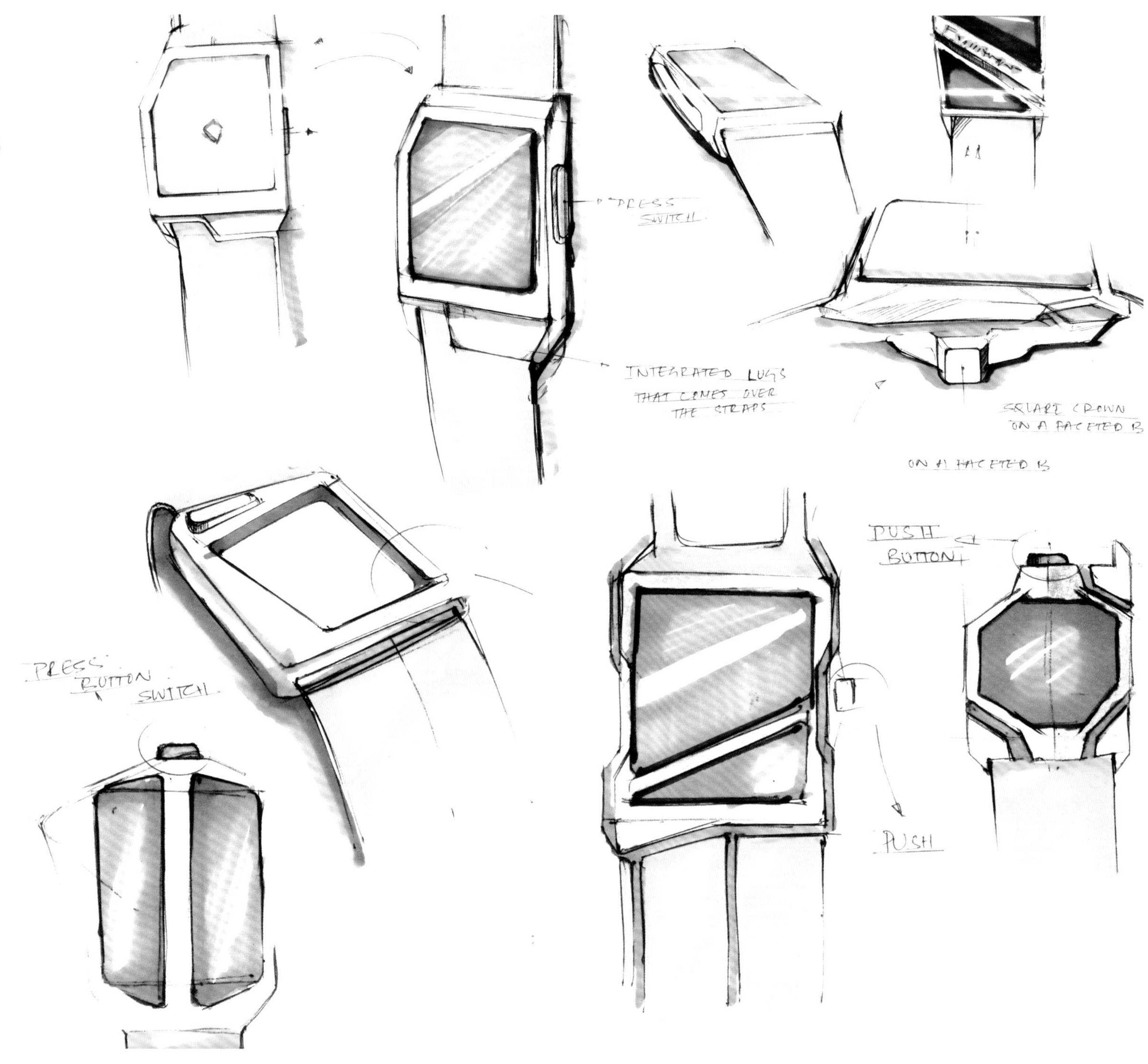

设计师：尤瓦尔·阿南德（Ujjwal Anand）
客户：Fastrack公司，非真实项目

这是基于人体工程学的设计理念为Fastrack品牌打造的一款概念手表，其设定时间的方式非比寻常。设计师从上百张草图中选定了几个方案，并最终对其中一个方案进行了深化设计。

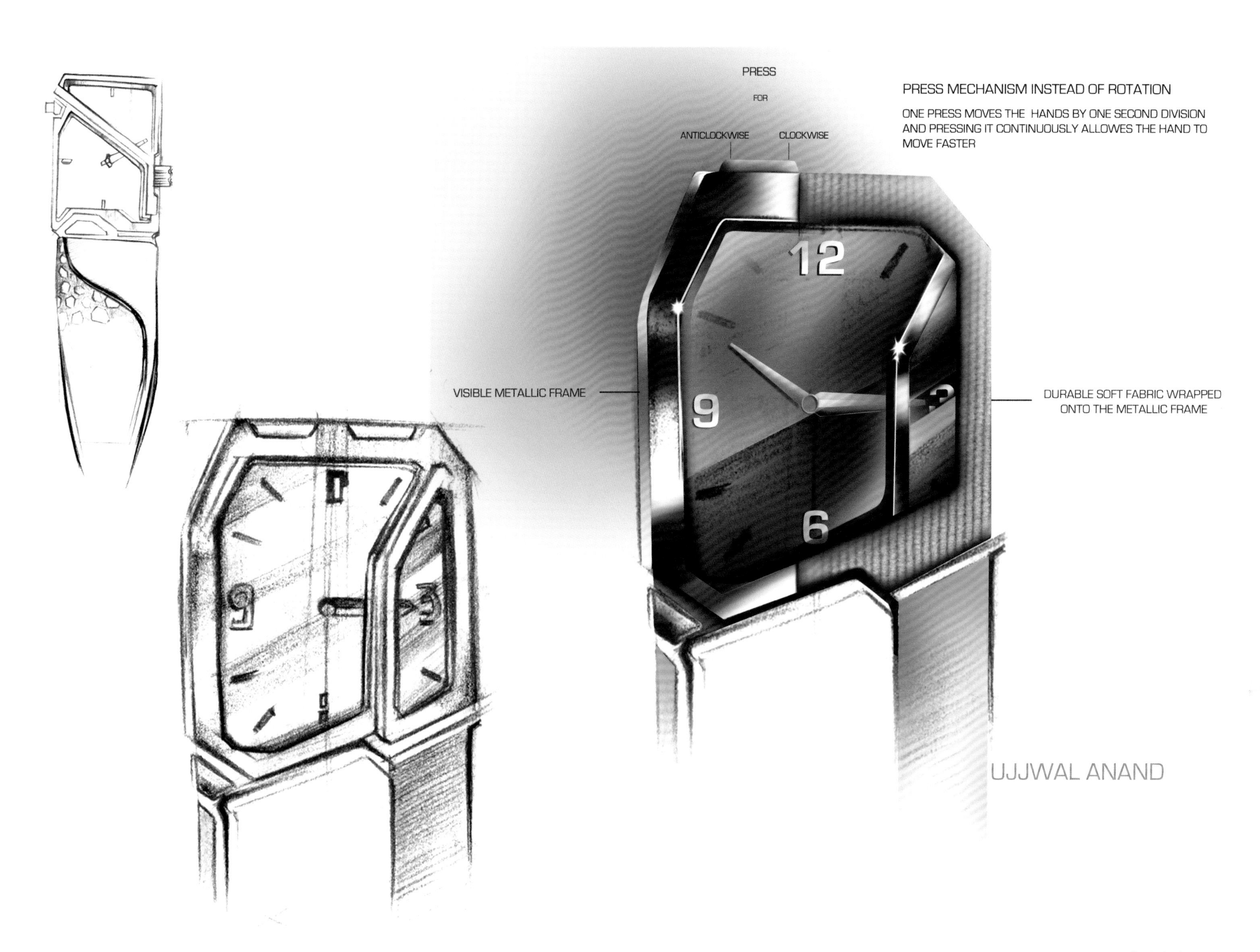
PRESS
FOR
ANTICLOCKWISE
CLOCKWISE
PRESS MECHANISM INSTEAD OF ROTATION
ONE PRESS MOVES THE HANDS BY ONE SECOND DIVISION AND PRESSING IT CONTINUOUSLY ALLOWES THE HAND TO MOVE FASTER
VISIBLE METALLIC FRAME
DURABLE SOFT FABRIC WRAPPED ONTO THE METALLIC FRAME
12
9
6
UJJWAL ANAND

Fastrack Watch: CT-1

Fastrack CT-1
手表

设计师：尤瓦尔·阿南德
客户：Fastrack公司，非真实项目

在这款带有Porsche标志的机械手表的概念设计中，设计师从F1赛车中的方格旗获得灵感，寓意着运动精神，活力十足。最终的方案简单而具有说服力——方格旗的元素体现在表带和表盘的设计上。

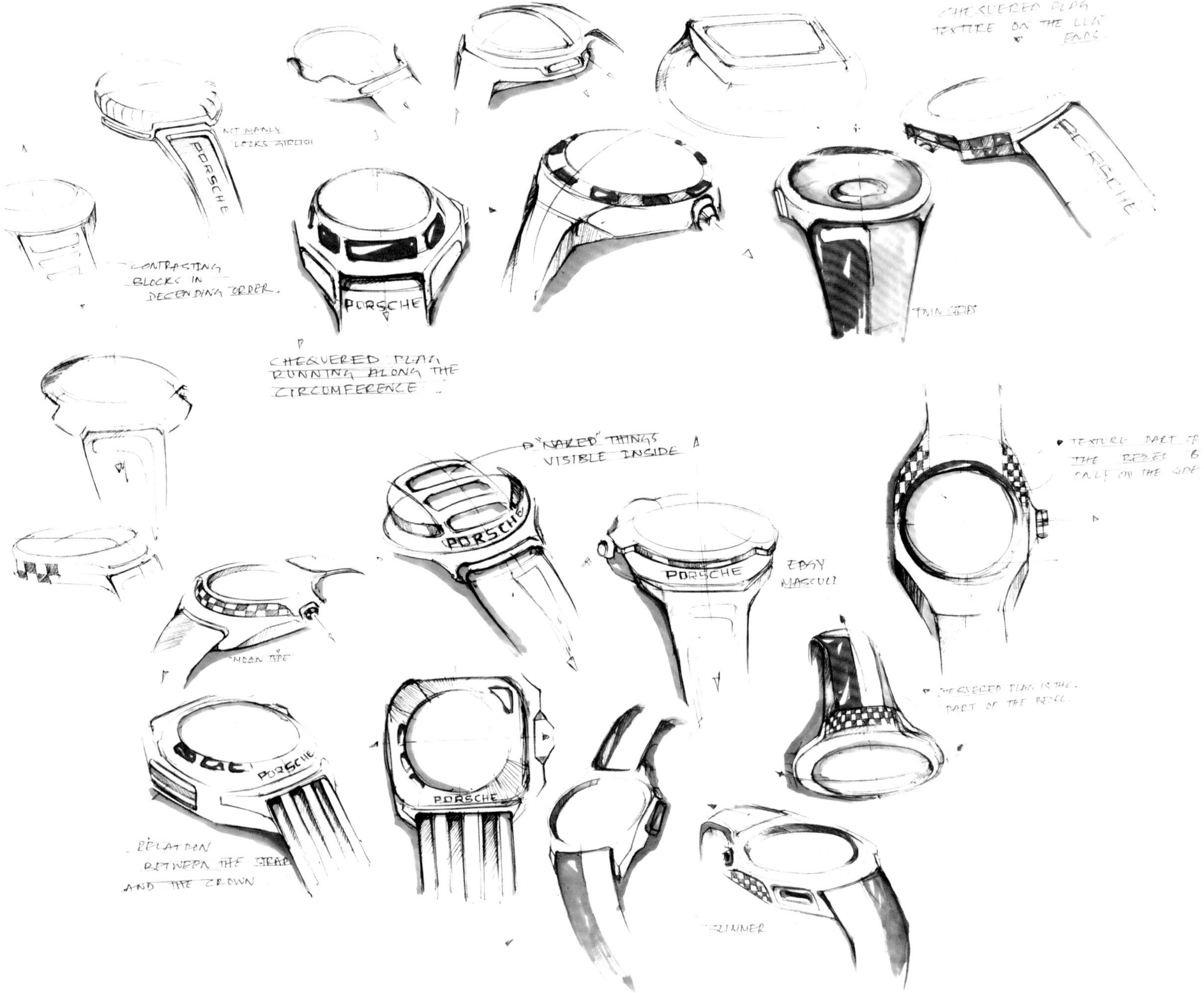

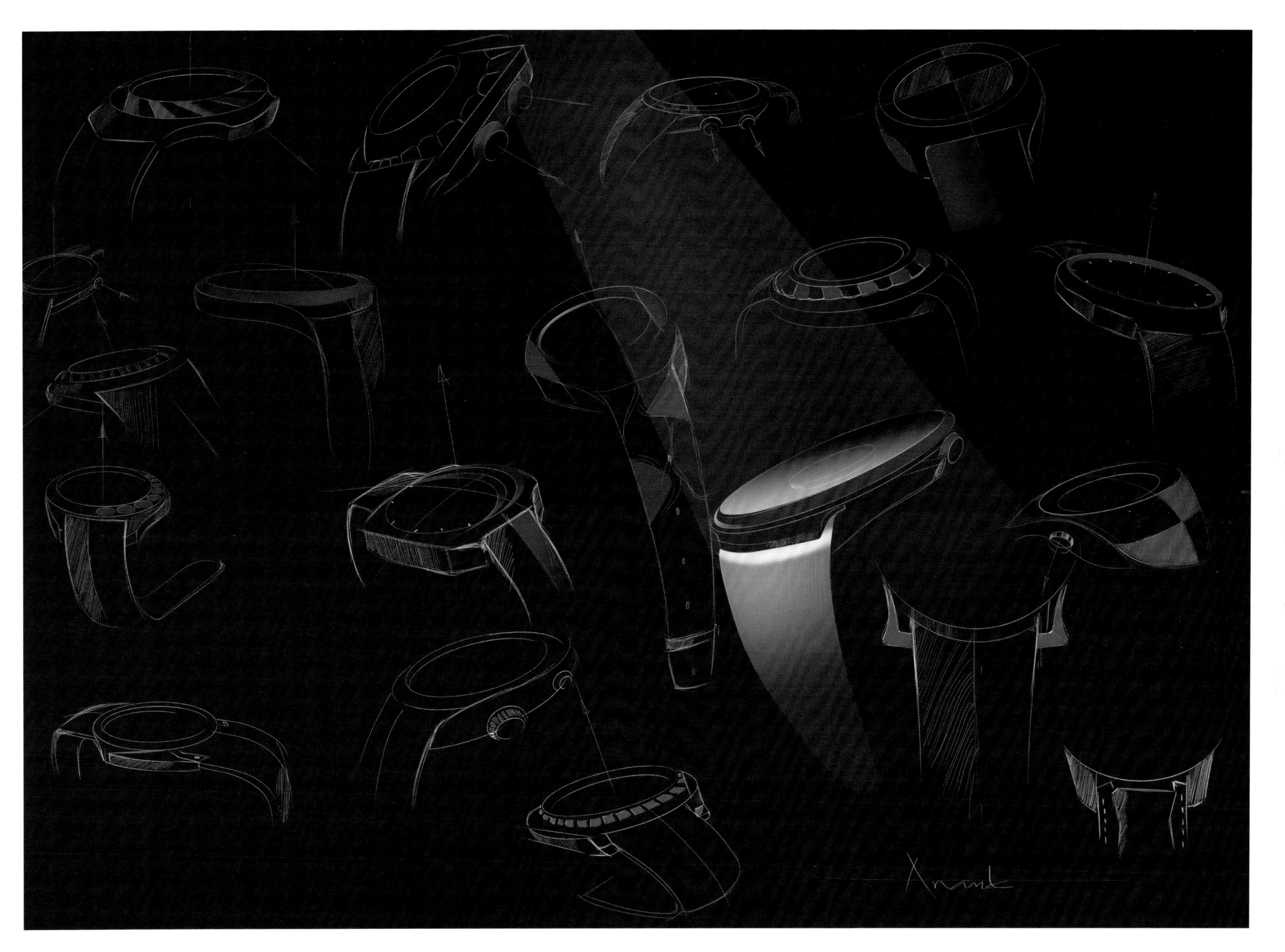
Anand

Grand Tourer De Ville Watch

Grand Tourer De Ville手表

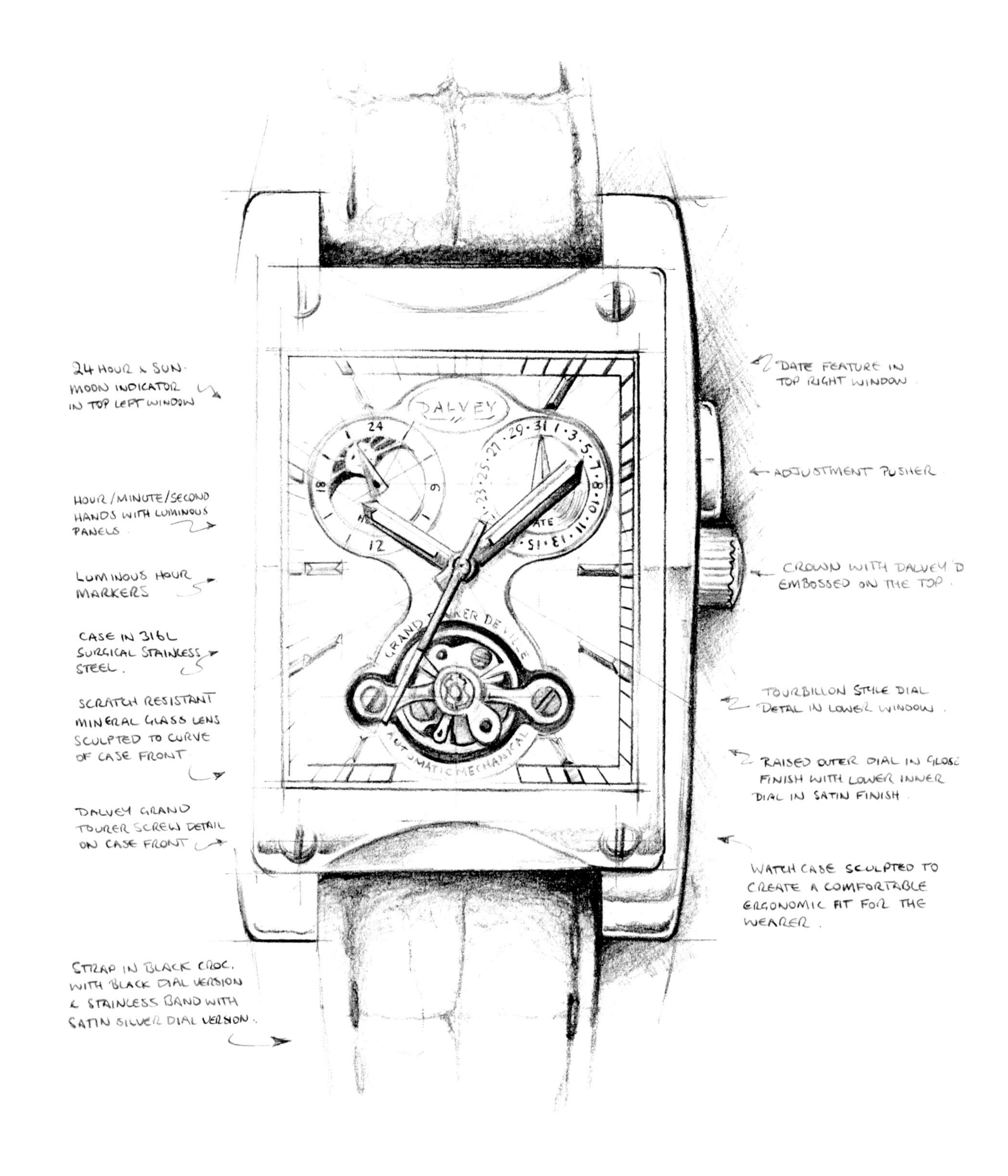

设计师：保罗·J.斯塔利斯基（Paul J. Stariski）

客户：Dalvey公司

这是为Dalvey品牌打造的男士奢华腕表，其矩形的造型也与Dalvey品牌的Grand Tourer系列产品保持了一致的产品形象。设计师用铅笔精细地表现出了手表的矩形不锈钢表盘。

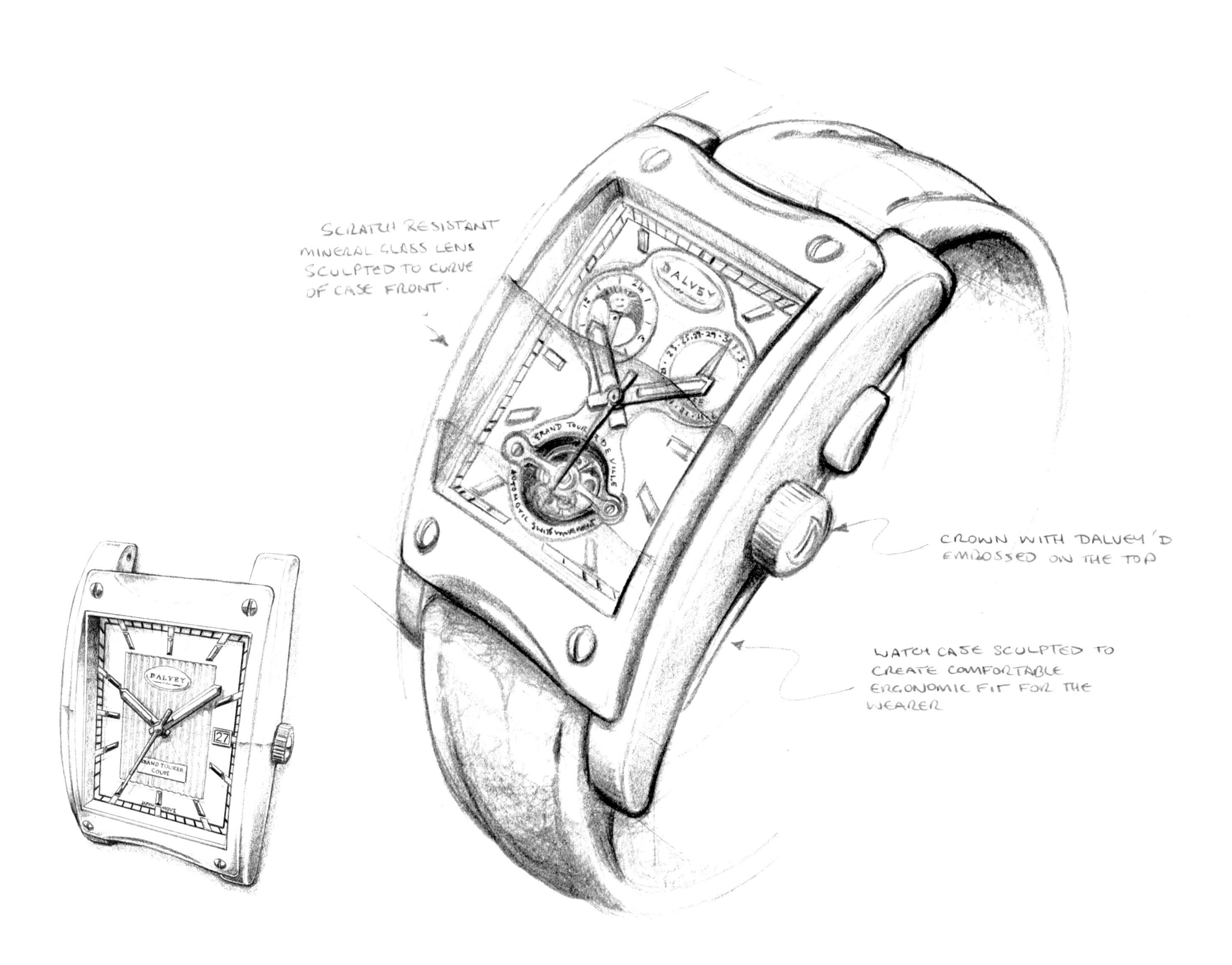
SCRATCH RESISTANT MINERAL GLASS LENS SCULPTED TO CURVE OF CASE FRONT.
CROWN WITH DALVEY 'D EMBOSSED ON THE TOP
WATCH CASE SCULPTED TO CREATE COMFORTABLE ERGONOMIC FIT FOR THE WEARER
DALVEY
GRAND TOURER COUPE

New Bold

Bold全键盘智能手机

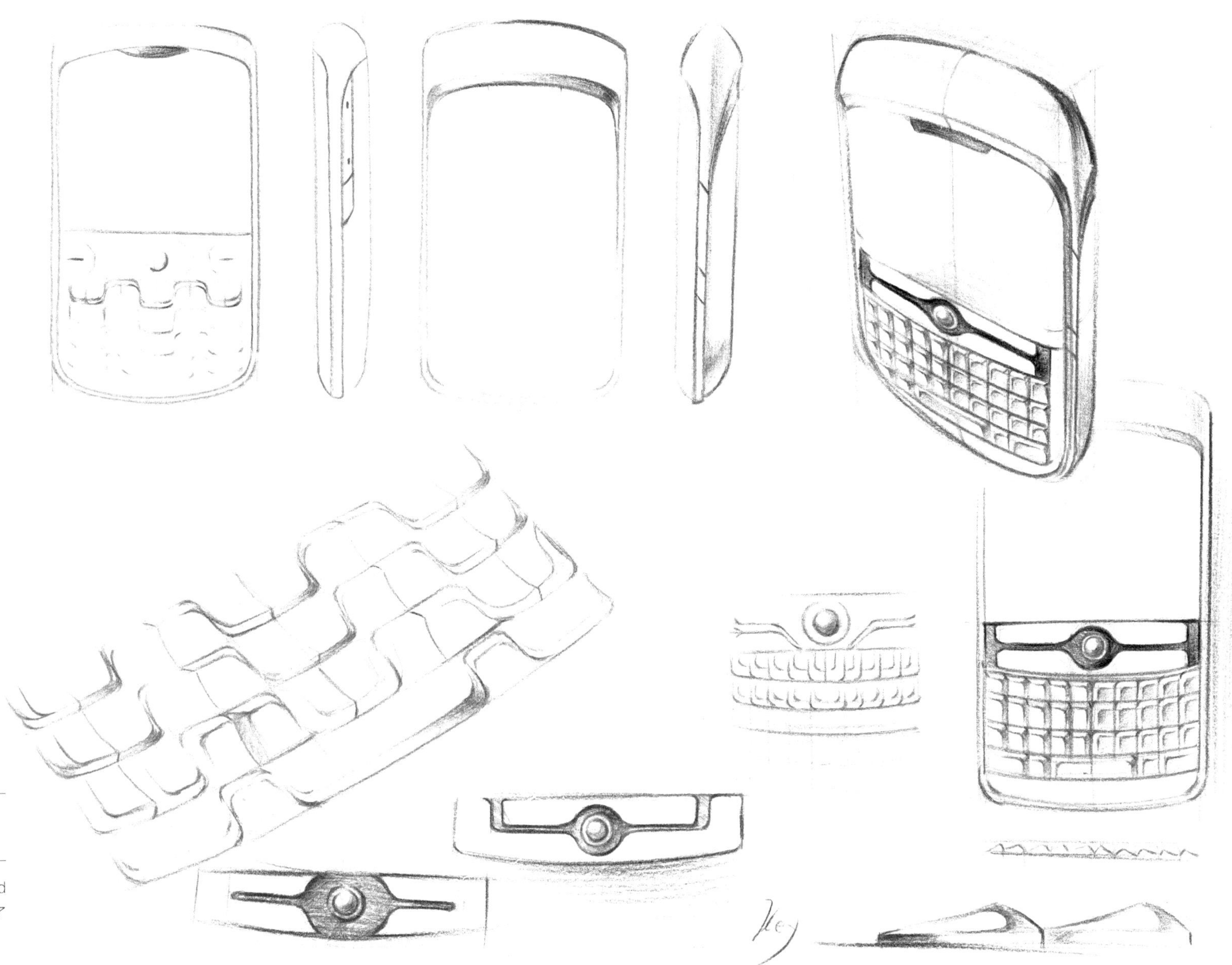

设计师：赵珠明（Jeongmin Jo）
客户：Blackberry公司

该全键盘智能手机的设计旨在表现出Bold系列手机年轻、柔和的产品形象，使用了红色及流线型的设计元素。

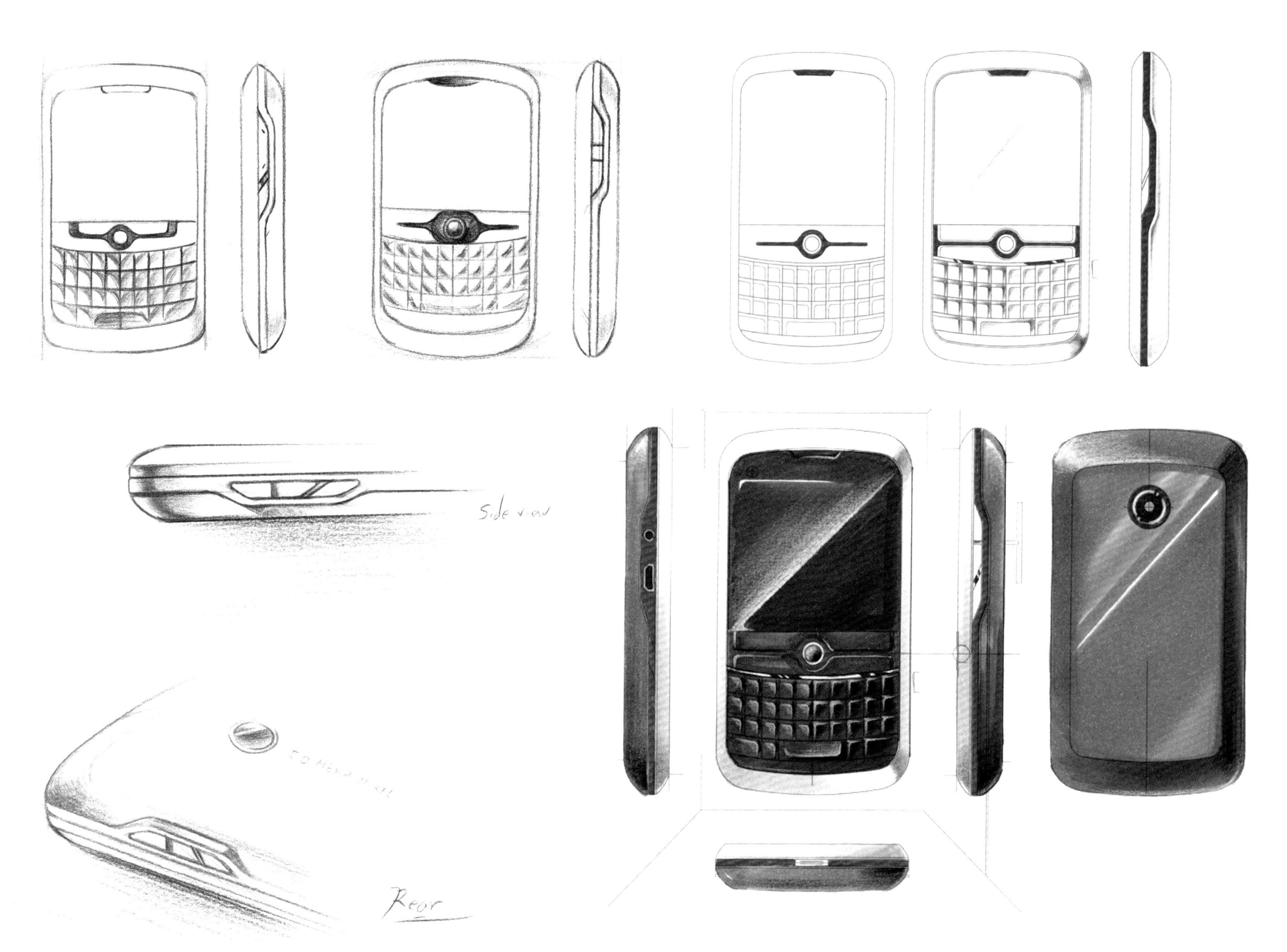
Side view
Rear

C3 Project
C3手机方案

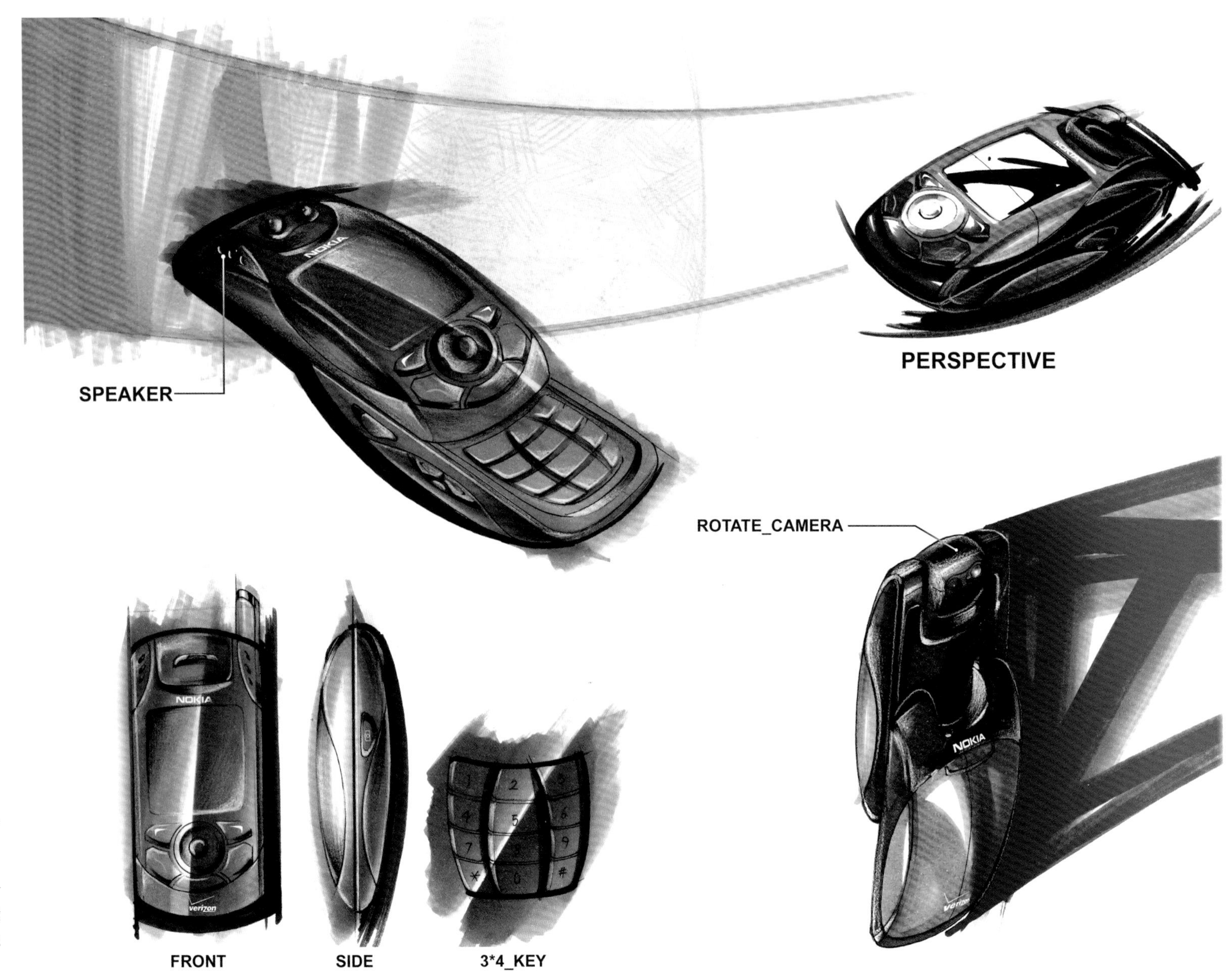

设计师：赵珠明
客户：Nokia公司

这是Nokia为年轻人打造的滑盖手机，设计理念凸显产品的动感与活力，并带有独特的可旋转手机镜头。

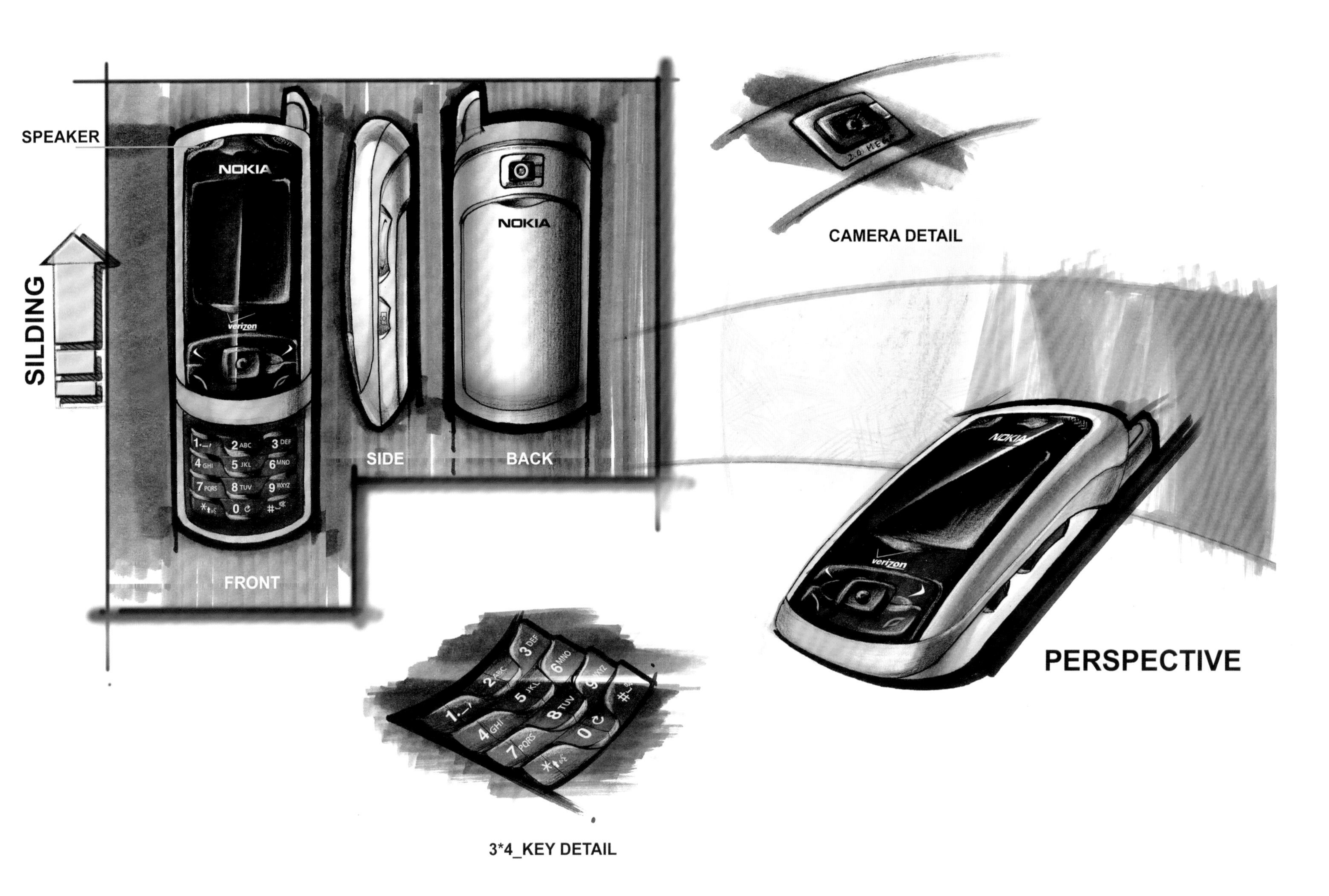
SPEAKER
SILDING
NOKIA
verizon
FRONT
SIDE
BACK
CAMERA DETAIL
PERSPECTIVE
3*4_KEY DETAIL

Tábula

Tábula平板电脑

设计师：卡洛斯·希门尼斯（Carlos Jiménez）

客户：延雪平大学、Yellon公司

该项目旨在设计一款可在建筑施工环境中使用的平板电脑，其主要功能为便于工人参看建筑设计图纸，从而代替纸质图纸。最初的产品造型设计方案主要采用矩形轮廓，之后又考虑改为圆形，这是为了减少产品损坏的几率。而最终的设计融合了这两种设计理念——整体为矩形，并带有柔和的边缘和倒角。同时，对于产品操作界面的多功能化也进行了深入推敲。

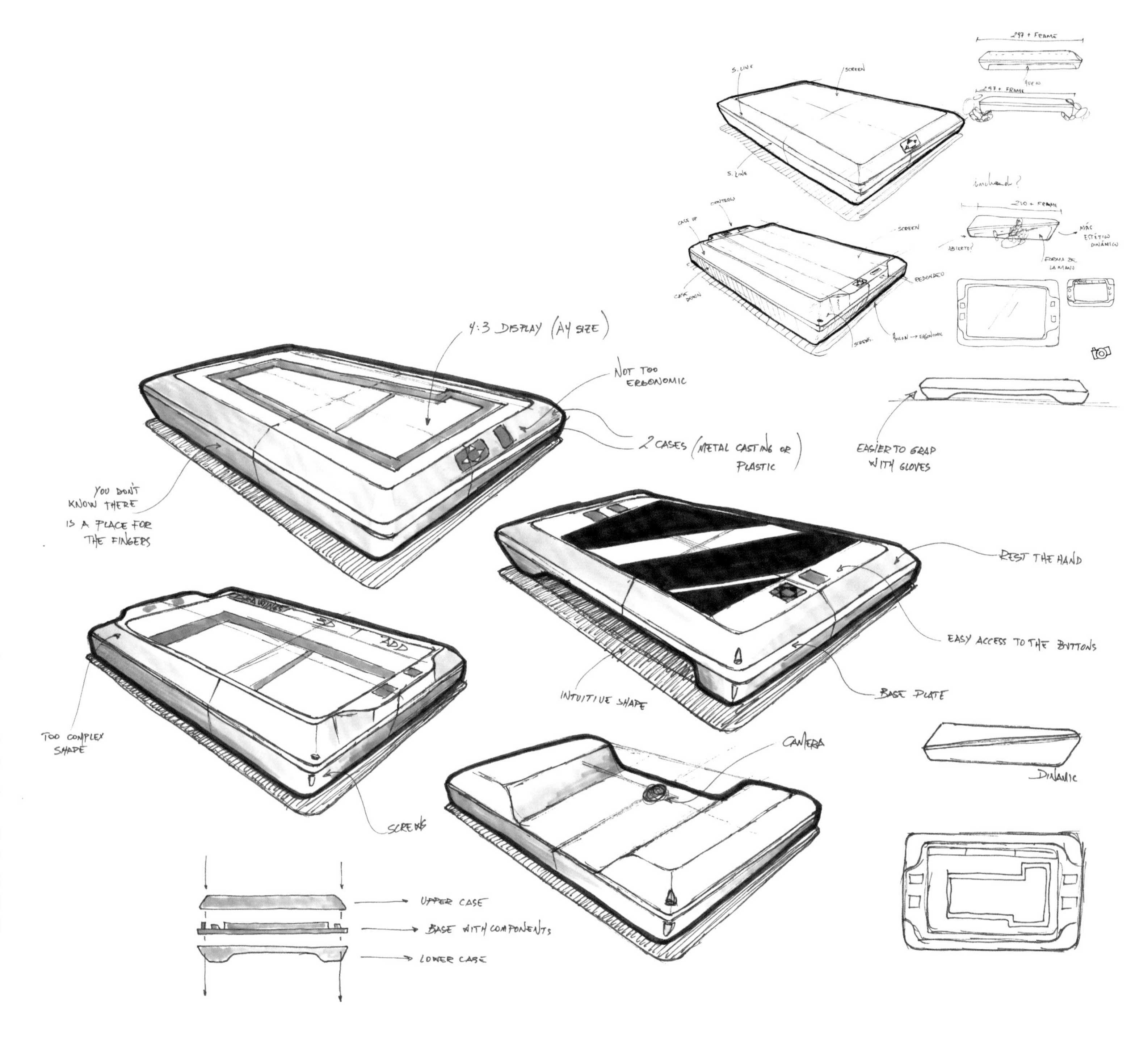

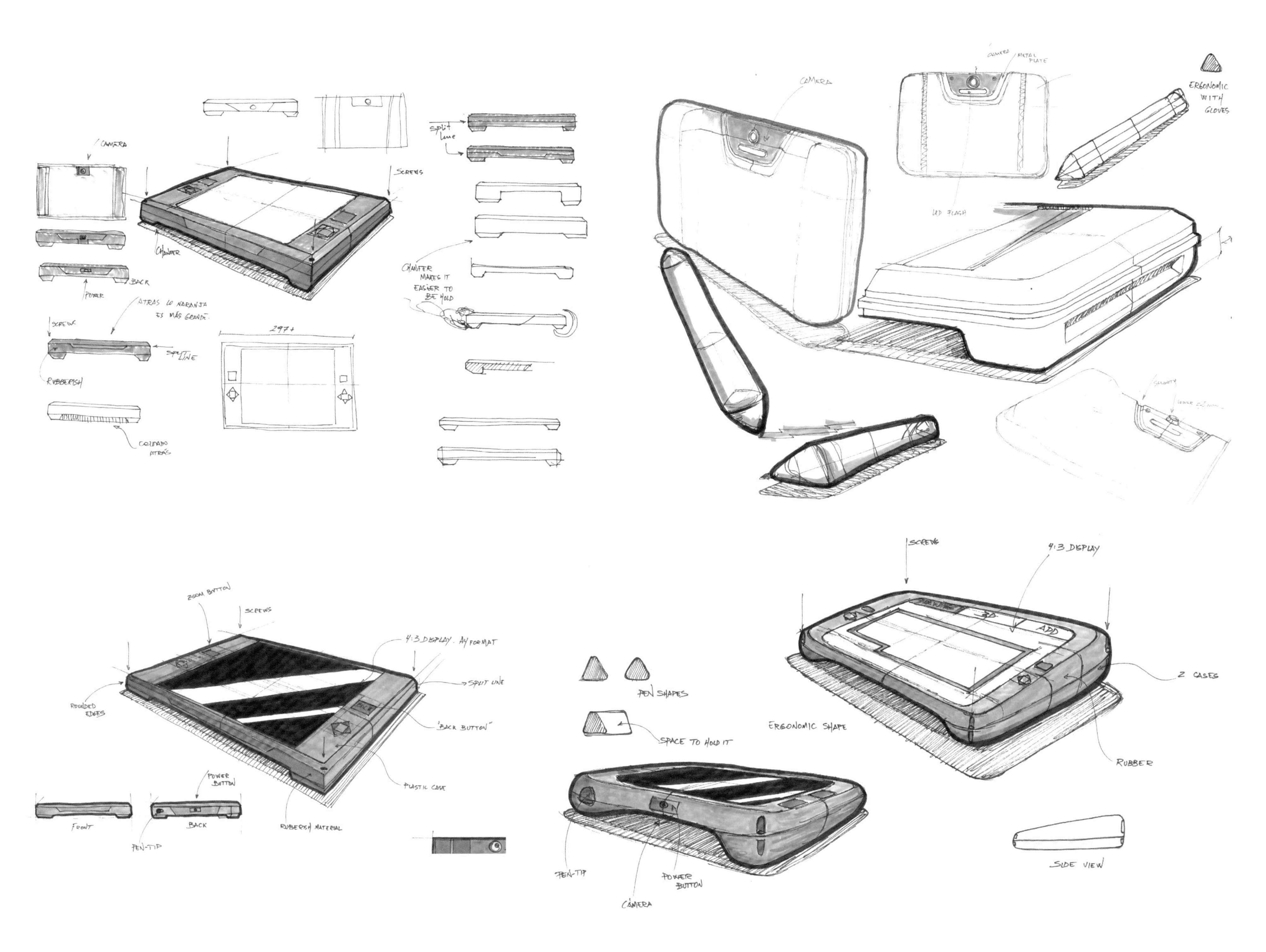
CAMERA
SCREWS
CHAMFER
BACK
POWER
ATRAS LO NARANJA ES MÁS GRANDE
SCREWS
SPLIT LINE
RUBBERISH
CORTADO ATRÁS
297+
SPLIT LINE
CHAMFER MAKES IT EASIER TO BE HOLD
CAMERA
METAL PLATE
ERGONOMIC WITH GLOVES
LED FLASH
ZOOM BUTTON
SCREWS
4:3 DISPLAY. A4 FORMAT
SPLIT LINE
ROUNDED EDGES
"BACK BUTTON"
PLASTIC CASE
POWER BUTTON
FRONT
BACK
PEN-TIP
RUBBERISH MATERIAL
SCREWS
4:3 DISPLAY
2 CASES
PEN SHAPES
SPACE TO HOLD IT
ERGONOMIC SHAPE
RUBBER
PEN-TIP
POWER BUTTON
CAMERA
SIDE VIEW

Ripple - Nettop Mini PC

Ripple
迷你台式电脑

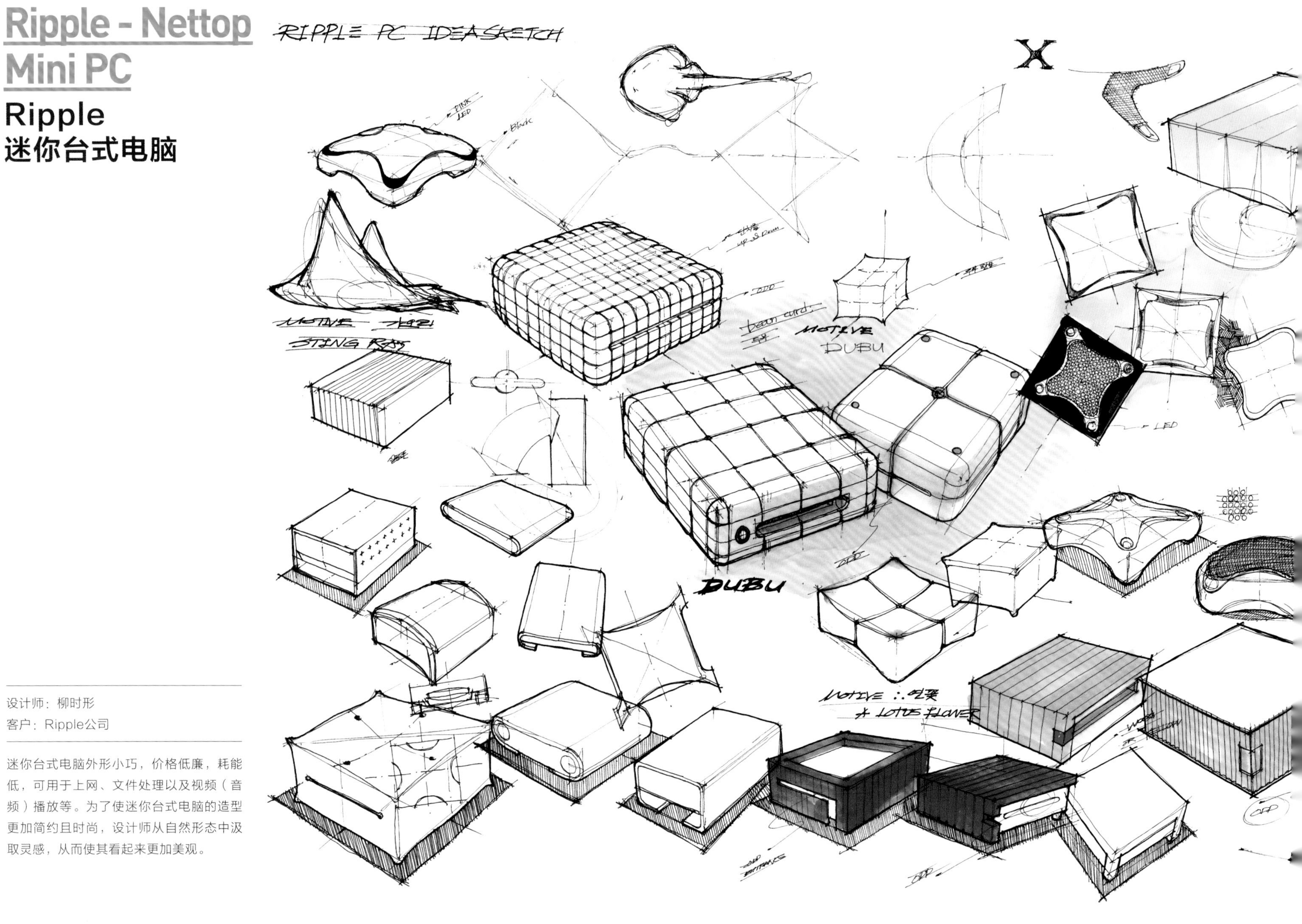

设计师：柳时彤

客户：Ripple公司

迷你台式电脑外形小巧，价格低廉，耗能低，可用于上网、文件处理以及视频（音频）播放等。为了使迷你台式电脑的造型更加简约且时尚，设计师从自然形态中汲取灵感，从而使其看起来更加美观。

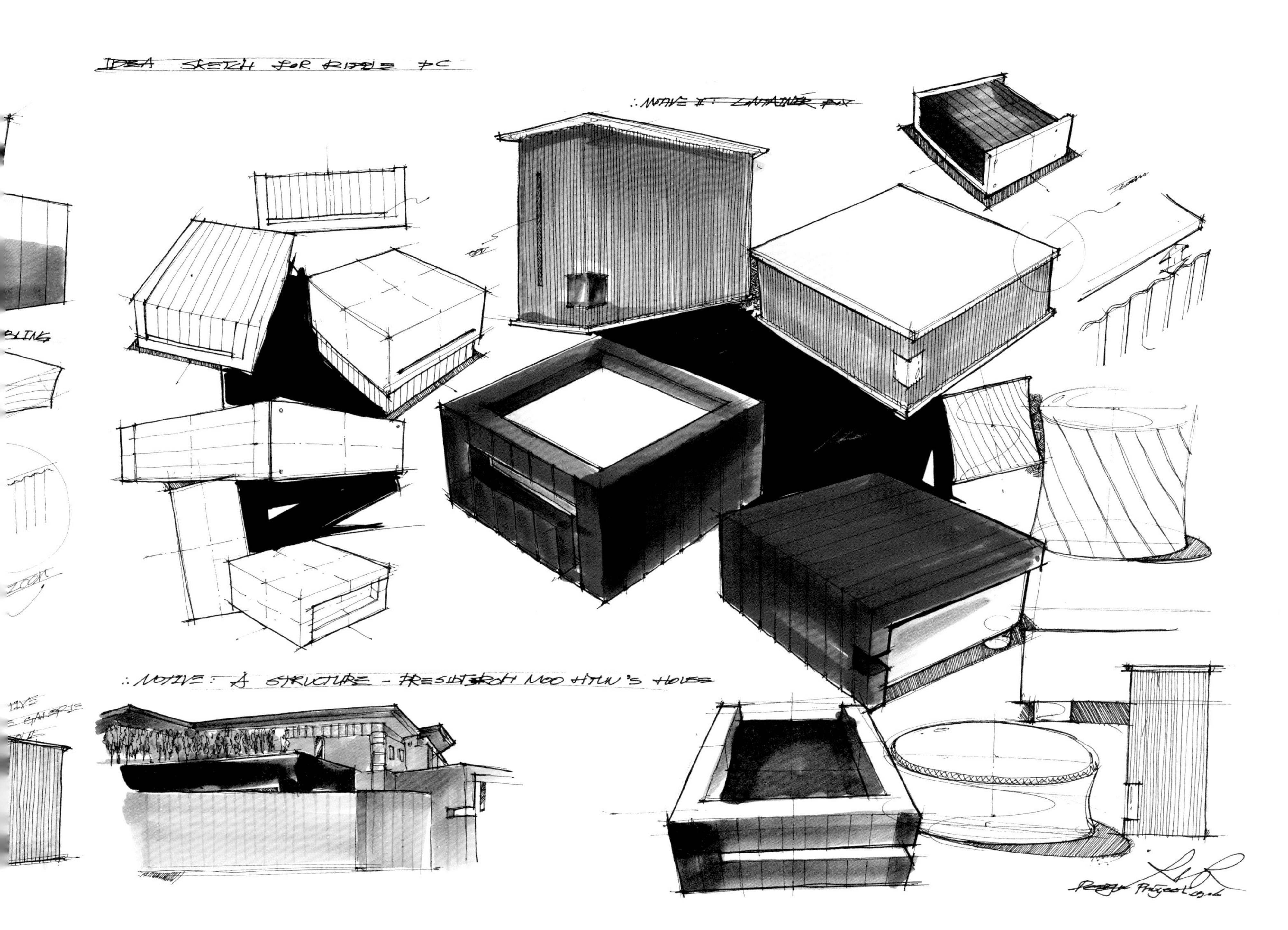
IDEA SKETCH FOR RIPPLE PC

IDEA SKETCH FOR RIPPLE PC

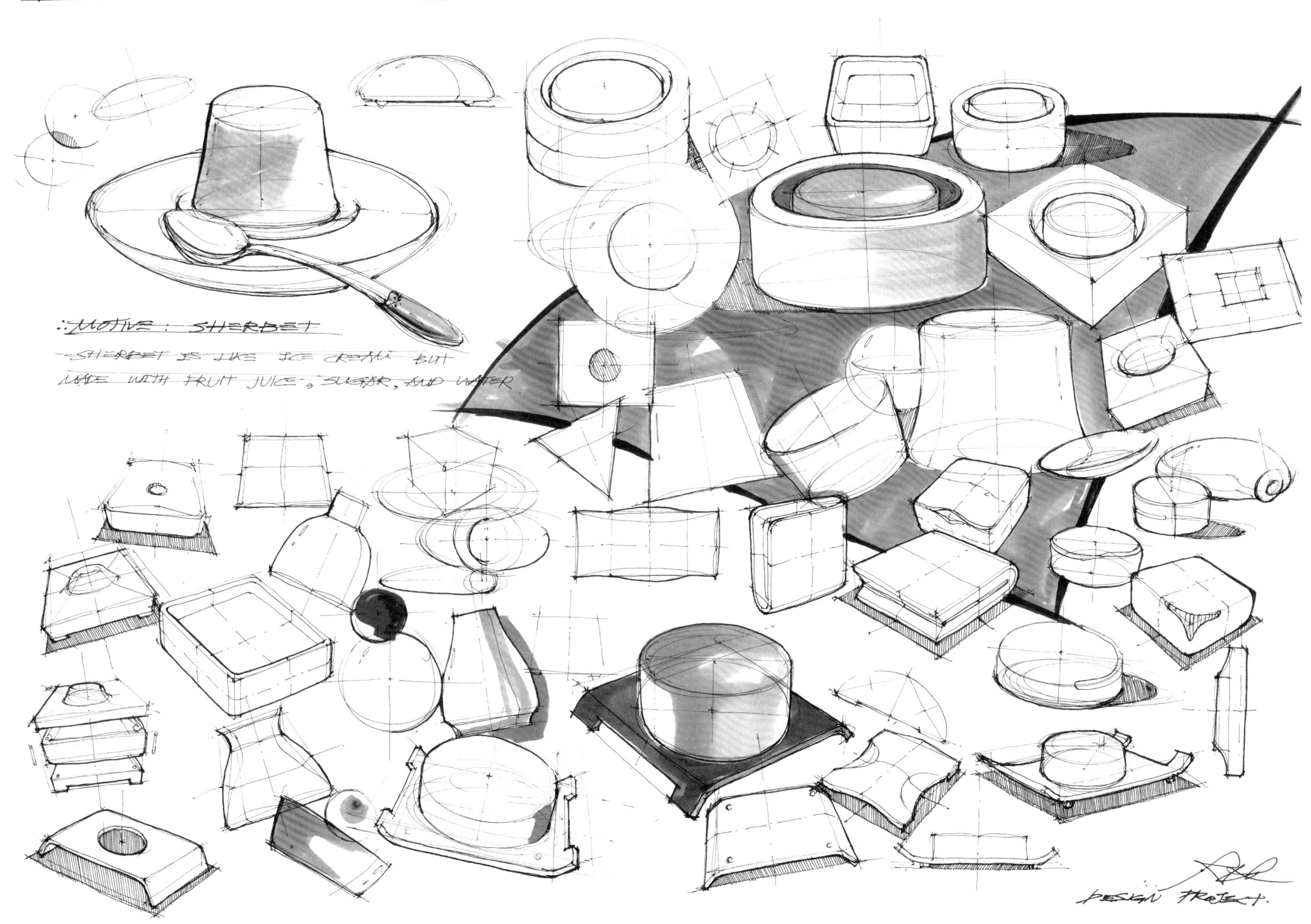
::MOTIVE : SHERBET
SHERBET IS LIKE ICE CREAM BUT
MADE WITH FRUIT JUICE, SUGAR, AND WATER.
DESIGN PROJECT.

IDEA SKETCH FOR RIPPLE PC
ANIMAL CHARACTER
SQUIRREL
COW
SNAKE
CROCODILE
HIPPOPOTAMUS
SIDE VIEW
FRONT VIEW
팬더곰의 얼굴모양을 형상화한다
PC의 받침부분을 팬더의 4다리로
표현
MOTIVE: PANDA
A PANDA IS A LARGE ANIMAL RATHER LIKE
A BEAR, WHICH HAS BLACK AND WHITE FUR
AND LIVES IN THE BAMBOO FOREST OF CHINA
DESIGN PROJECT
SIDE VIEW
ODD

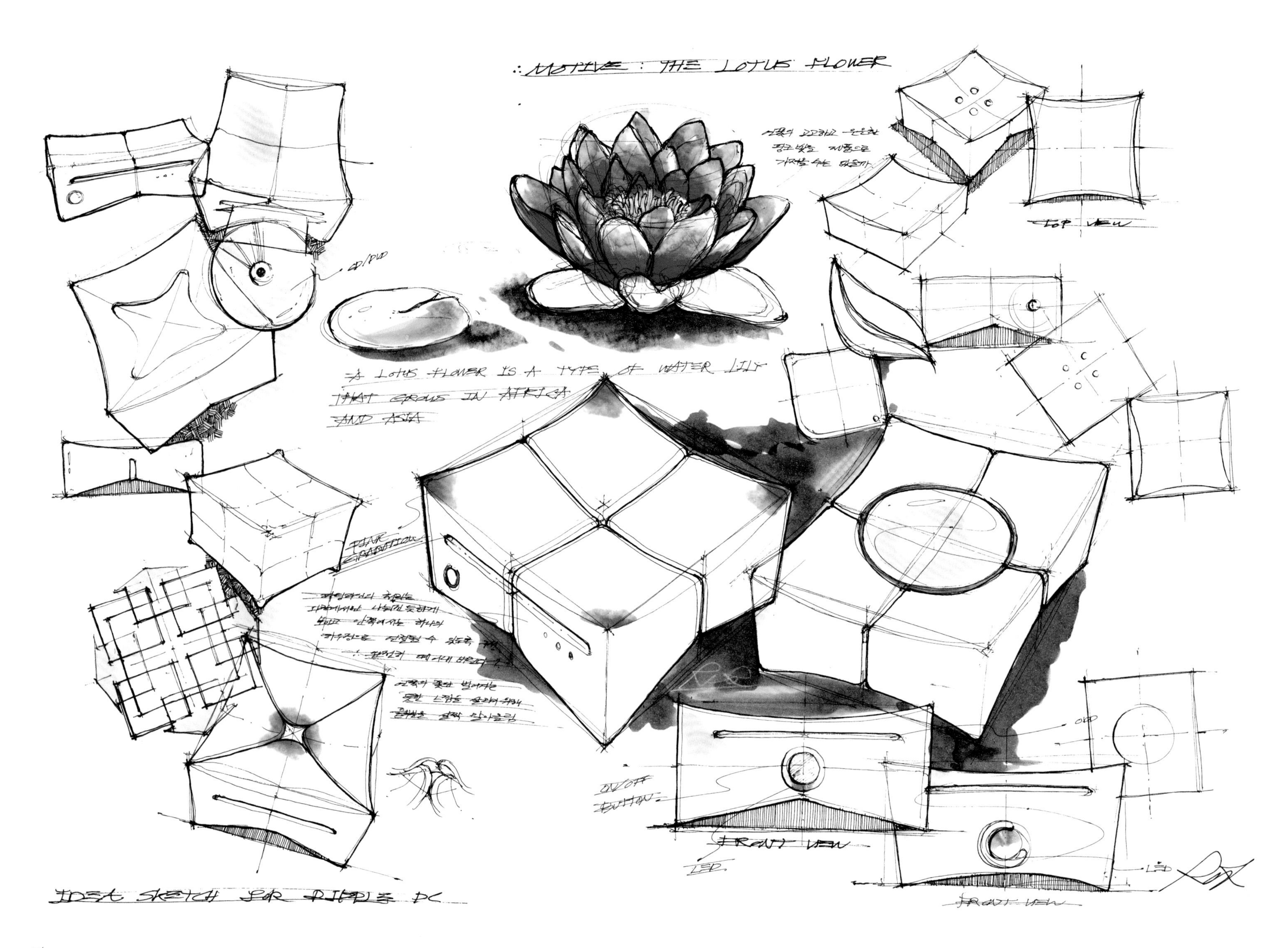

MOTIVE : THE LOTUS FLOWER
A LOTUS FLOWER IS A TYPE OF WATER LILY
THAT GROWS IN AFRICA
AND ASIA
TOP VIEW
CD/DVD
ON/OFF
BUTTON
FRONT VIEW
LED
FRONT VIEW

IDEA SKETCH FOR RIPPLE P
VOLCANO
FRONT VIEW
TOP VIEW
:: MOTIVE : VOLCANO
A VOLCANO IS A MOUNTAIN
FROM WHICH HOT MELTED
ROCK, GAS, STEAM, AND ASH
FROM INSIDE THE EARTH
SOMETIMES
BURST.

Automatic Hand Sanitizer
自动洗手液器

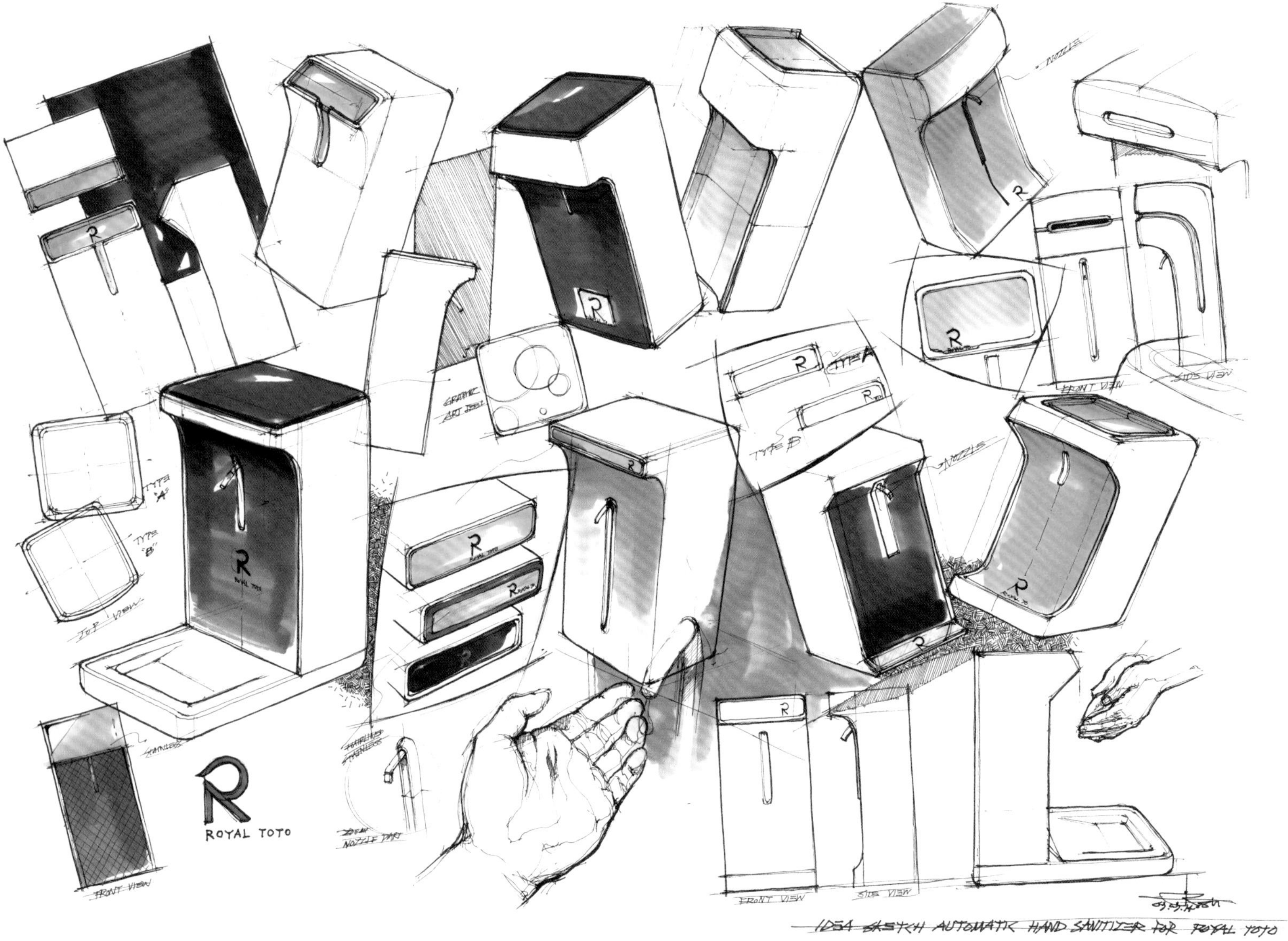

设计师：柳时彤
客户：TS JAVA公司

在卫生间或公共场合，自动洗手液器更加卫生，因此越来越受欢迎。研究表明，在公共卫生间，50%的人希望在无接触的环境中洗手。因此，在卫生间中安装自动洗手液器能够带来很多好处，例如可减少细菌交叉感染的风险、提升员工和顾客的满意度等。该产品的造型设计趋于简洁，功能上体现出卫生易用，便于清洗。

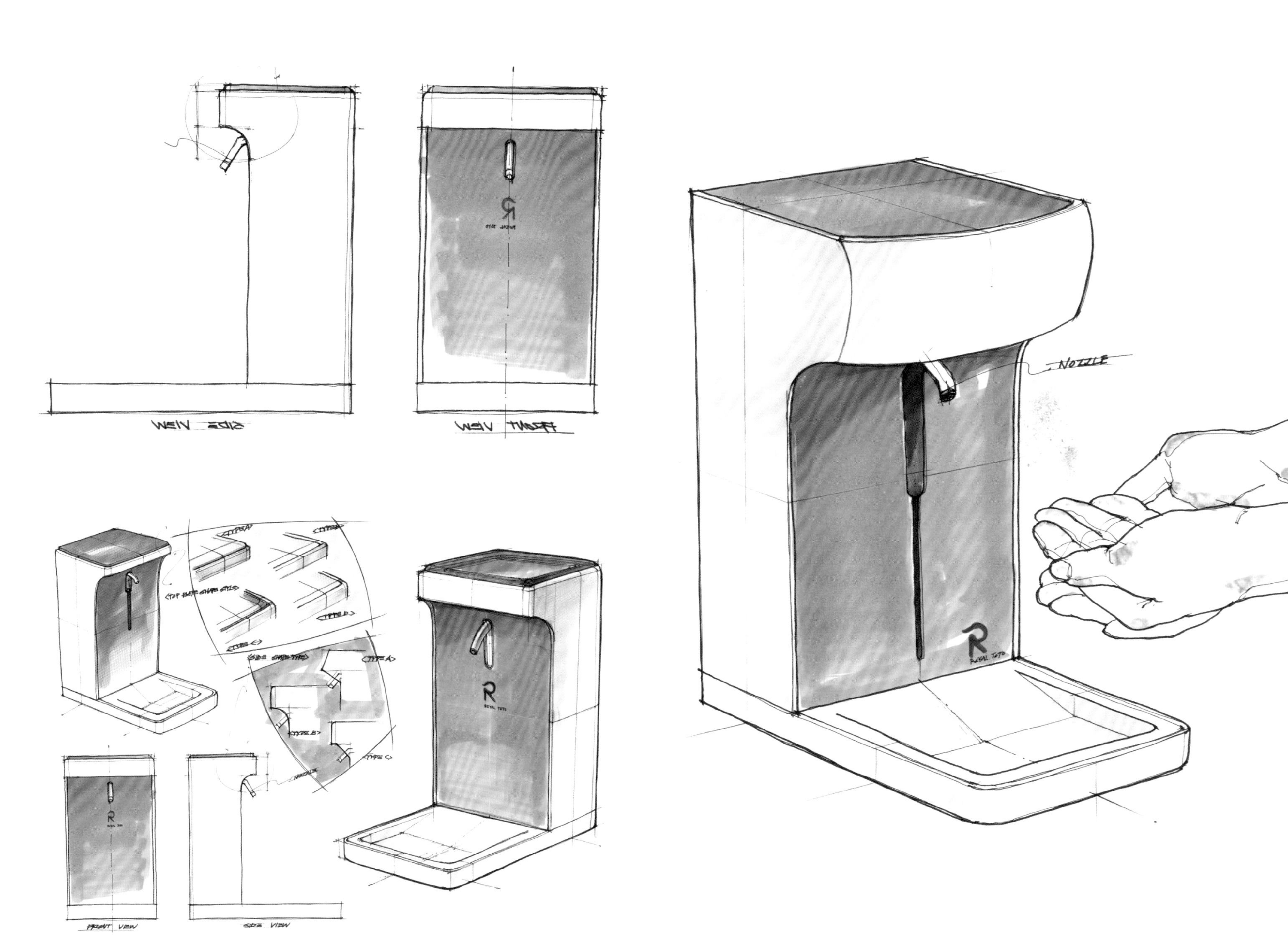
NOZZLE
ROYAL TOTO
FRONT VIEW
SIDE VIEW

Guitar
吉他

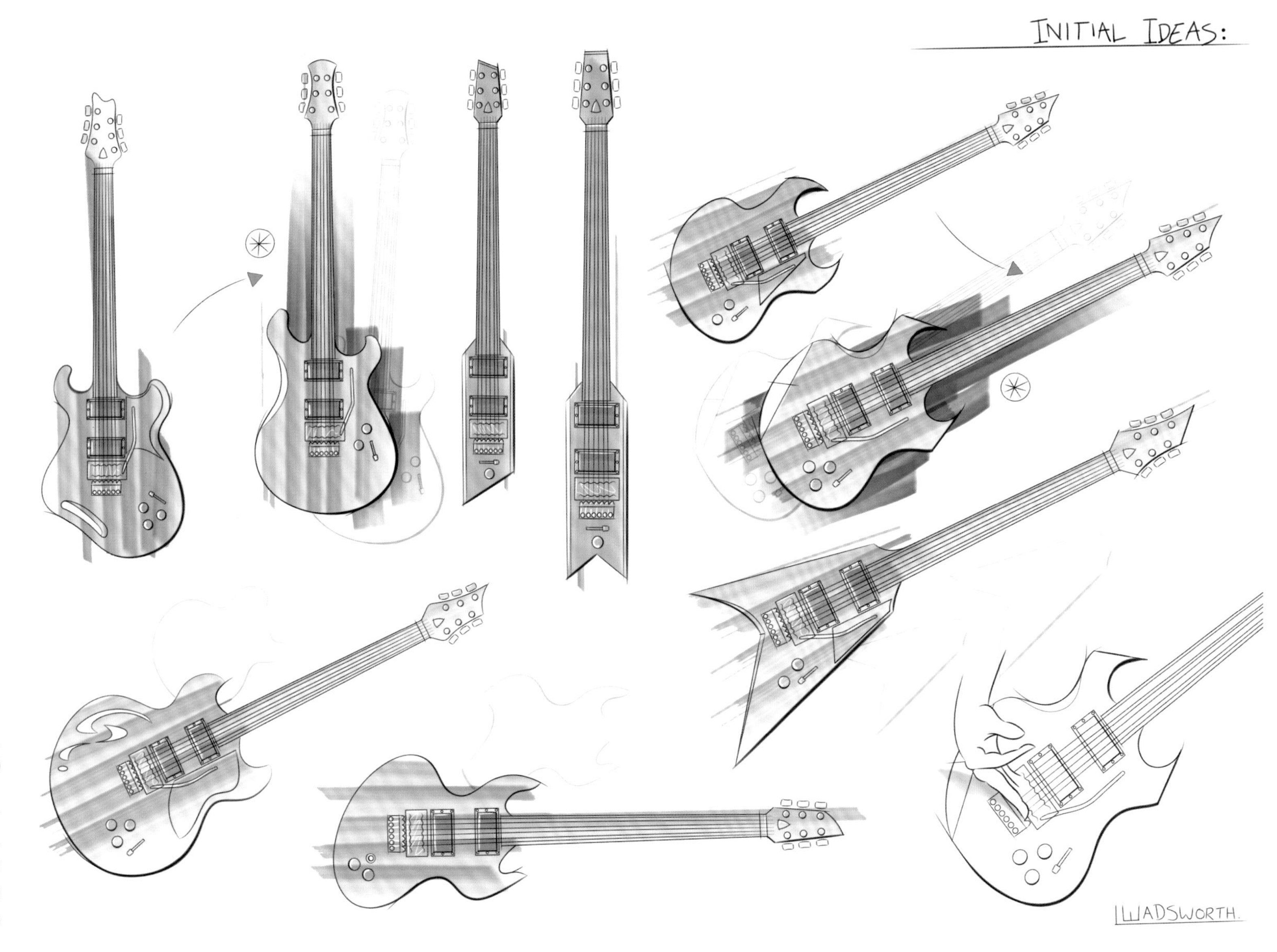

设计师：卢克·沃兹沃思（Luke Wadsworth）
客户：个人作品

弹吉他是设计师的个人爱好。设计师认为，最佳的设计草图是能够突出细节，注重展示产品颜色和材质。这些草图是使用计算机软件绘制完成的，这样能够轻松地改变线条粗细，从而创造出完美的曲线。

CONCEPT DEVELOPMENT.
FLOYD ROSE
LOCKING
TREMOLO.

ARTICLES FOR DAILY USE
生活家居

Body Engine Toy

Body Engine
儿童玩具车

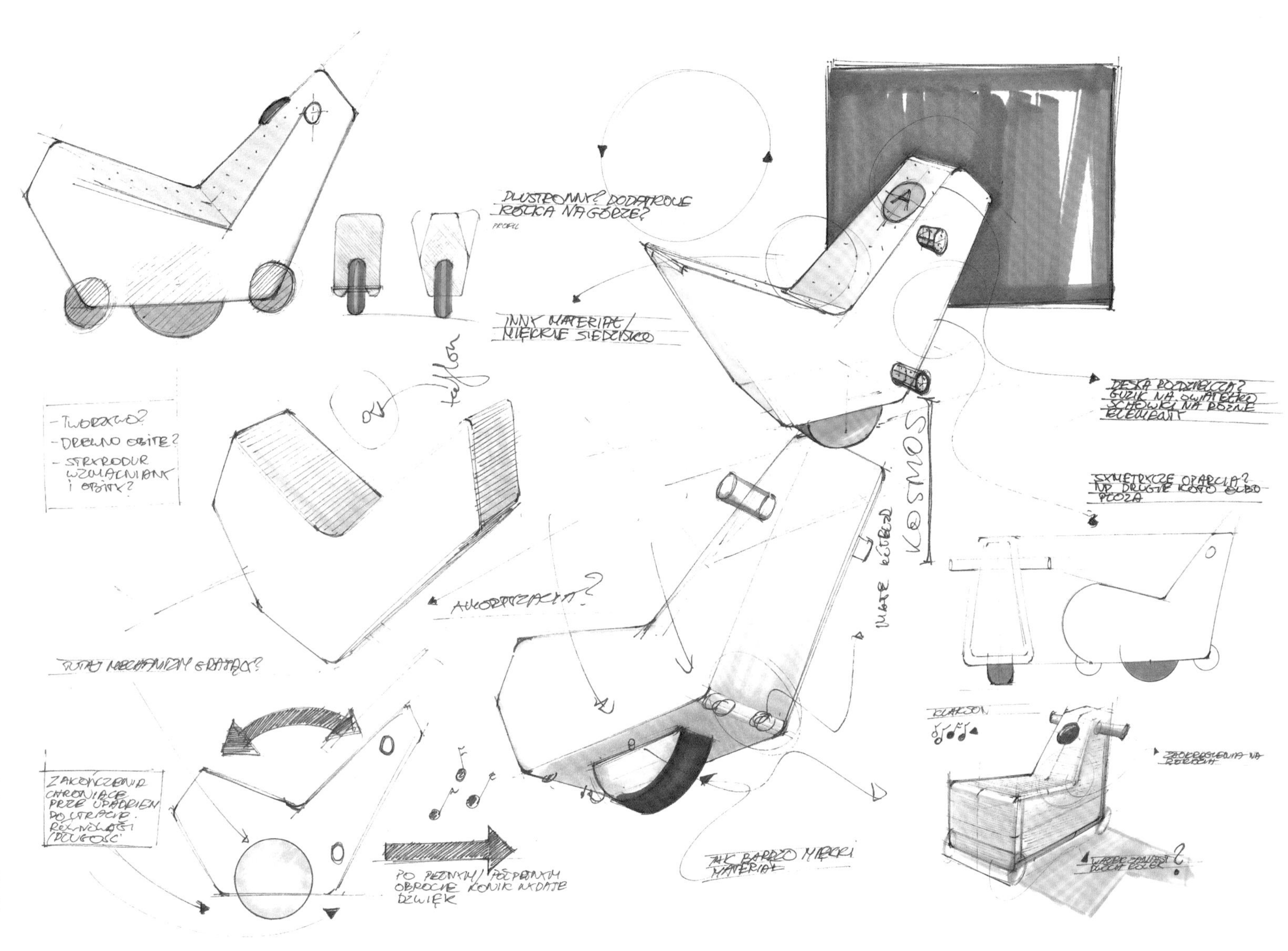

设计师：迈克尔·马尔凯维奇（Michal Markiewicz）
客户：学校项目

这是一款简单的儿童玩具车，供两至三岁儿童使用。其主要设计目标是凸显趣味性，锻炼孩子的平衡力和想象力。

最初的草图是对基础造型的研究。设计的目标是体现趣味与快乐，因此在设计的过程中享受手绘带来的快乐同样显得十分重要。在对人体工程学的相关数据进行分析之后，设计师开始选择最佳的产品造型。同时，完整地阐述产品的功能也很重要。图文结合的草图诠释了产品和用户之间的关系和比例。最终草图呈现了产品的材质细节与使用方式。

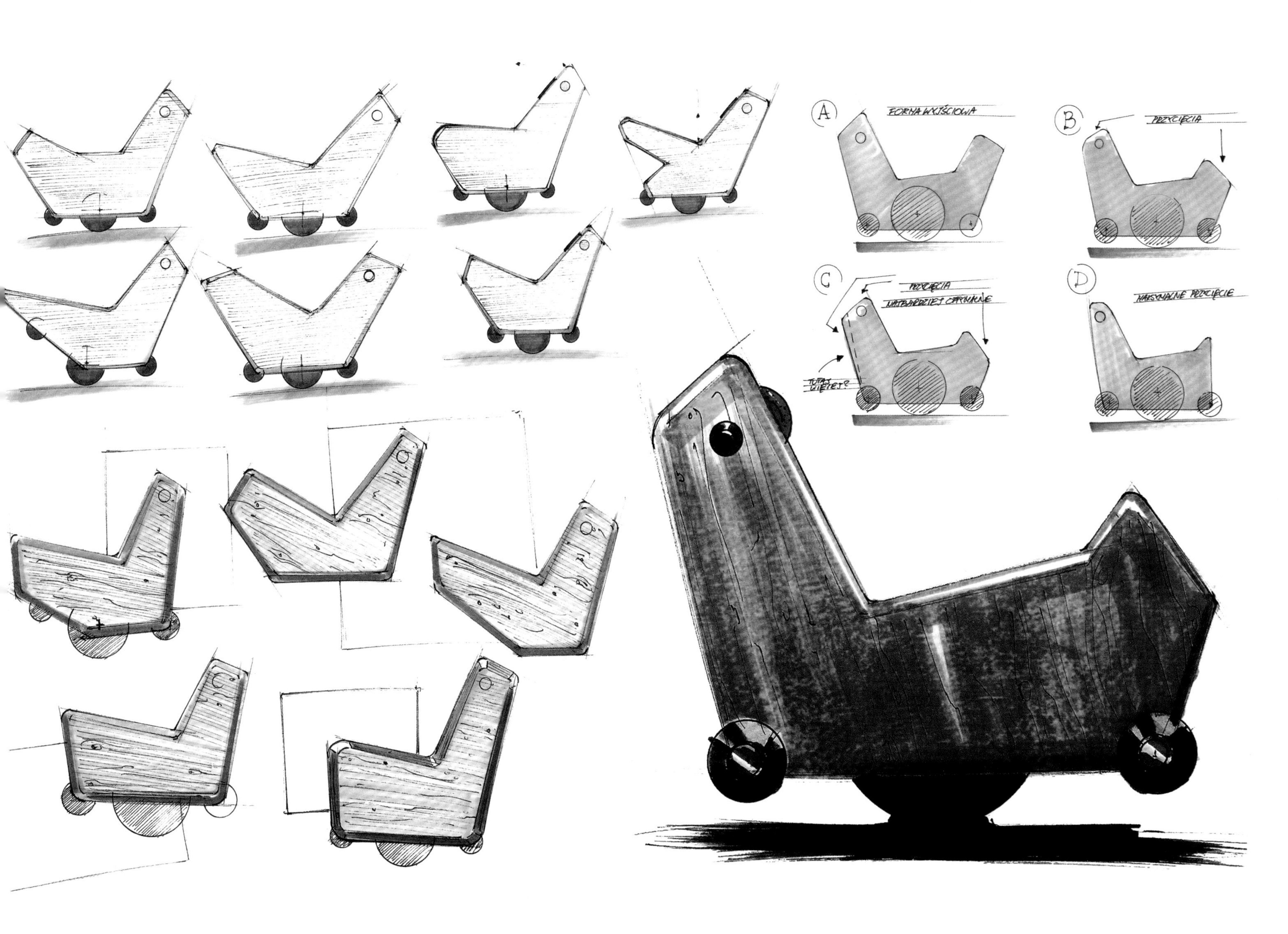
A
FORMA WYJŚCIOWA
B
PRZYCIĘCIA
C
PRZYCIĘCIA
TUTAJ WIĘCEJ?
D
MAKSYMALNE PRZYCIĘCIE

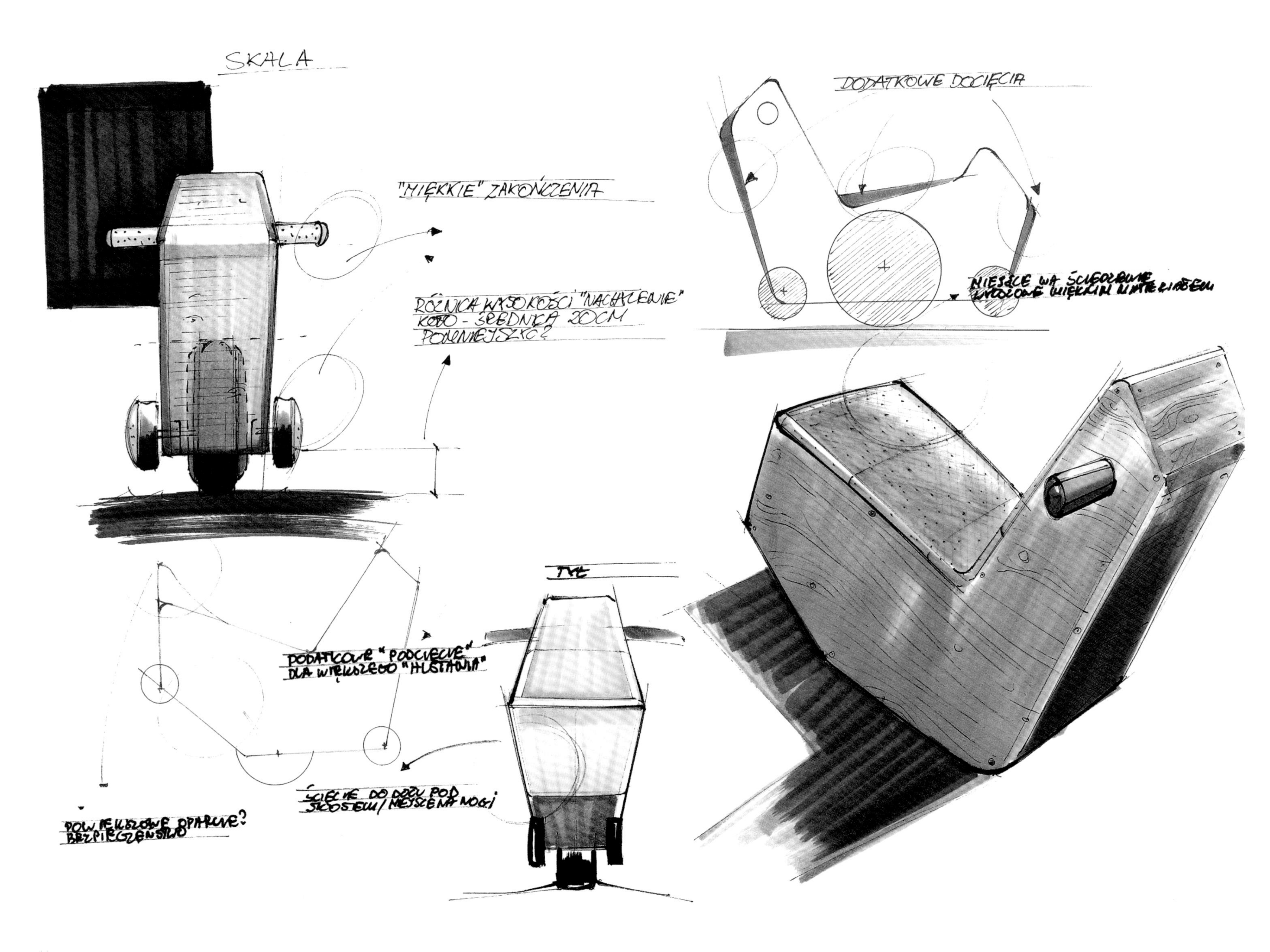

SKALA
"MIĘKKIE" ZAKOŃCZENIA
DODATKOWE DOCIĘCIA

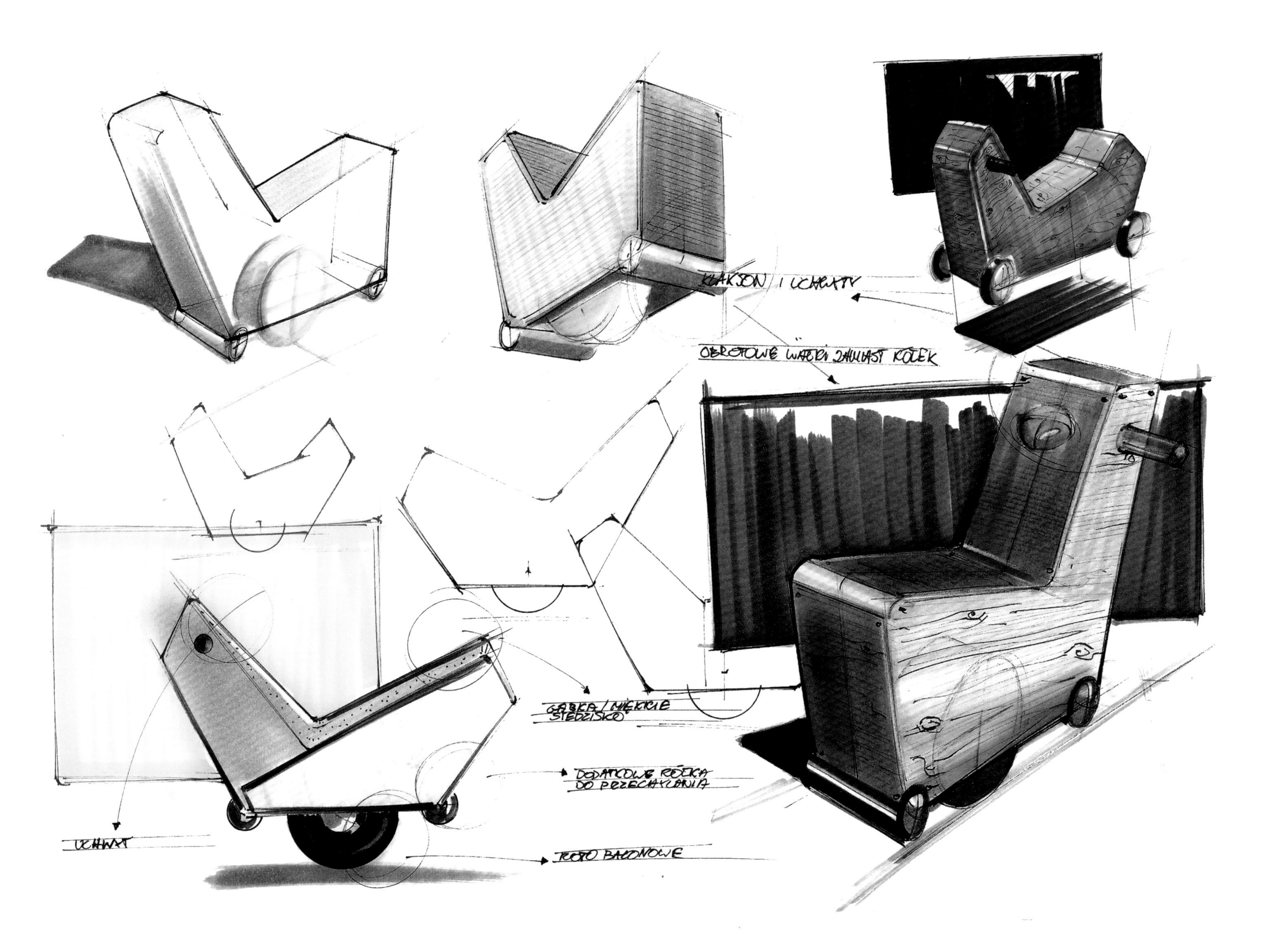
KLAKSON I UCHWYTY
UCHWYT

Tea Kettle
茶壶

设计师：迈克·塞拉芬（Mike Serafin）、托尼·罗斯（Tony Ruth）、钟丹（Dan Chung）、斯蒂夫·阿布瓦（Steve Abueva）
客户：World Kitchen公司

这款茶壶集成了加热的功能，产品主体为一个耐高温的蓝白色玻璃瓶，其表面设计借鉴了中国传统图案装饰。选择热电铬（热激活）的材料设计，可以起到提示水温的作用，当水温足够高时便可启动。设计草图围绕产品的使用方式展开，体现了产品简约且典雅的设计风格。

Clear Sky Perfume
Clear Sky
香水瓶

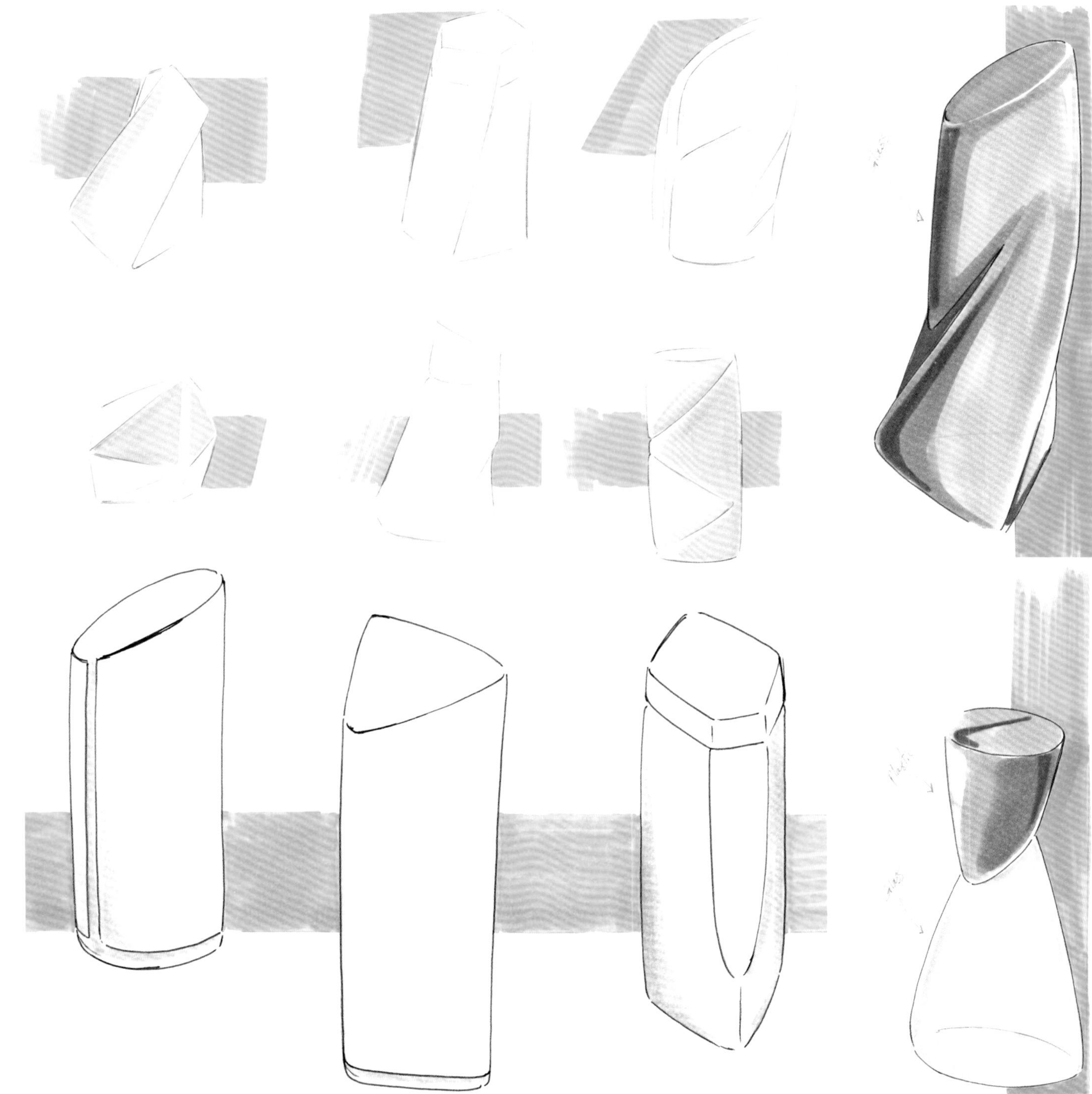

设计师：赛义夫 · A. 艾尔 · 卡米斯（Saif A. Al Khamissi）
客户：学校项目，温德斯海姆应用科学大学

设计师尝试将闻香水时的记忆和感觉转换成一种产品形态。设计的第一步从闻某一种香水开始，随后确定自己的感受。第二步将这些记忆和感觉转换成不同的草图和设计语言。最终的产品造型需要凸显纯净和简约的感觉。

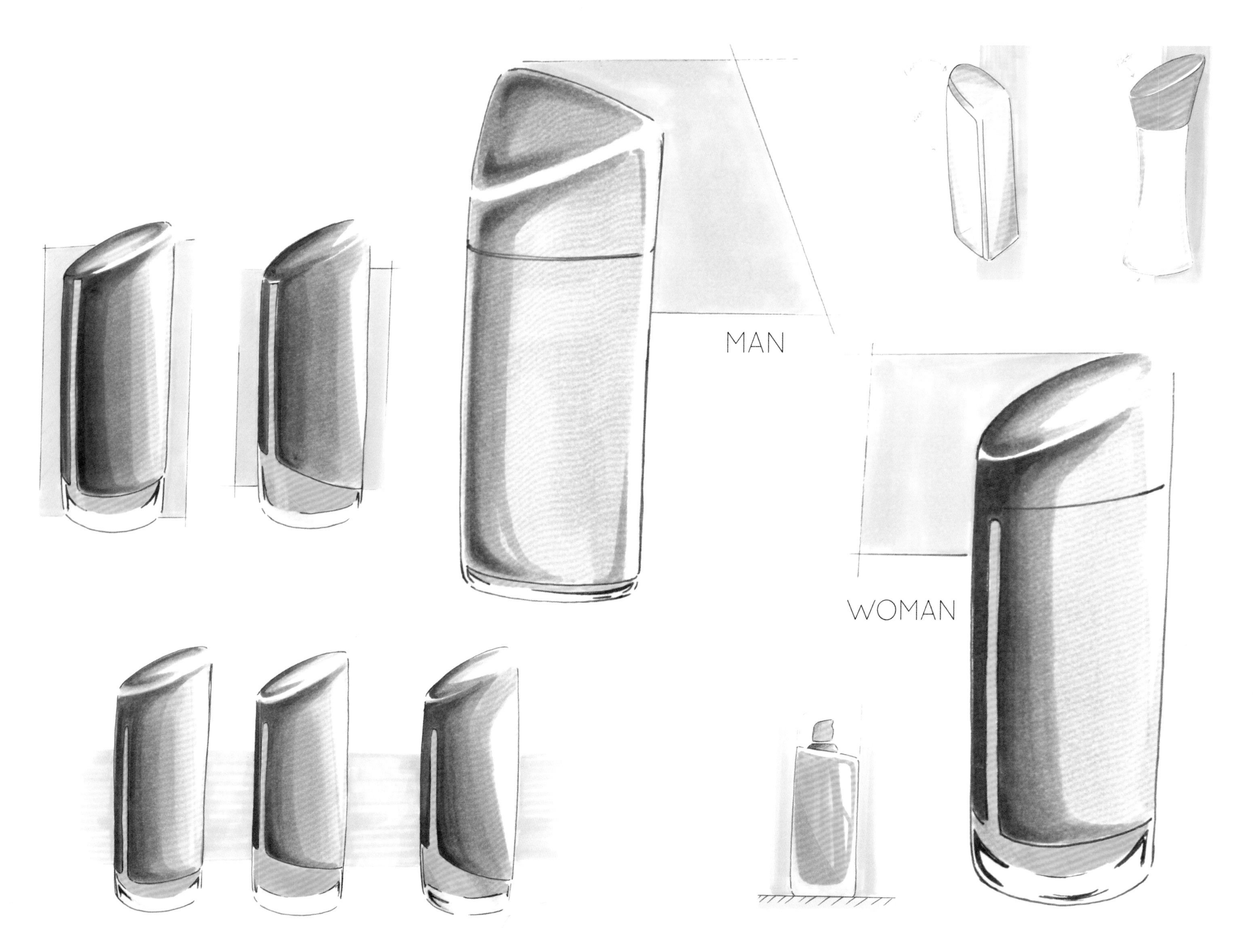
MAN
WOMAN

Beer with a Twist

扭动的啤酒罐

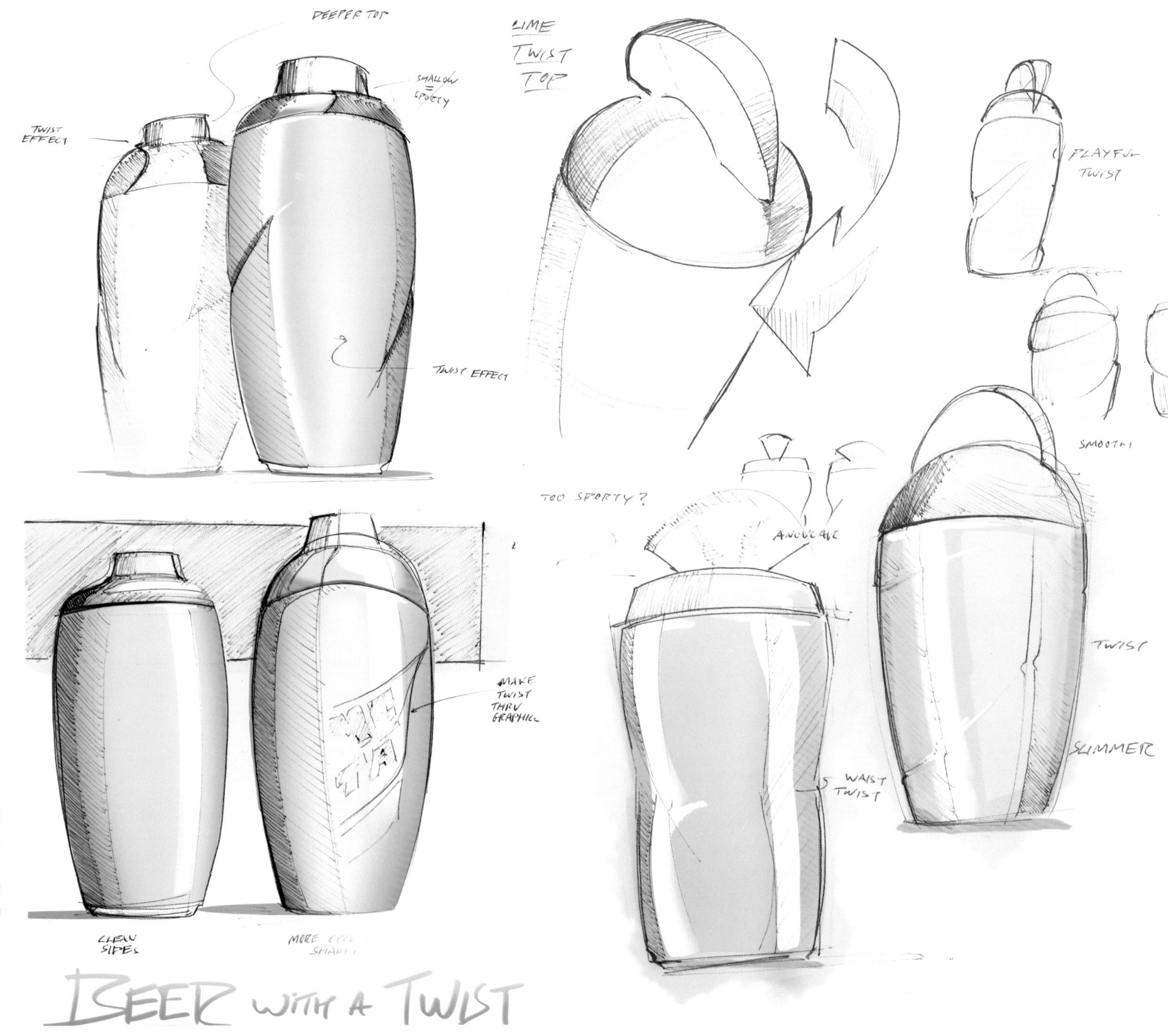

设计师：查利·兰斯康姆（Charlie Ranscombe）
客户：InnoCentive公司

这里展示了设计一款全新的啤酒罐的所有草图，旨在诠释“扭动的啤酒罐”的理念。设计师尝试将“Twist”（扭动）这个单词通过不同的啤酒罐造型表现出来。同时，设计师还借鉴了传统鸡尾酒瓶的设计，这着重体现在瓶身侧面以及瓶口处的设计上。

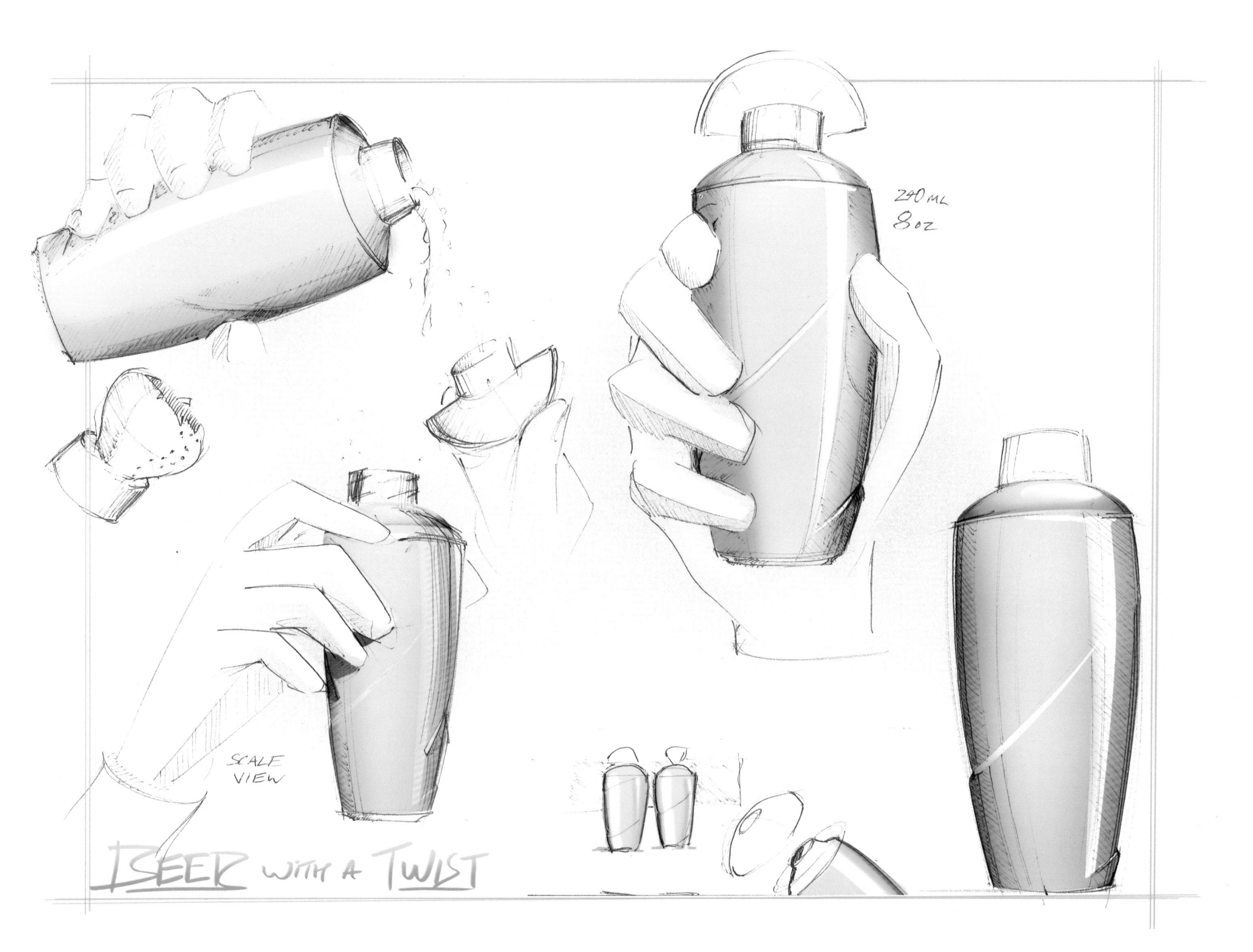
240 ML
8 oz
SCALE
VIEW
BEER WITH A TWIST

Kids Tumbler
儿童饮水器

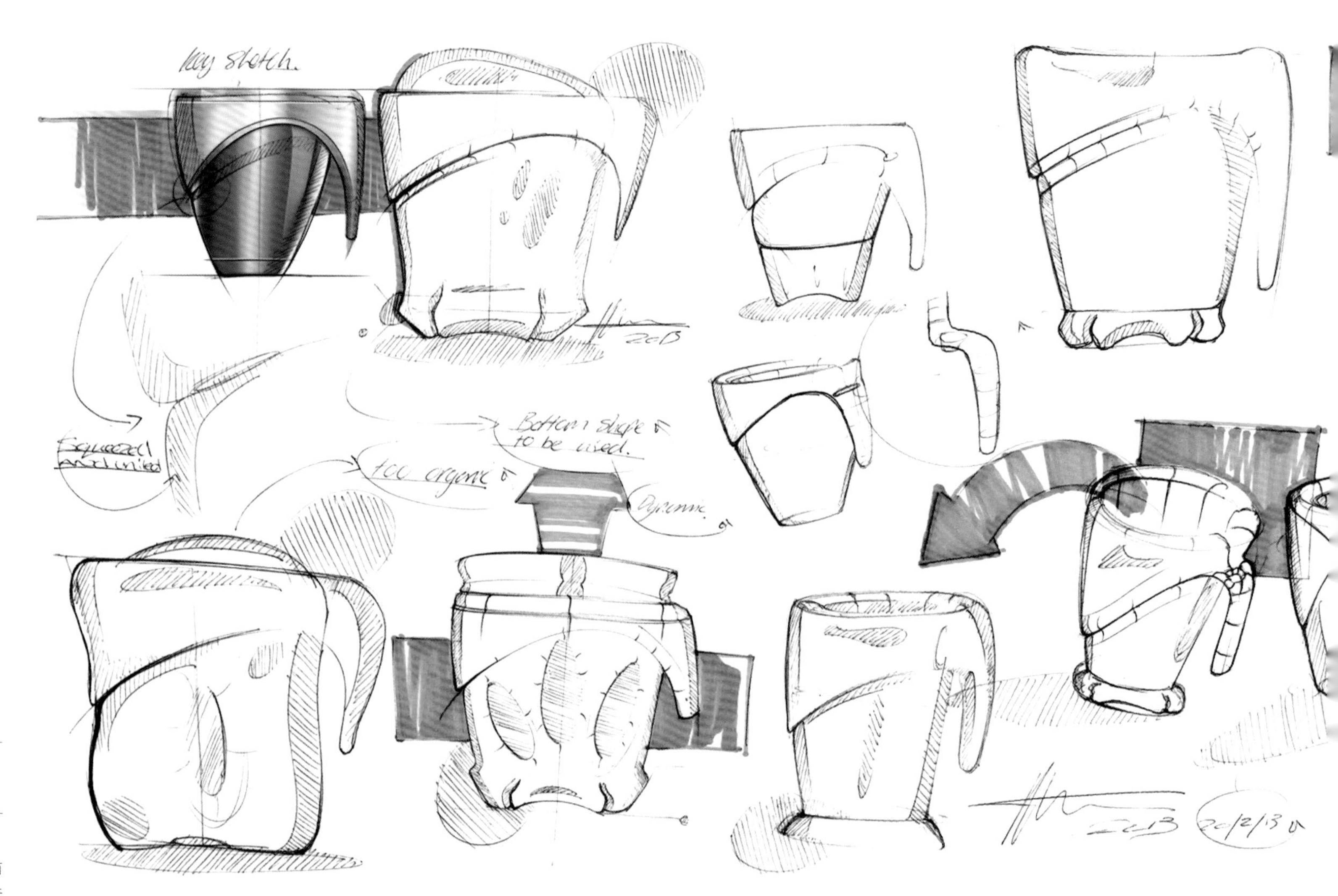

设计师：丹尼尔 · 法默
客户：CobaltNiche Design公司

该设计旨在打造不同形态的儿童饮水器，造型奇妙而带有趣味性。构思过程围绕有趣的线条和色彩展开，设计师在草图中运用橙黄色营造出了引人注目的视觉效果。

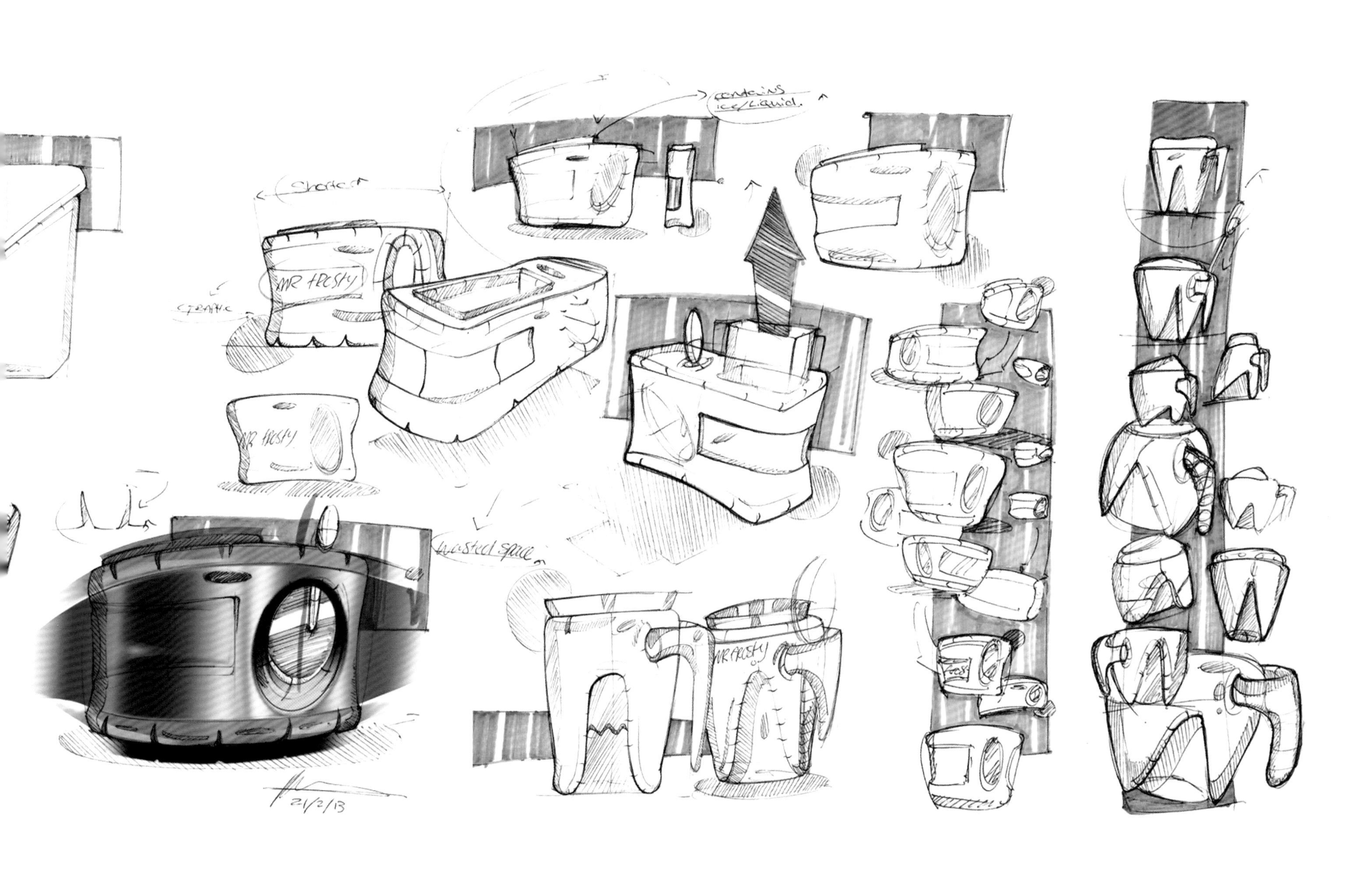
contains ice/liquid.
Shorter
MR FROSTY
GRAPHIC
MR FROSTY
wasted space
MR FROSTY
21/2/13

Fastdraw Water Bottle
便携水瓶

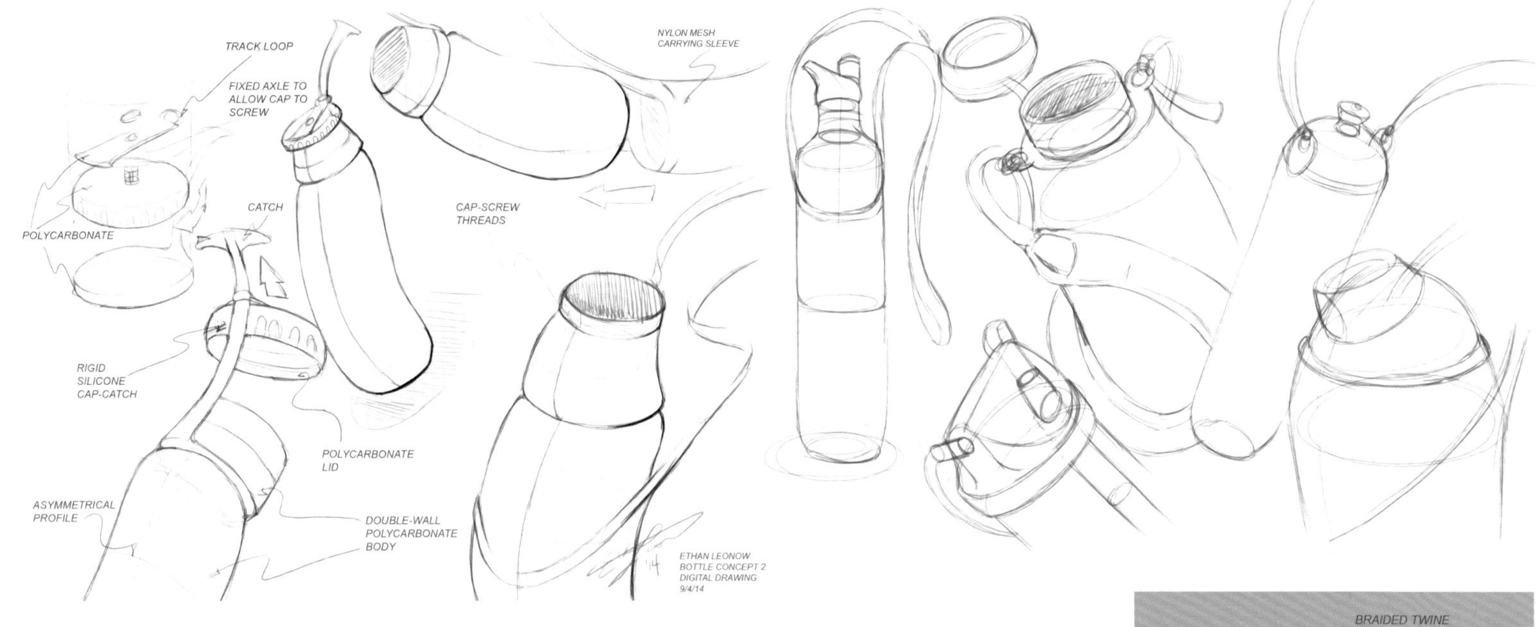

设计师：伊桑·利奥诺（Ethan Leonow）
客户：个人作品

突出便携性是这款水瓶的设计初衷，要求遵循传统的使用方式，同时便于携带和开启瓶盖。

这一设计的所有草图都是在Photoshop中描绘完成的。所有设计方案都是从基本形态圆柱体发展而来，然后在此基础上增添瓶盖等细节。随着设计的深入，许多细节也愈加完善，最终的设计草图还着重表现了水瓶的使用方式。

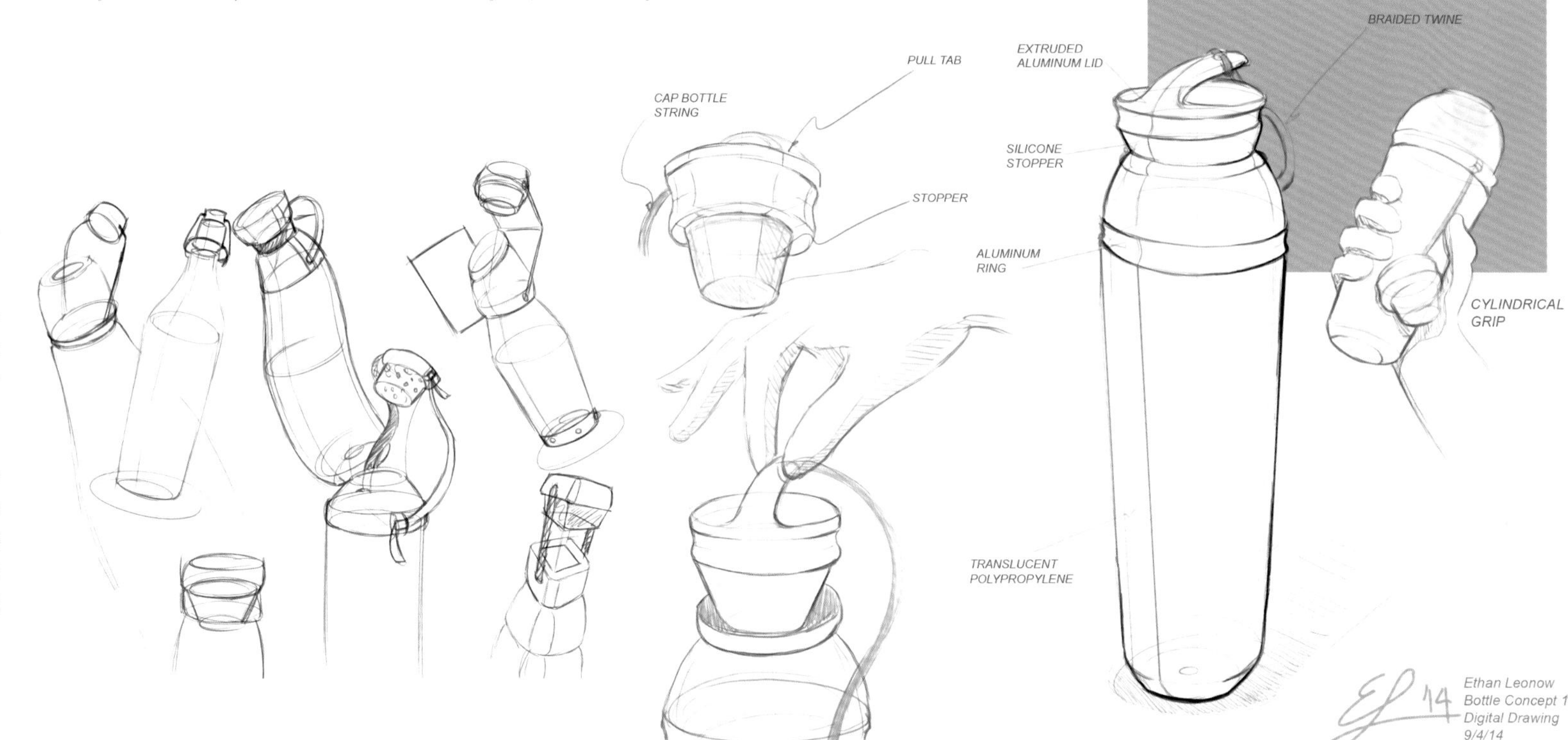

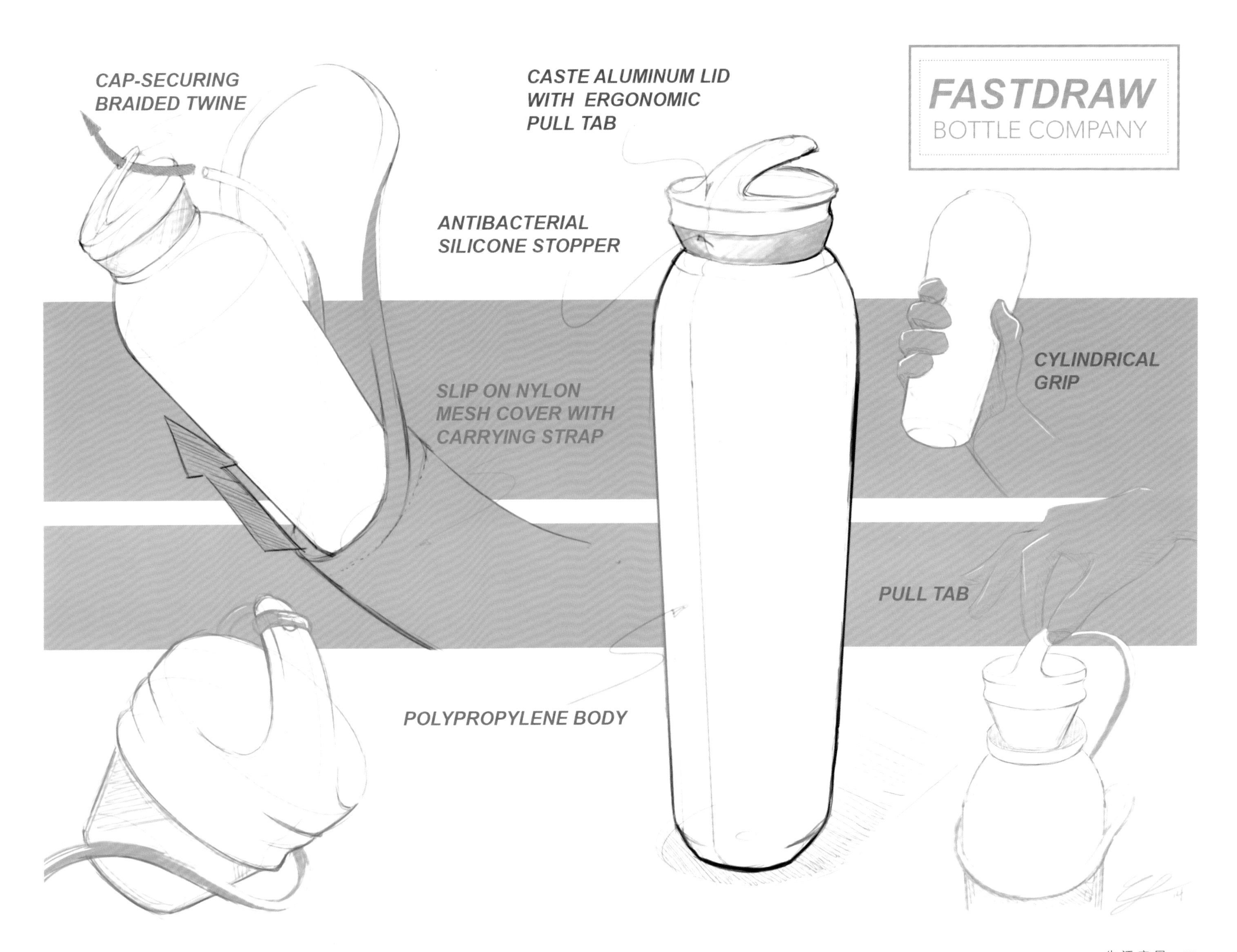
CAP-SECURING
BRAIDED TWINE
CASTE ALUMINUM LID
WITH ERGONOMIC
PULL TAB
FASTDRAW
BOTTLE COMPANY
ANTIBACTERIAL
SILICONE STOPPER
CYLINDRICAL
GRIP
SLIP ON NYLON
MESH COVER WITH
CARRYING STRAP
PULL TAB
POLYPROPYLENE BODY

Bedroom Furniture
卧室家具

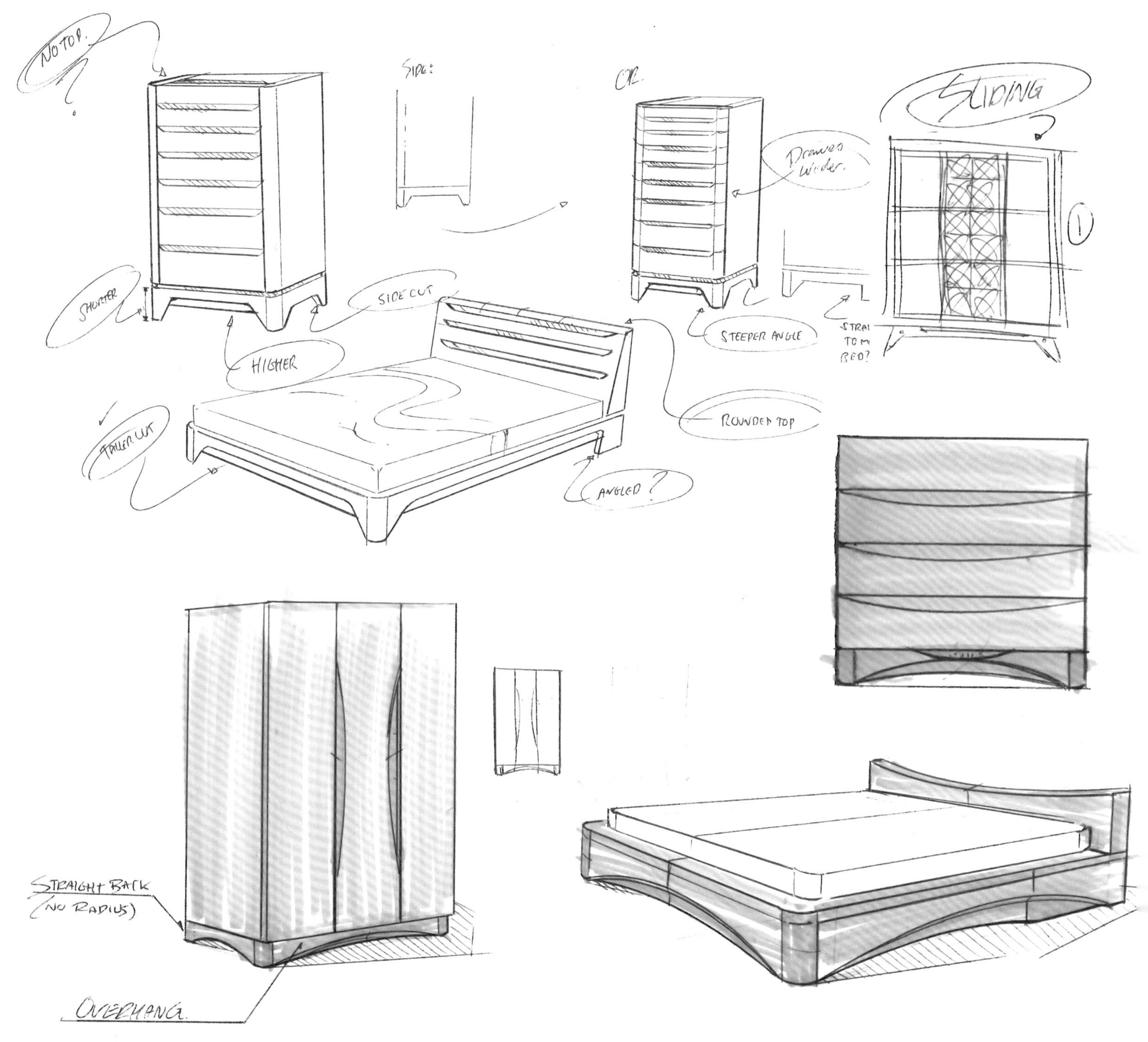

设计师：卢克 · 沃兹沃思

客户：个人作品

设计师决定采用现代日式风格进行一组卧室家具设计。设计师首先使用黑色圆珠笔勾勒出产品草图，之后用马克笔来表现家具的颜色及材质。最后，可以利用添加阴影的方法，提升画面的纵深感以及突出某一个设计方案。

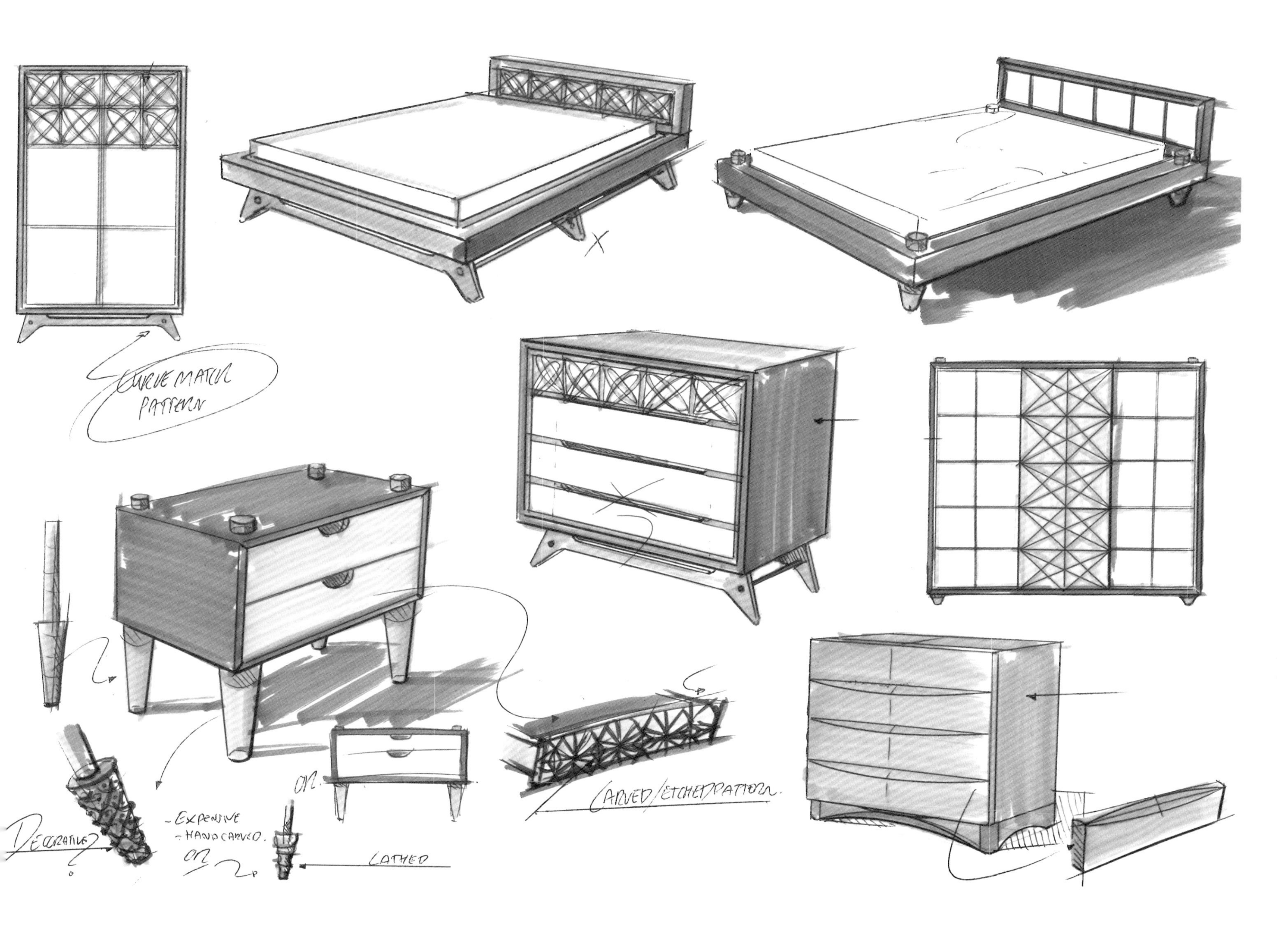
CURVE MATCH PATTERN
DECORATIVE?
- EXPENSIVE
- HAND CARVED.
OR
OR.
LATHED
CARVED/ETCHED PATTERN.

Oben Muebles Collection

Oben Muebles
系列家具

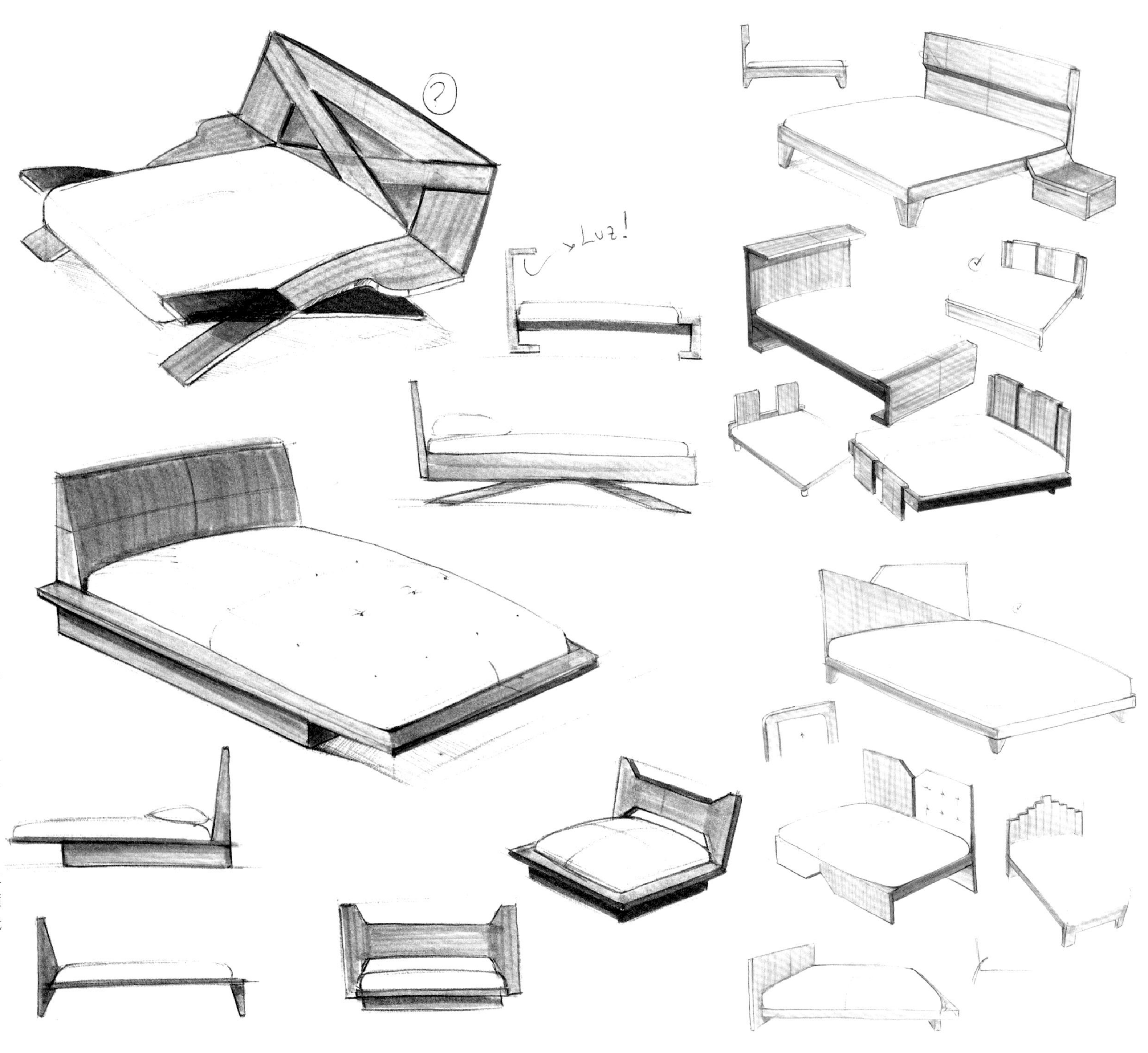

设计师：毛里西奥·萨宁·M.（Mauricio Sanin M.）
客户：Oben Muebles公司

这是为Oben Muebles家具品牌打造的一系列概念设计，其中包括床和床头柜等，旨在呈现卧室家具的特点。

在哥伦比亚，木材是备受钟爱的材料之一，因此设计师决定打造能够凸显木材质感的家具设计，并在其中融合传统与现代的设计语言。

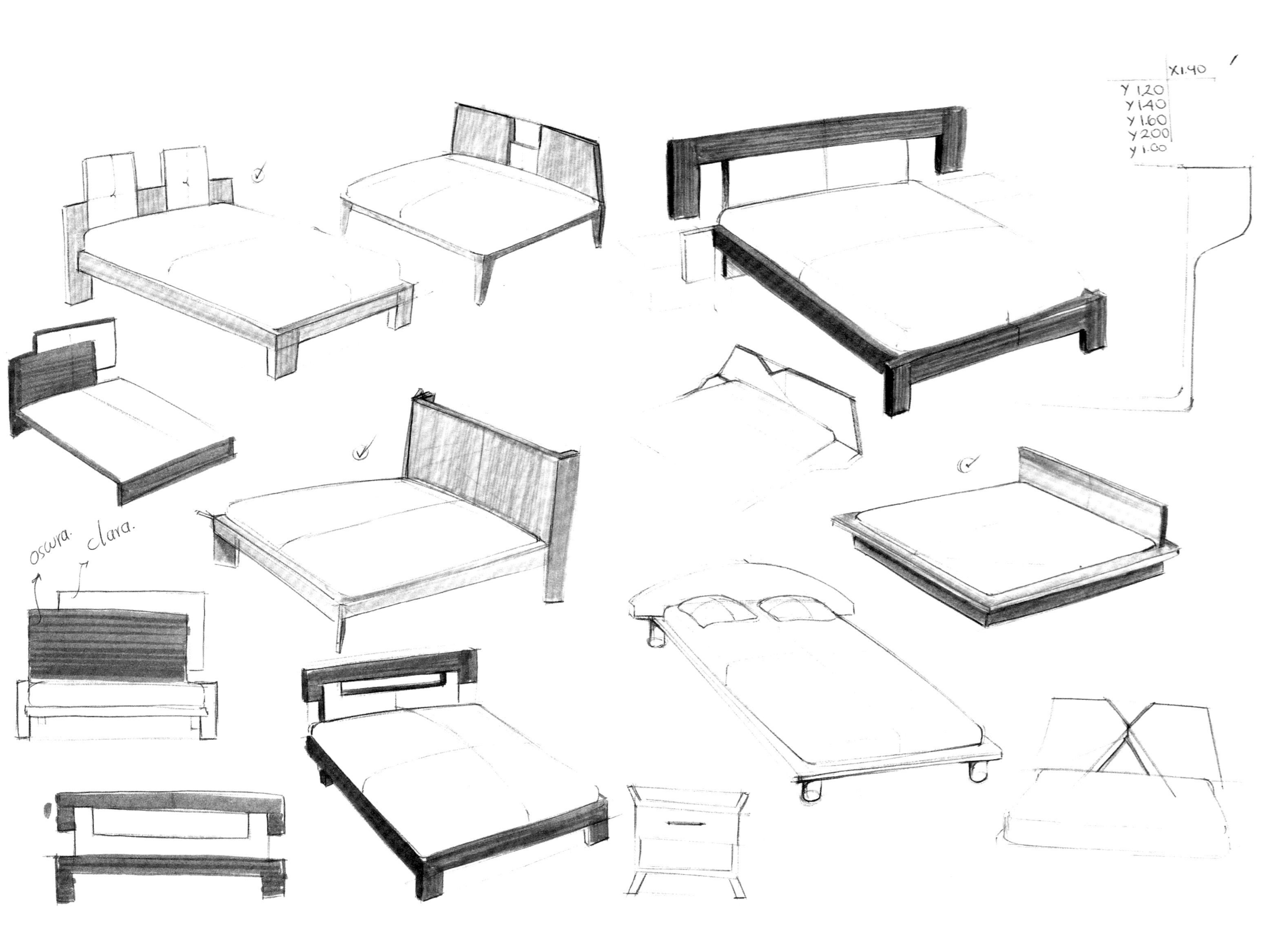
X1.40
Y 1.20
Y 1.40
Y 1.60
Y 2.00
Y 1.00
oscura.
clara.

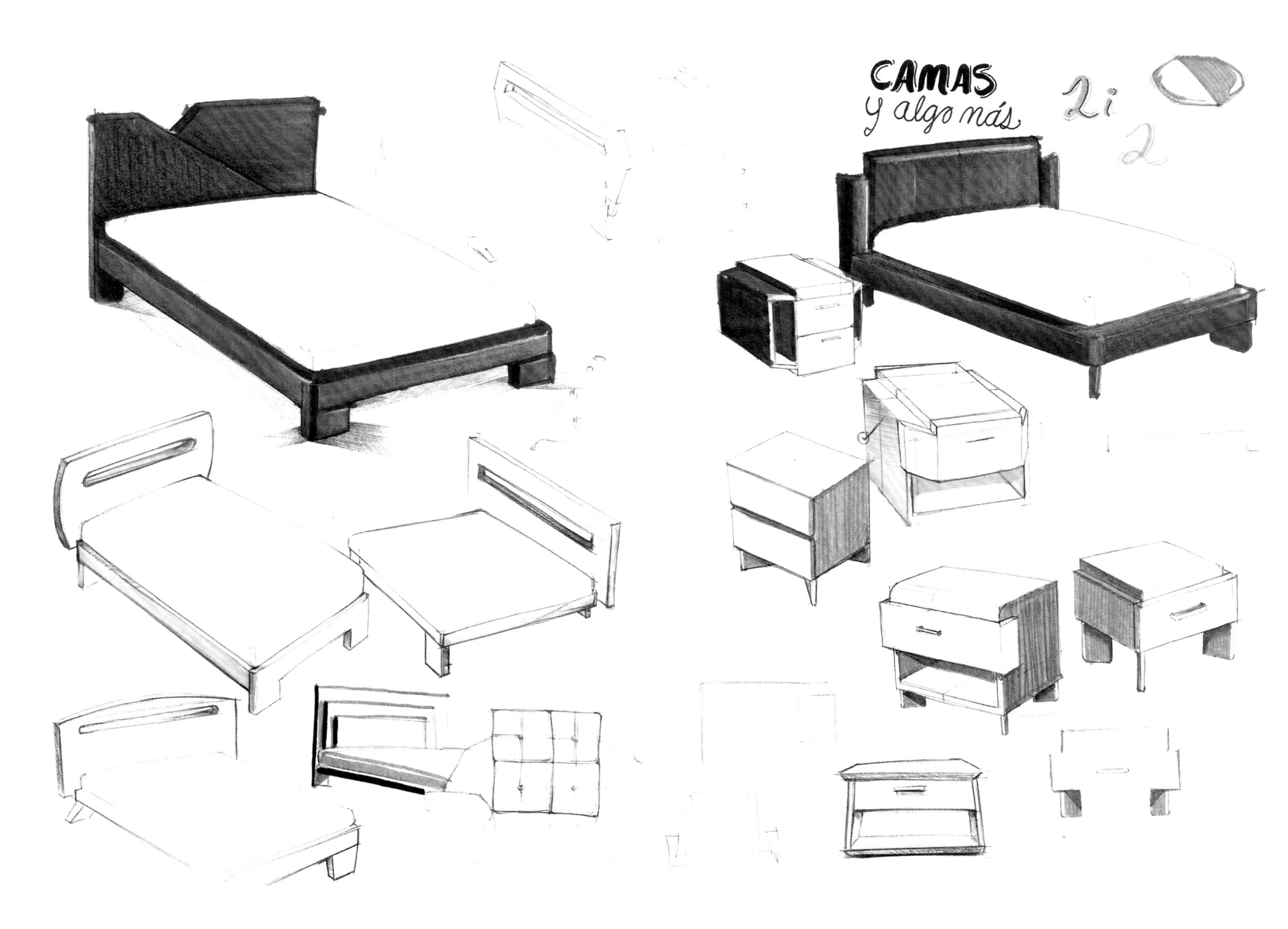
CAMAS
y algo más

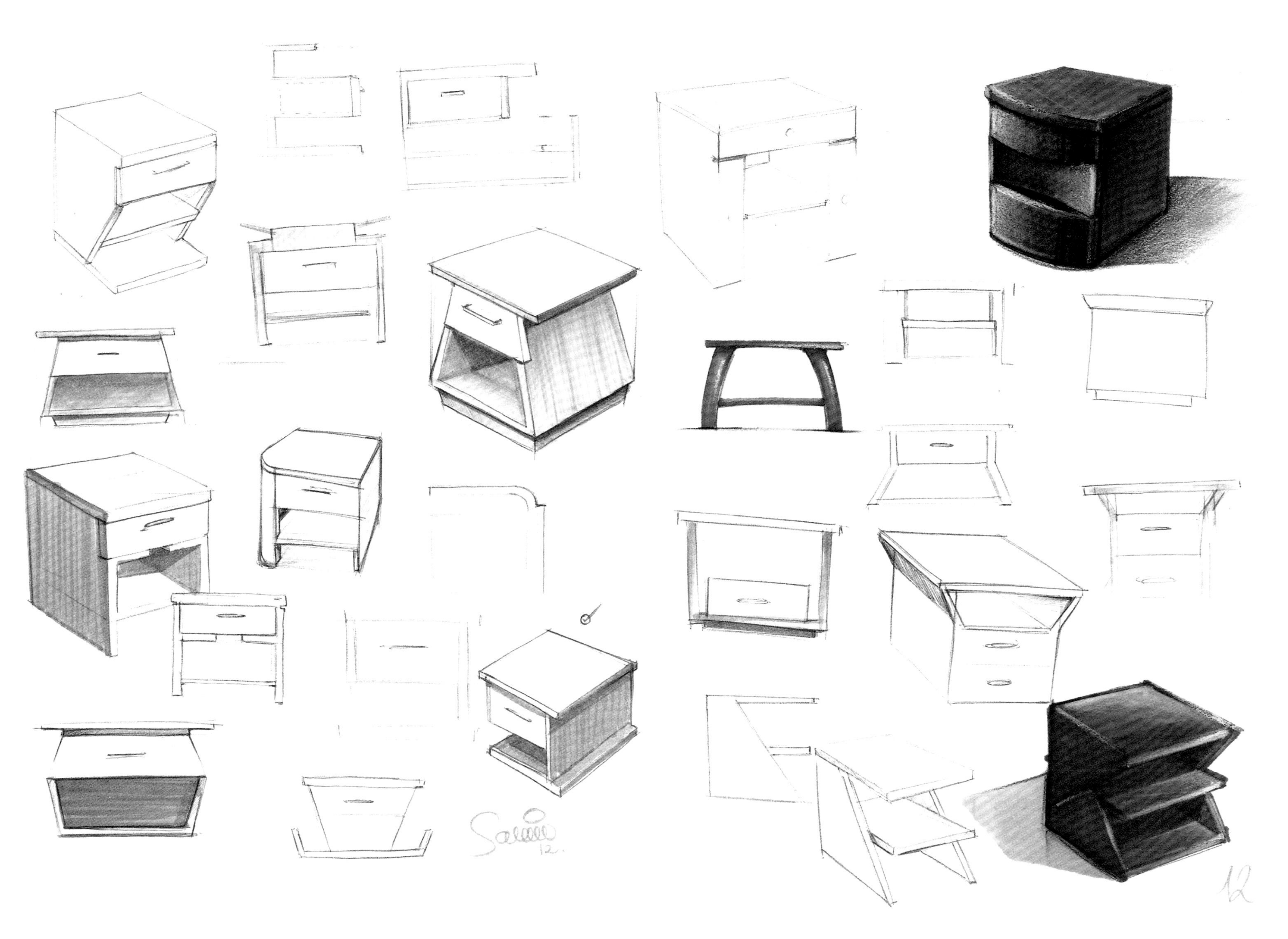

The Depilatator

剃须器

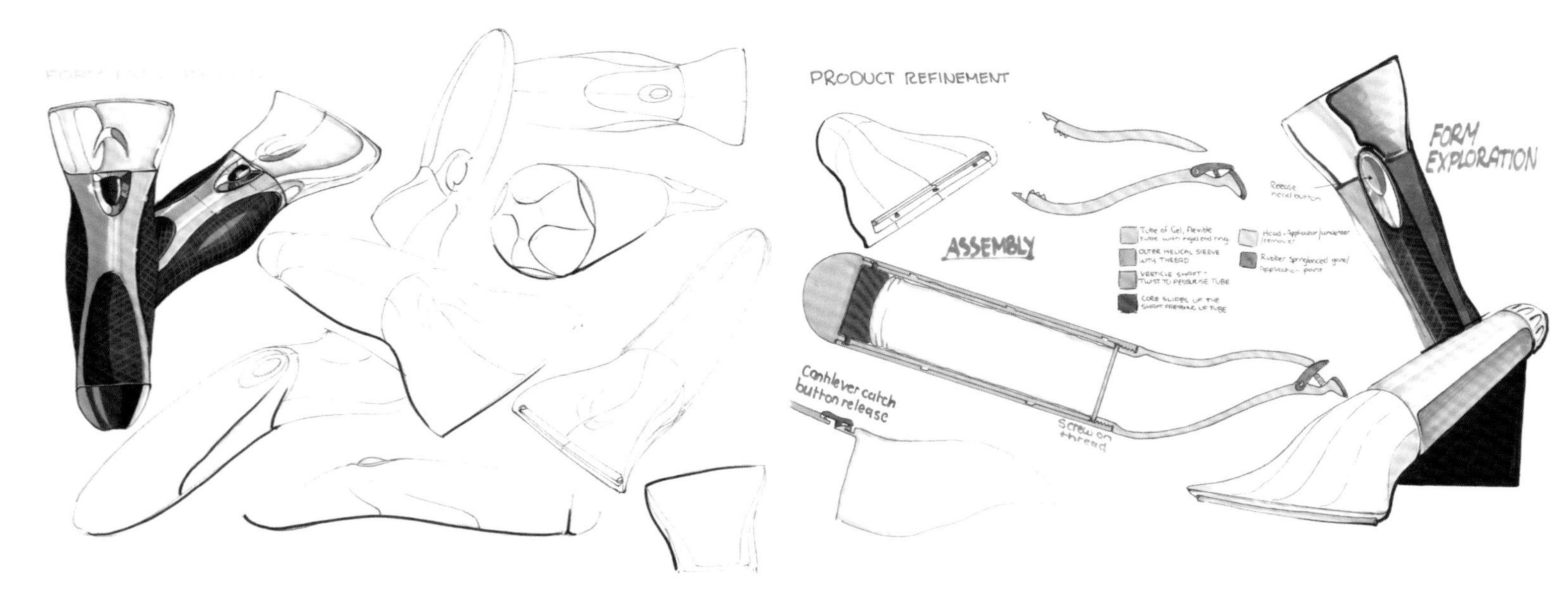

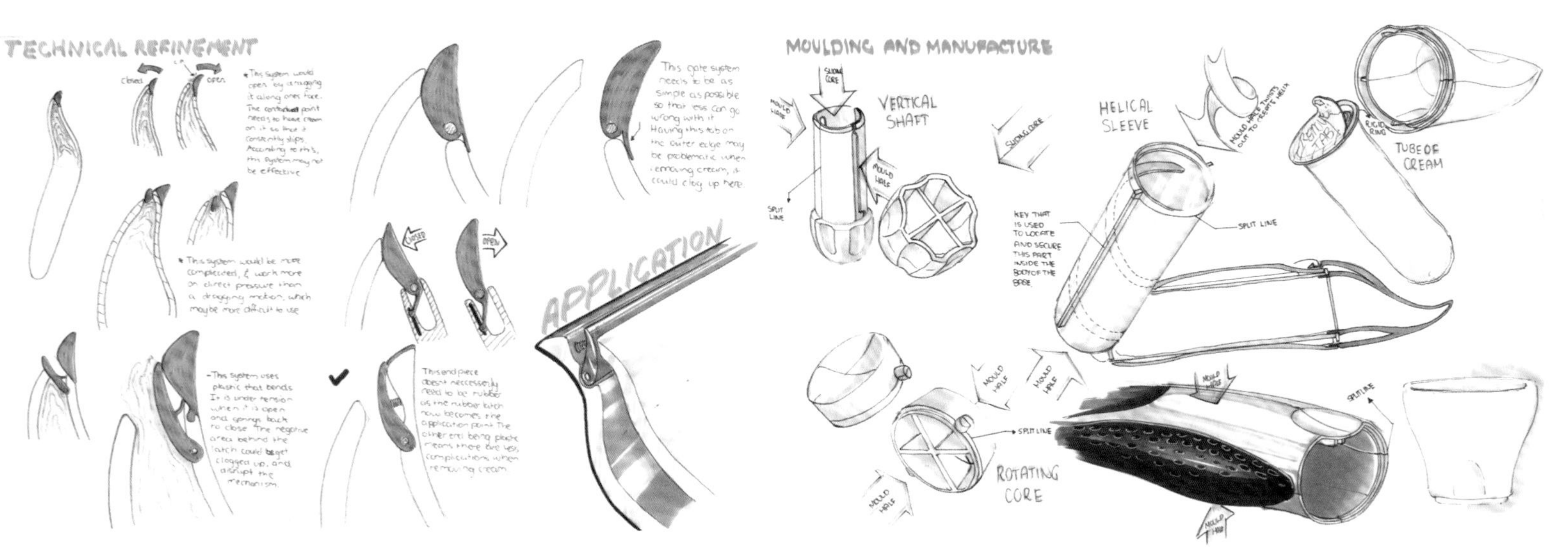

设计师：默里·夏普

客户：学校项目，约翰内斯堡大学

根据2012年“博朗国际工业设计大赛”（BraunPrize）的要求，设计师需要设计一款能够让生活更美好的产品。这一产品必须与众不同，而且目前还未上市。

剃须，对于大多数男士来说已经成为了繁琐的事情。随着都市性感男士形象的不断涌现，拥有一个干净的面庞已经成为了必须要做的事情。这款剃须器可以将剃须膏均匀地涂抹于面颊两侧，然后用顶部的刀头将剃须膏及胡须刮干净。此外，特制的剃须膏可被装在剃须器内部，如果每两周使用一次，可以持续使用四个月。

BRAUNPRIZE

FORM EXPLORATION

THIS DESIGN IS BASED ON THE PRINCIPLES OF GOOD DESIGN BY DIETER RAMS. BECAUSE THE COMPETITION IS HELD BY BRAUN, I HAVE DESIGNED A WELL ENGINEERED, SIMPLE AND FUNCTIONAL PRODUCT.

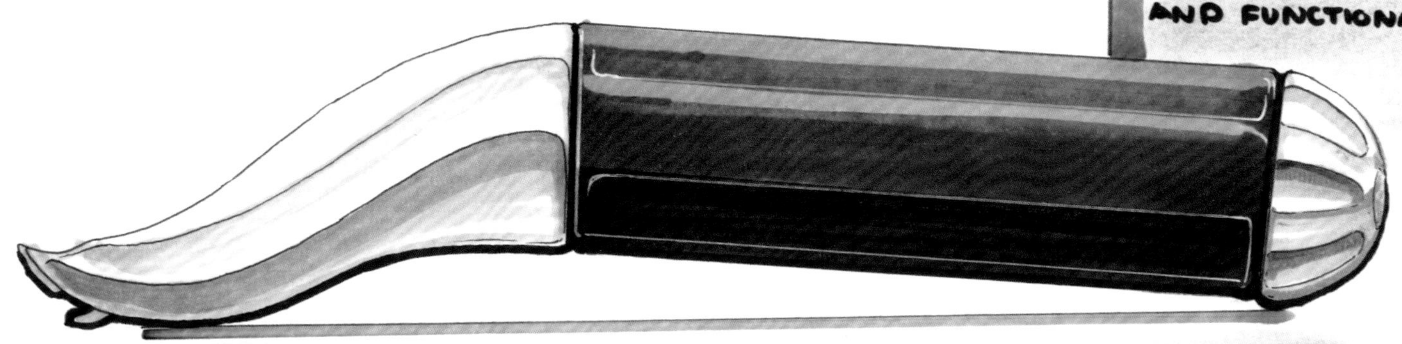

Sweet Potato Screwdriver

“甘薯”螺丝刀

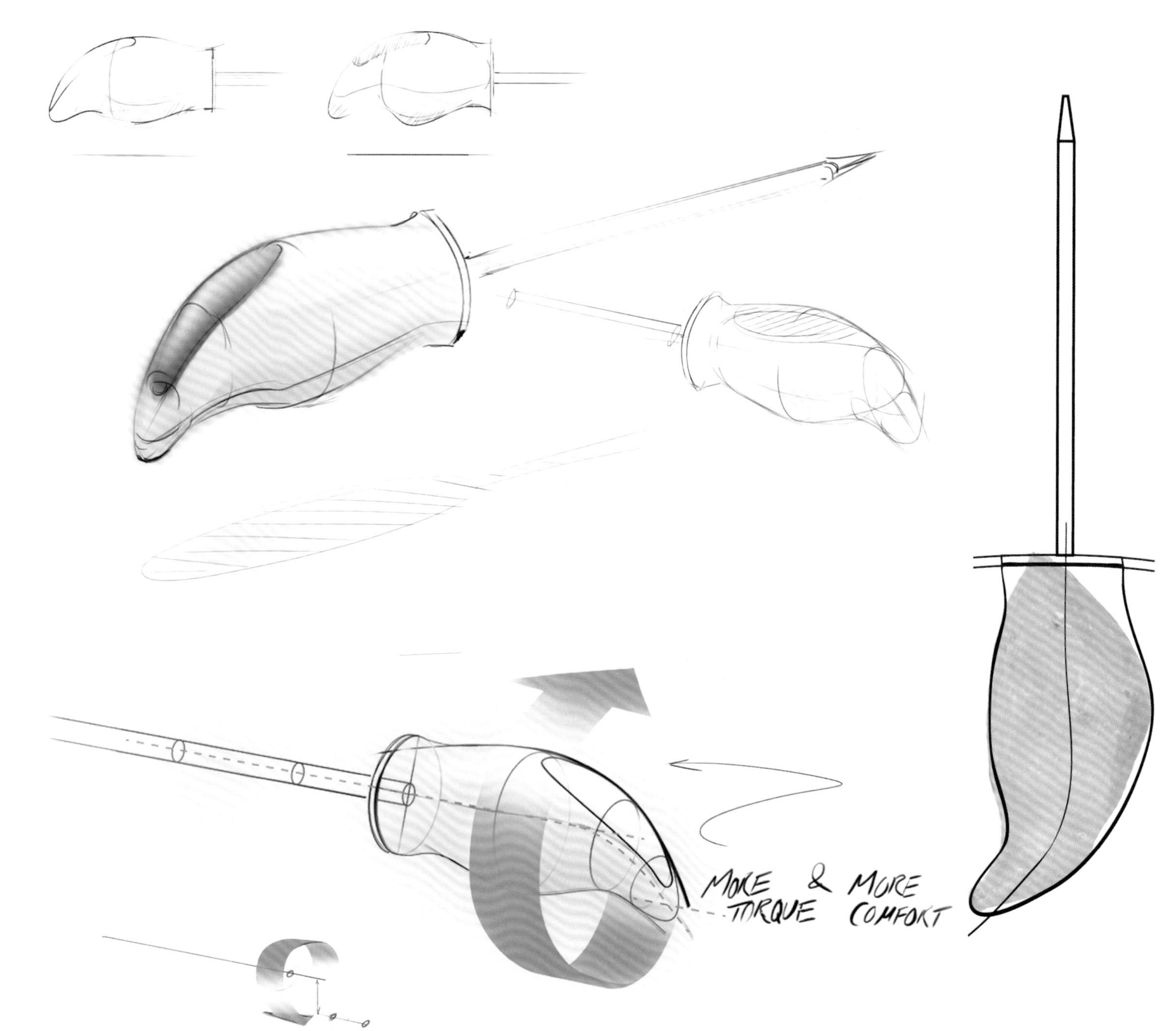

设计师：查利 · 兰斯康姆

客户：个人作品

灵感往往来自于奇怪的形态，例如蔬菜。设计师正是在烹饪甘薯的时候受到了启发——其截面形状非常适于手握。而传统螺丝刀的手柄使用起来并不方便，旋转的时候一直感觉握不住。“甘薯”形态的手柄正符合了产品人体工程学的需要。

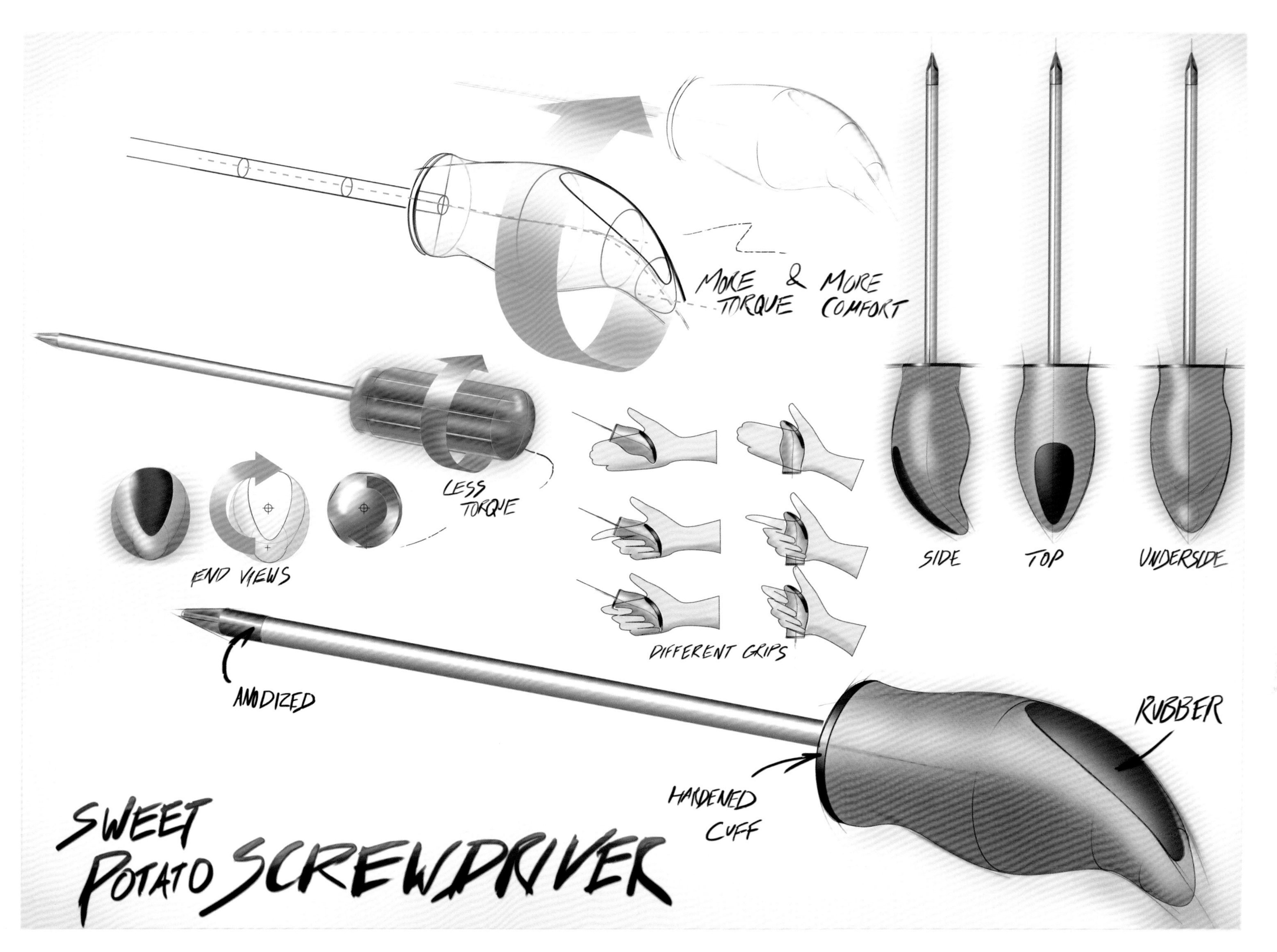
MORE TORQUE & MORE COMFORT
LESS TORQUE
END VIEWS
DIFFERENT GRIPS
SIDE
TOP
UNDERSIDE
ANODIZED
RUBBER
HARDENED CUFF
SWEET POTATO SCREWDRIVER

The Nosey Piggy Timer
“小猪”定时器

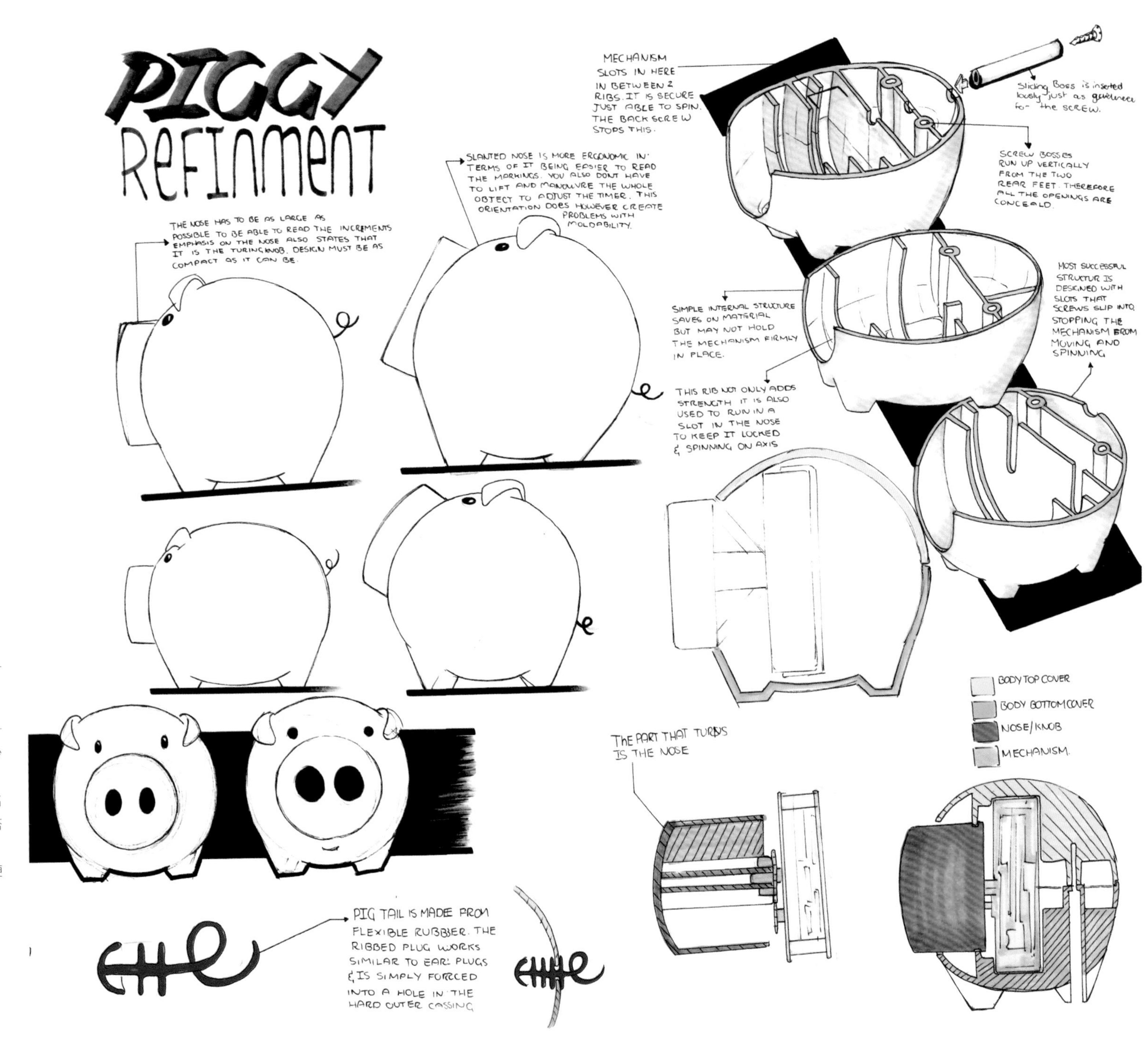

PIGGY REFINMENT
THE NOSE HAS TO BE AS LARGE AS POSSIBLE TO BE ABLE TO READ THE INCREMENTS EMPHASIS ON THE NOSE ALSO STATES THAT IT IS THE TURING KNOB. DESIGN MUST BE AS COMPACT AS IT CAN BE.
SLANTED NOSE IS MORE ERGONOMIC IN TERMS OF IT BEING EASIER TO READ THE MARKINGS. YOU ALSO DON'T HAVE TO LIFT AND MANOUVRE THE WHOLE OBJECT TO ADJUST THE TIMER. THIS ORIENTATION DOES HOWEVER CREATE PROBLEMS WITH MOLDABILITY.
MECHANISM SLOTS IN HERE IN BETWEEN 2 RIBS. IT IS SECURE JUST ABLE TO SPIN. THE BACK SCREW STOPS THIS.
Sliding Boss is inserted loosely just as guidance for the screw.
SCREW BOSSES RUN UP VERTICALLY FROM THE TWO REAR FEET. THEREFORE ALL THE OPENINGS ARE CONCEALD.
SIMPLE INTERNAL STRUCTURE SAVES ON MATERIAL BUT MAY NOT HOLD THE MECHANISM FIRMLY IN PLACE.
MOST SUCCESSFUL STRUCTUR IS DESIGNED WITH SLOTS THAT SCREWS SLIP INTO STOPPING THE MECHANISM FROM MOVING AND SPINNING
THIS RIB NOT ONLY ADDS STRENGTH IT IS ALSO USED TO RUN IN A SLOT IN THE NOSE TO KEEP IT LOCKED & SPINNING ON AXIS
BODY TOP COVER
BODY BOTTOM COVER
NOSE/KNOB
MECHANISM
THE PART THAT TURNS IS THE NOSE
PIG TAIL IS MADE FROM FLEXIBLE RUBBER. THE RIBBED PLUG WORKS SIMILAR TO EAR PLUGS & IS SIMPLY FORCED INTO A HOLE IN THE HARD OUTER CASSING

设计师：默里 · 夏普

客户：学校项目，约翰内斯堡大学

这是一款厨房定时器，其设计可以是奇妙的、繁复的，也可以是有趣的、古怪的。该产品设计的风格受到Alessi（世界著名的家居用品设计制造商）和菲利普 · 斯塔克（Philippe Starck，法国产品设计师）的影响。这一产品外形精灵古怪，同时便于使用。螺钉从小猪的眼睛和尾巴穿过，不影响视觉美感。倾斜的鼻子便于旋转，整体设计紧凑，且运用了较少的材料。

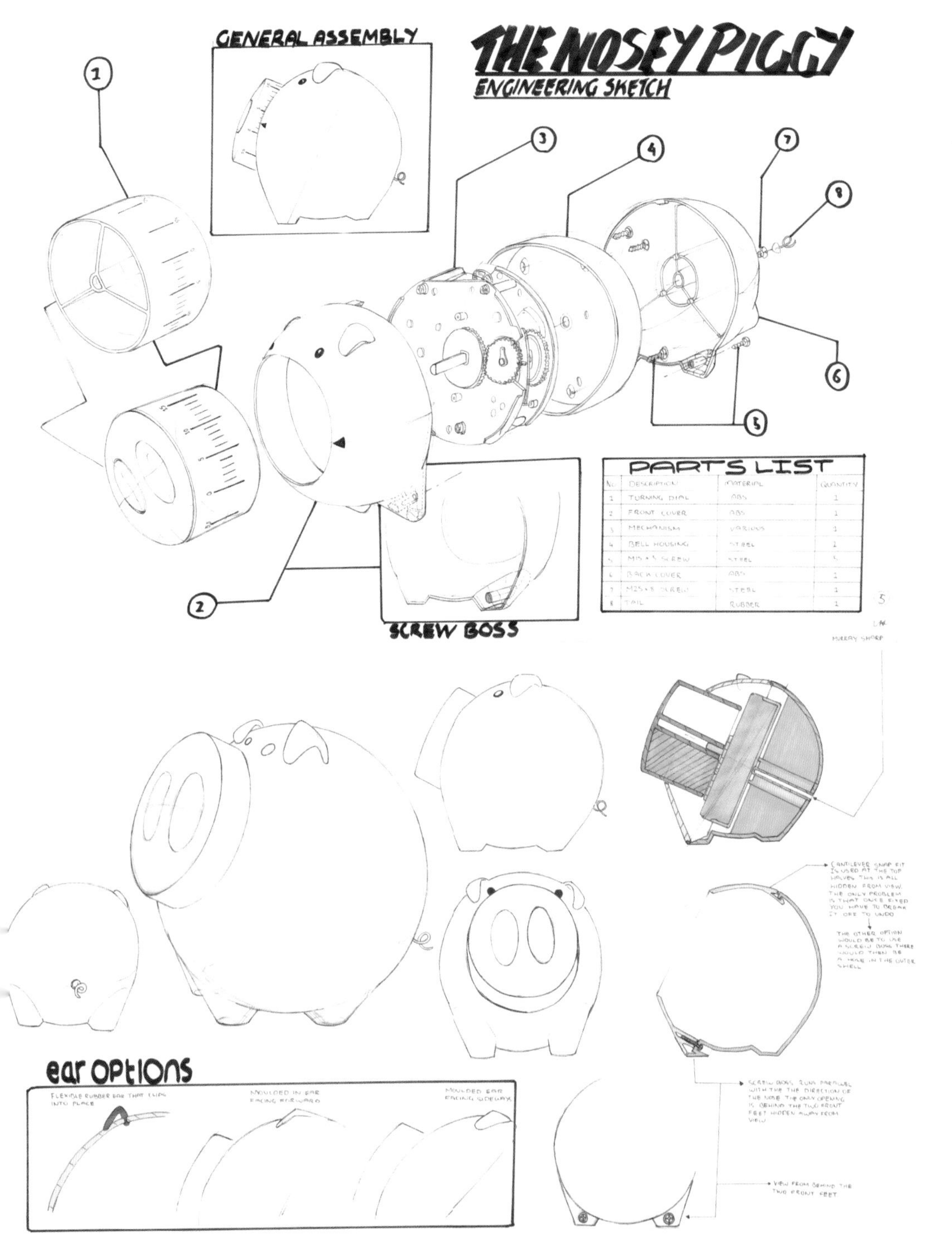
GENERAL ASSEMBLY
THE NOSEY PIGGY
ENGINEERING SKETCH
SCREW BOSS
PARTS LIST
No. DESCRIPTION MATERIAL QUANTITY
1 TURNING DIAL ABS 1
2 FRONT COVER ABS 1
3 MECHANISM VARIOUS 1
4 BELL HOUSING STEEL 1
5 M15×3 SCREW STEEL 5
6 BACK COVER ABS 1
7 M25×8 SCREW STEEL 1
8 TAIL RUBBER 1
MURRAY SHARP
ear options

Piggy Timer

Food Waste Recycling Unit

食品垃圾回收装置

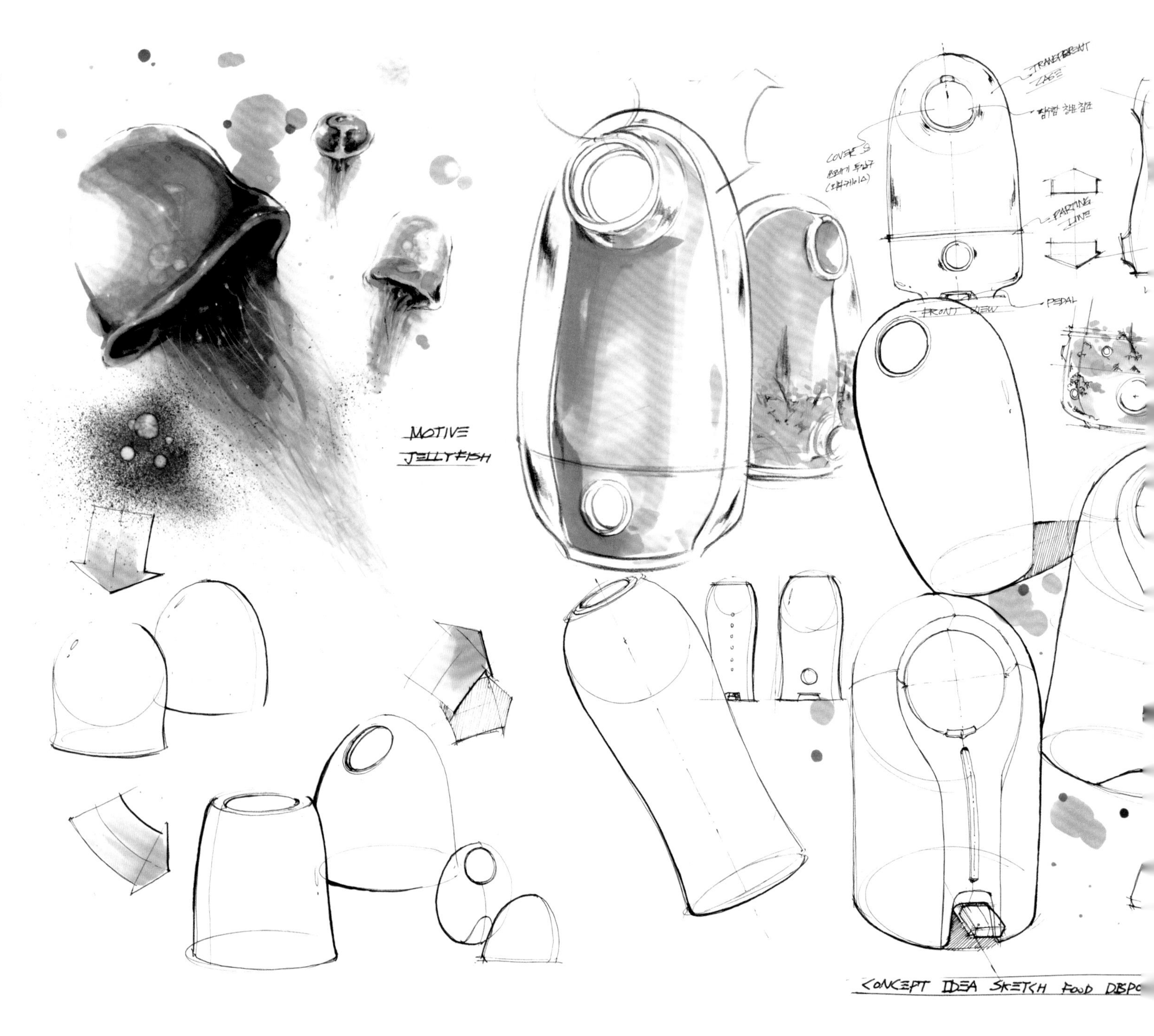

设计师：柳时形

客户：Loofen公司

这是一款在公共及商业场所使用的食品垃圾回收装置，且同样适用于家庭食品垃圾的回收和处理。食品垃圾脱水机是一个简单的处理装置，其能够使垃圾脱水成可直接回收的物质。很多家庭都在使用这种装置，但从调查及使用者的反馈来看，他们最头疼的是食品垃圾处理后的卫生状况。这一设计更注重用户体验及产品的形态与风格。设计师认为，经过处理的食品垃圾能够用做炉灶的燃料。

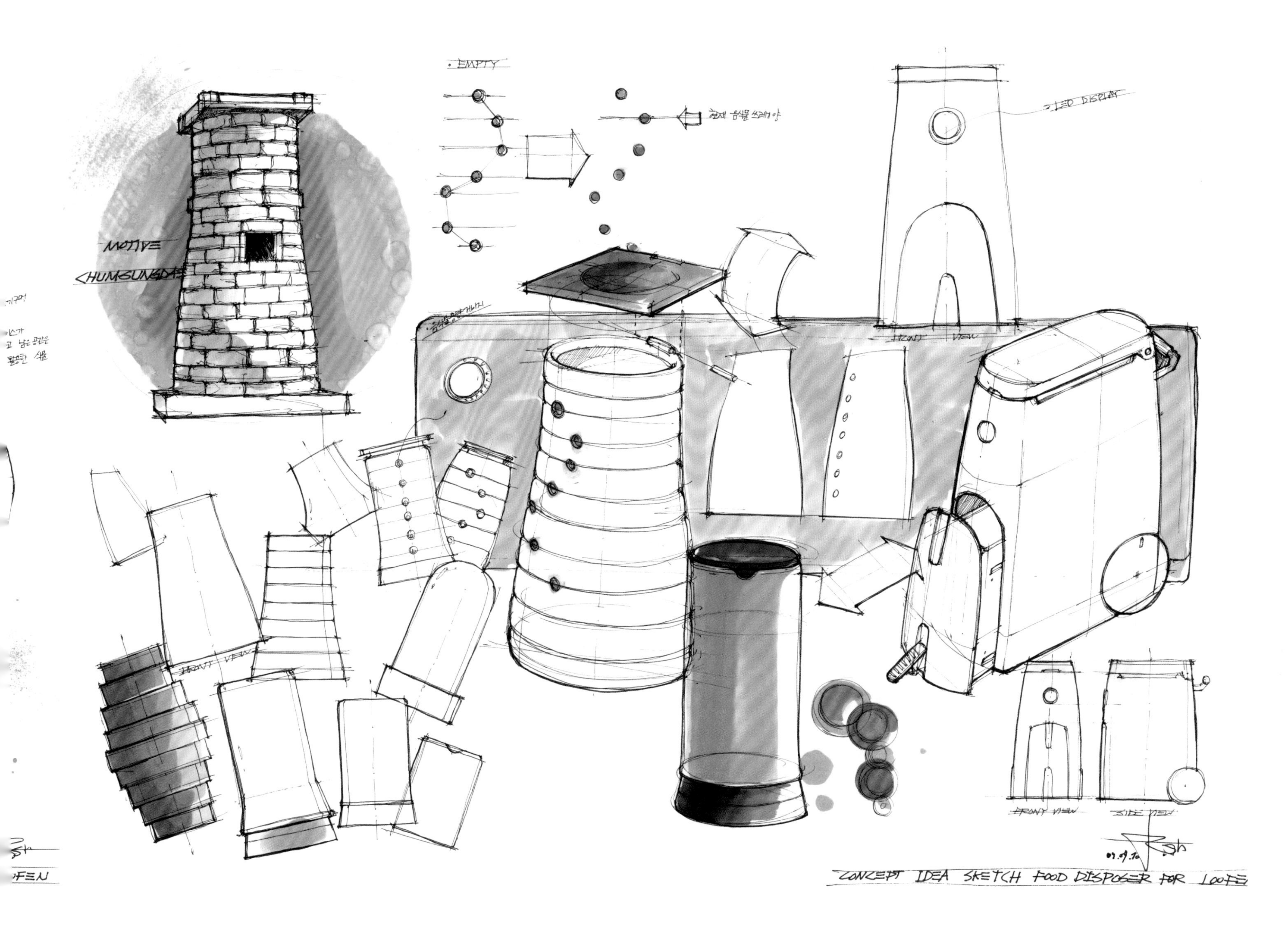

EMPTY
LED DISPLAY
MOTIVE
FRONT VIEW
SIDE VIEW
CONCEPT IDEA SKETCH FOOD DISPOSER FOR

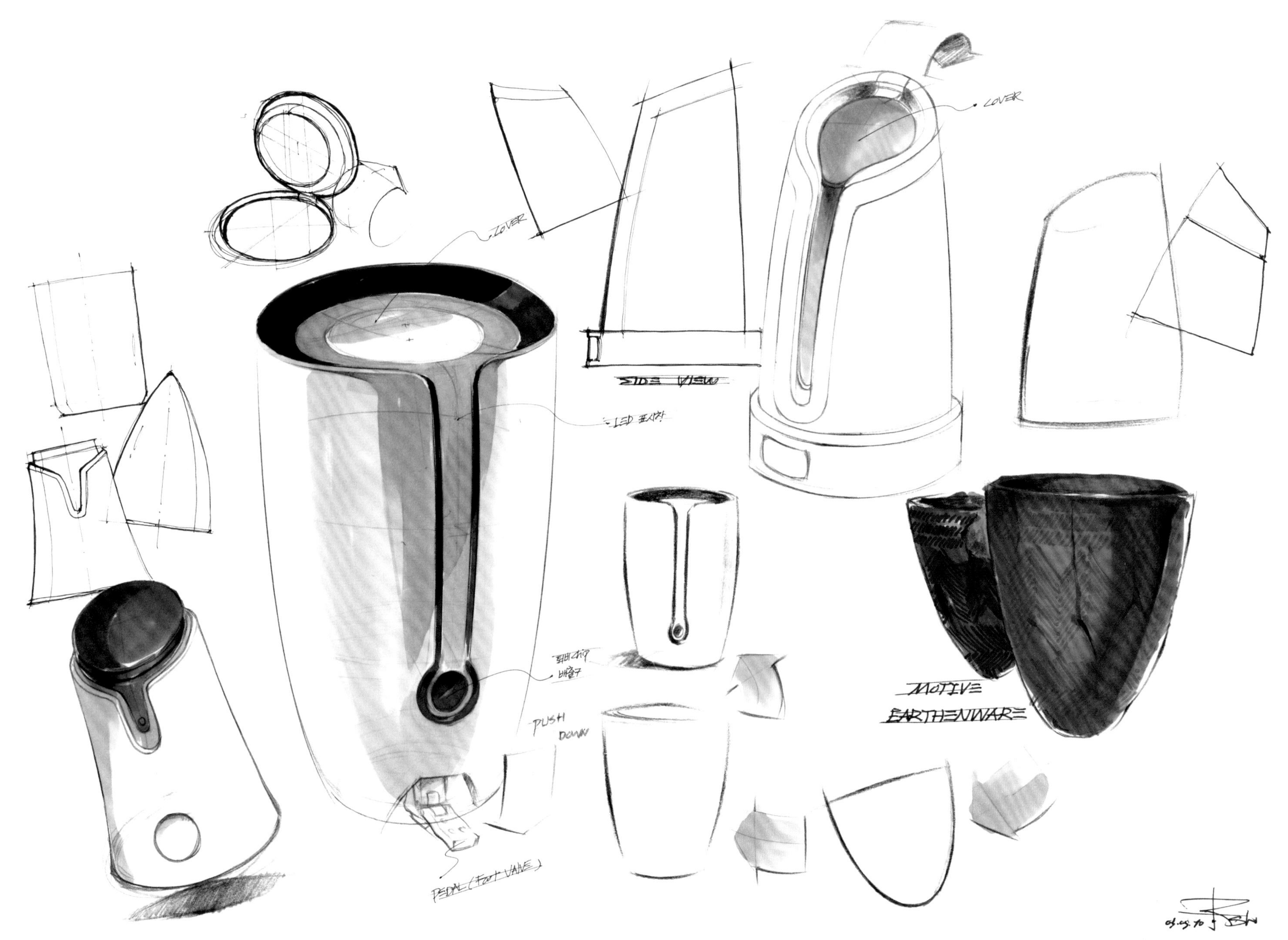
COVER
COVER
SIDE VIEW
COVER
LED
PUSH
DOWN
PEDAL (FOOT VALVE)
MOTIVE
EARTHENWARE
IDEA SKETCH FOOD DISPOSER FOR

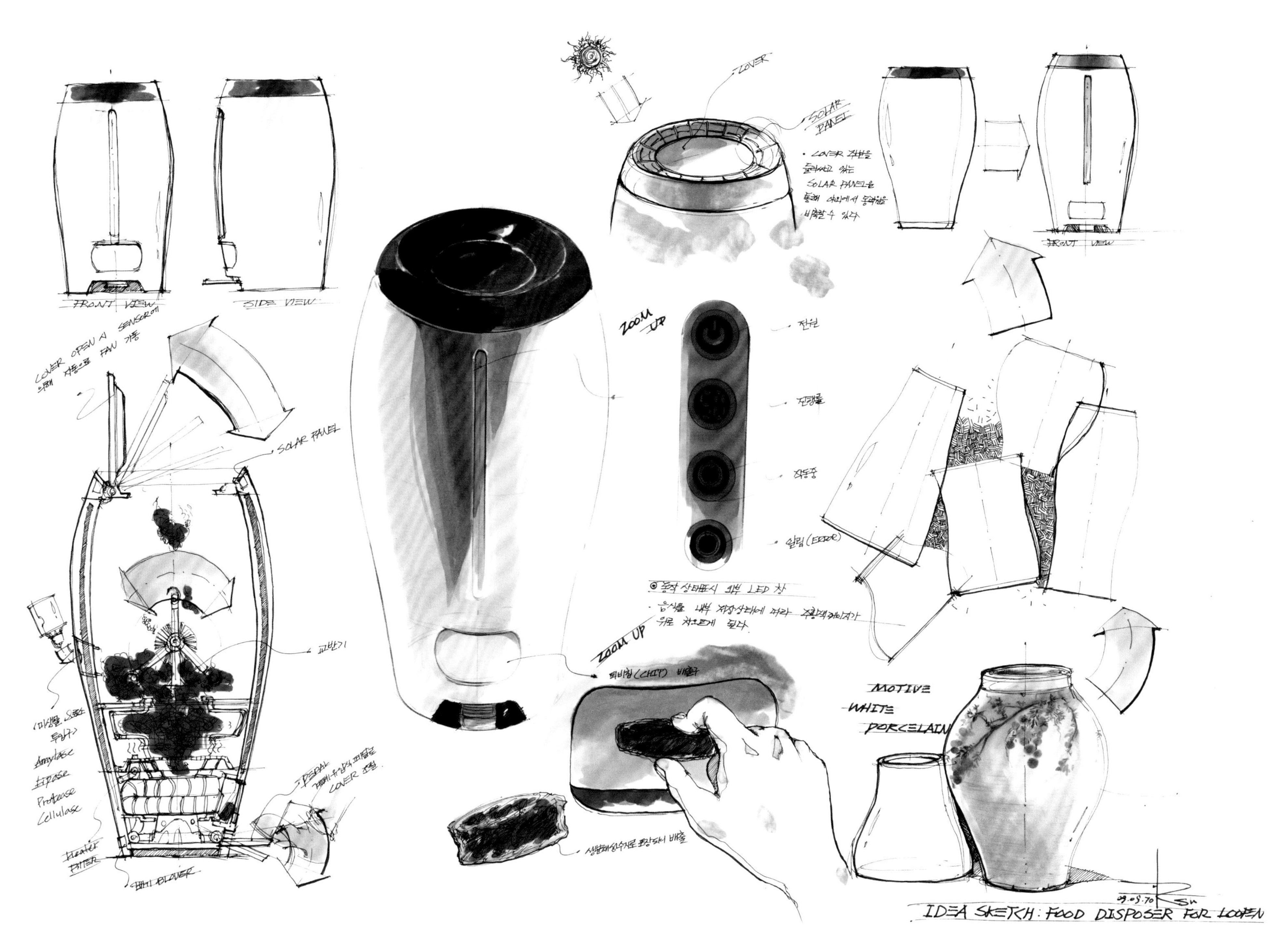
FRONT VIEW
SIDE VIEW
COVER
SOLAR PANEL
ZOOM UP
SOLAR PANEL
ZOOM UP
MOTIVE
WHITE
PORCELAIN
FRONT VIEW
IDEA SKETCH: FOOD DISPOSER FOR LOOPEN

Fruit Bowl
水果盘

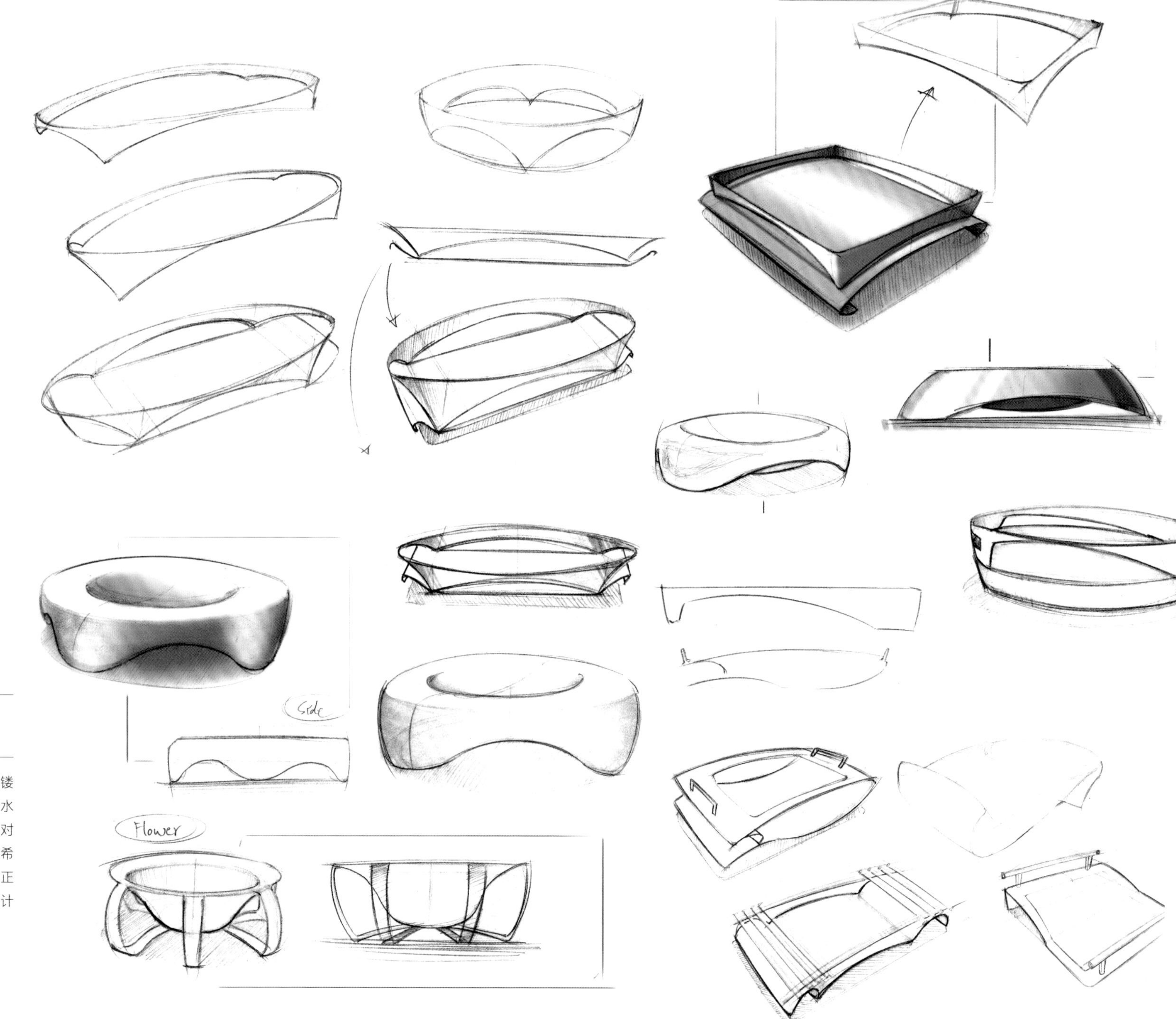

设计师：陈昶维（Chen Chang-Wei）
客户：个人作品

这一水果盘由不锈钢材料打造，边缘的镂空图案犹如树枝一般，让人感觉盘内的水果像是长在树上。在设计之初，设计师对很多日用水果盘的样式做了一些研究，希望让这一常见的产品看起来更加有趣。正是受到“长在树上的水果”的启发，设计师决定将树枝的影子作为主要设计元素。

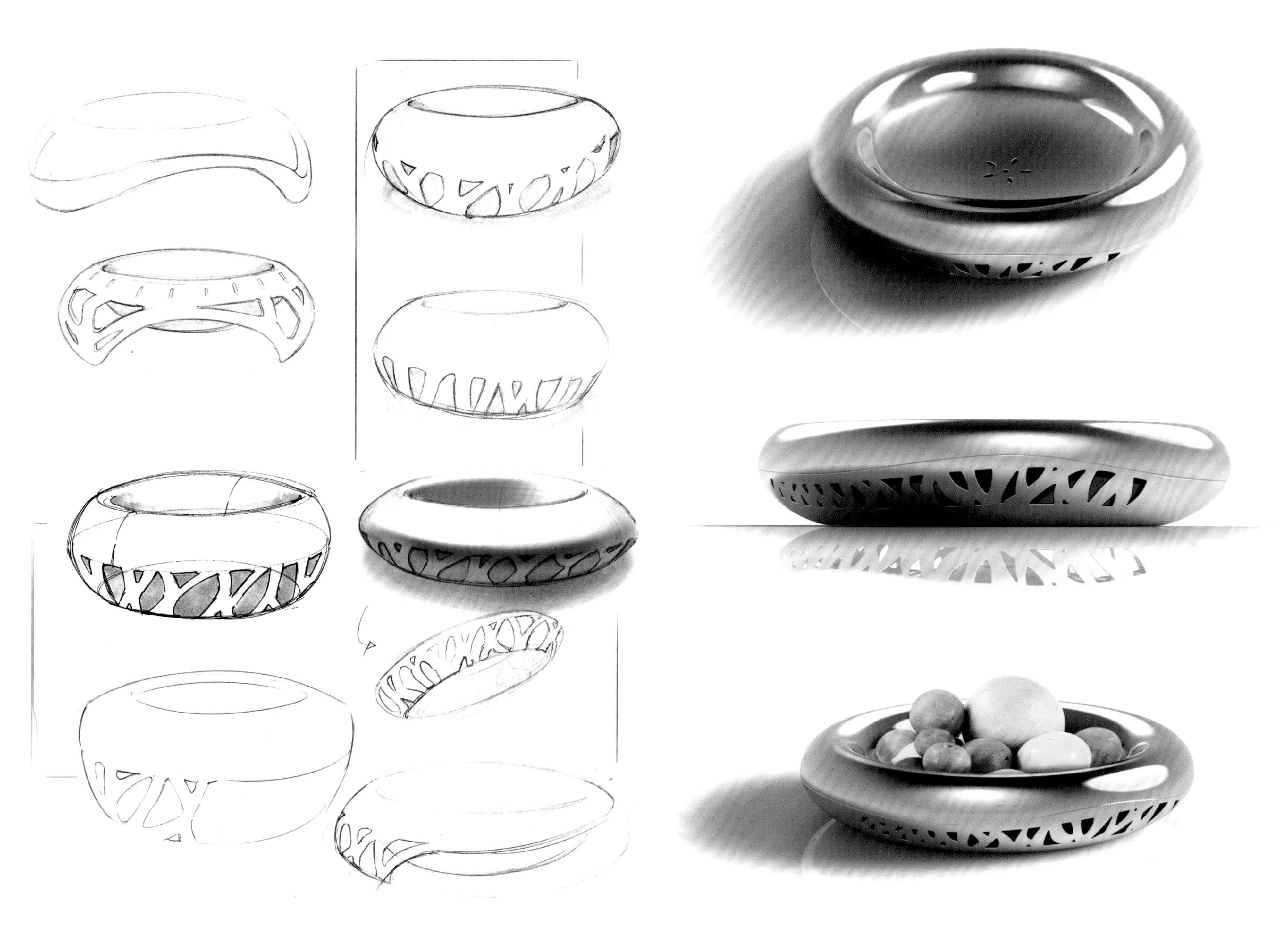

Baby Changing Station

换尿布台

设计师：柳时彤
客户：TS JAVA公司

如今，很多公共场所都设有换尿布台，市场上有很多种类似的产品，而TS JAVA公司则是一家专门从事卫生洁具研发生产的公司。设计师在设计时，着重将品牌的视觉形象与产品相融合。

JAVA
SIDE VIEW
STAND TYPE-A
STAND TYPE-B
B JAVA DIAPER CHANGER IDEA SKETCH

Sketching / Markers / Ideation

产品草图习作

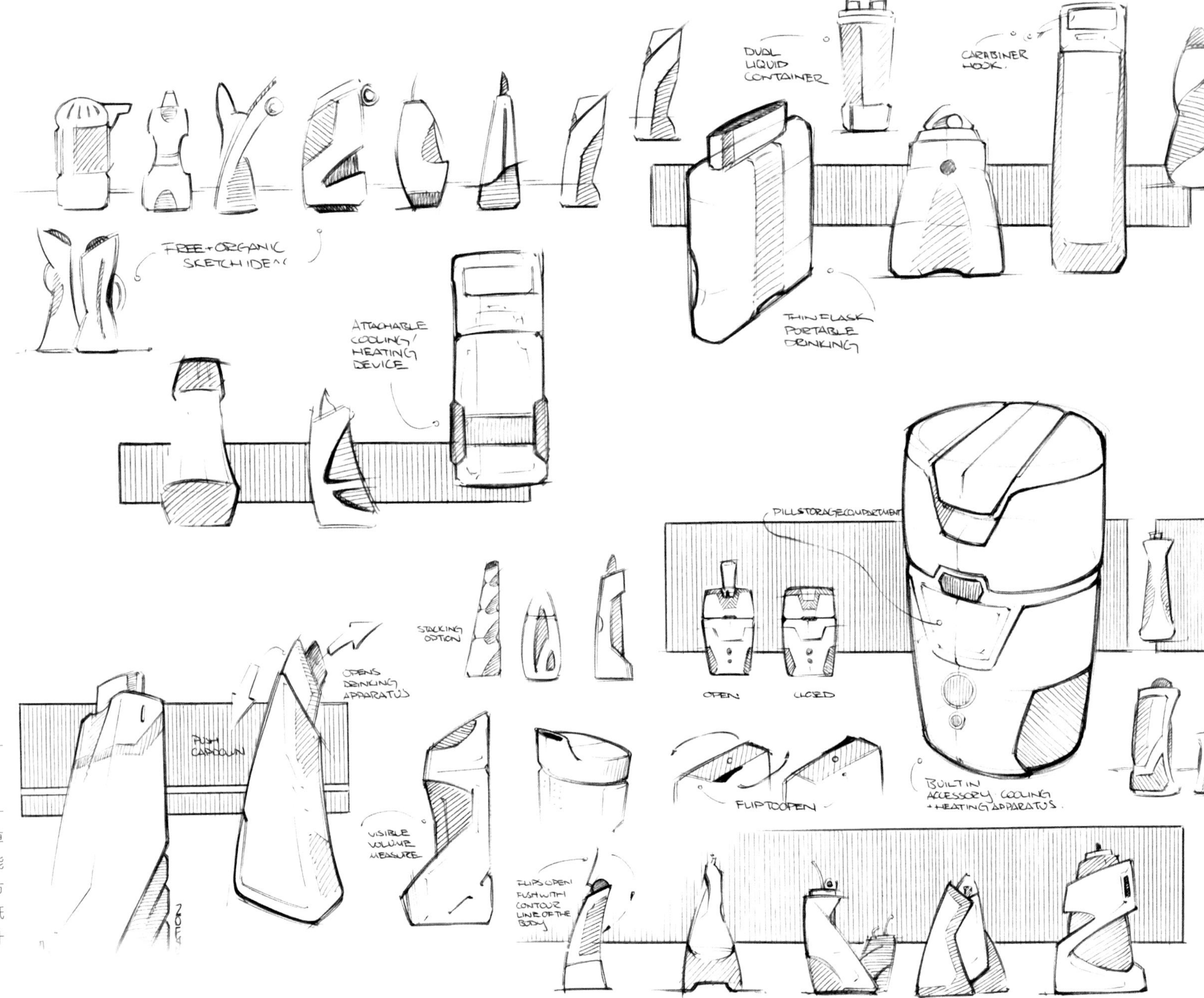

设计师：耿天亦（Tai TianyiGeng）

客户：个人作品

草图是设计师们交流的媒介，能够通过草图清晰、快速地表达想法是非常重要的能力。提升产品设计草图绘制水平的惟一方法便是坚持不懈的练习。不要害怕浪费纸张，不要担心涂改，学习用草图传达设计理念。

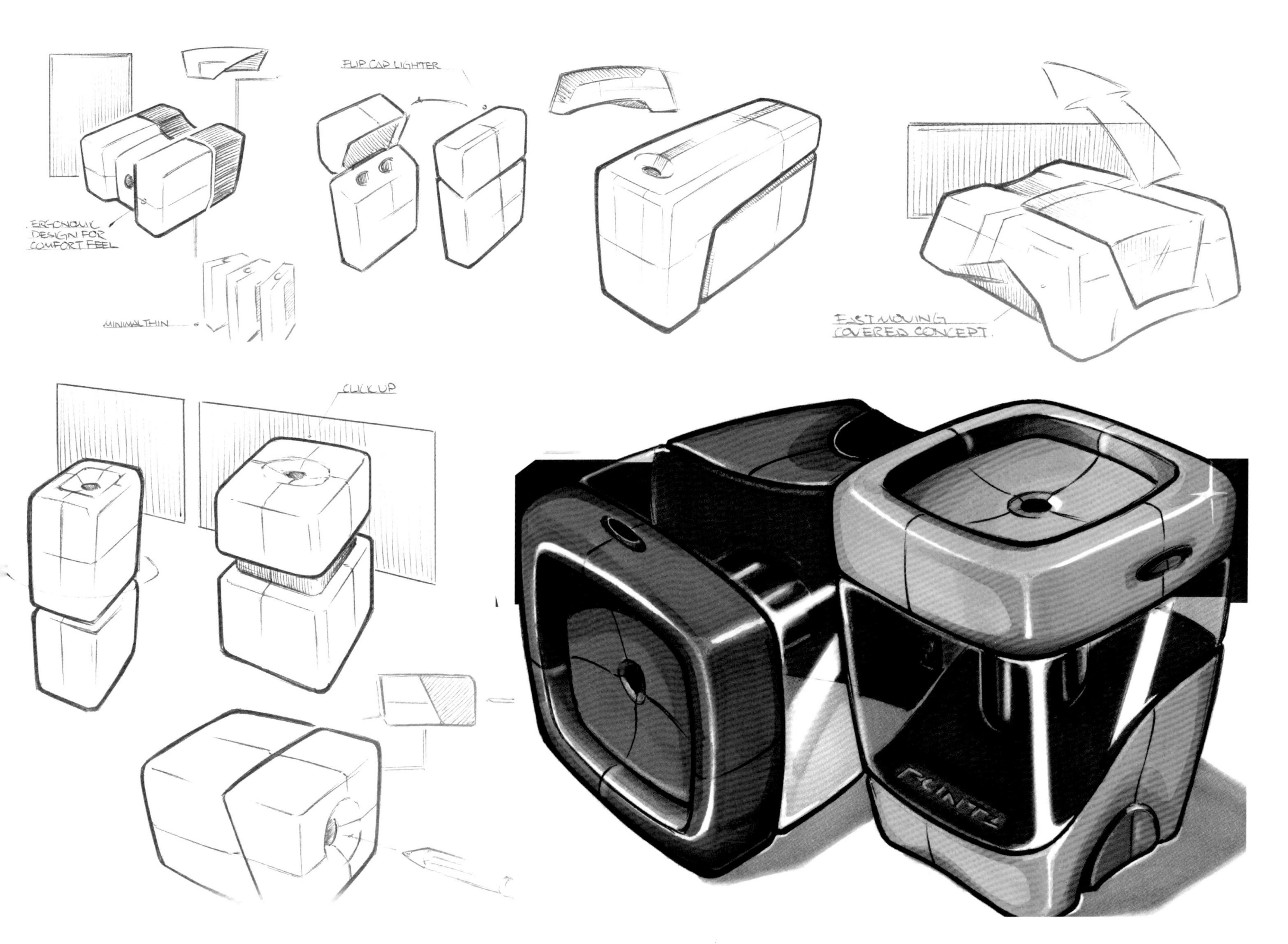

FLIP CAP LIGHTER
ERGONOMIC
DESIGN FOR
COMFORT FEEL
MINIMALTHIN
FAST MOVING
COVERED CONCEPT
CLICK UP

Frangit Alarm Clock
Frangit闹钟

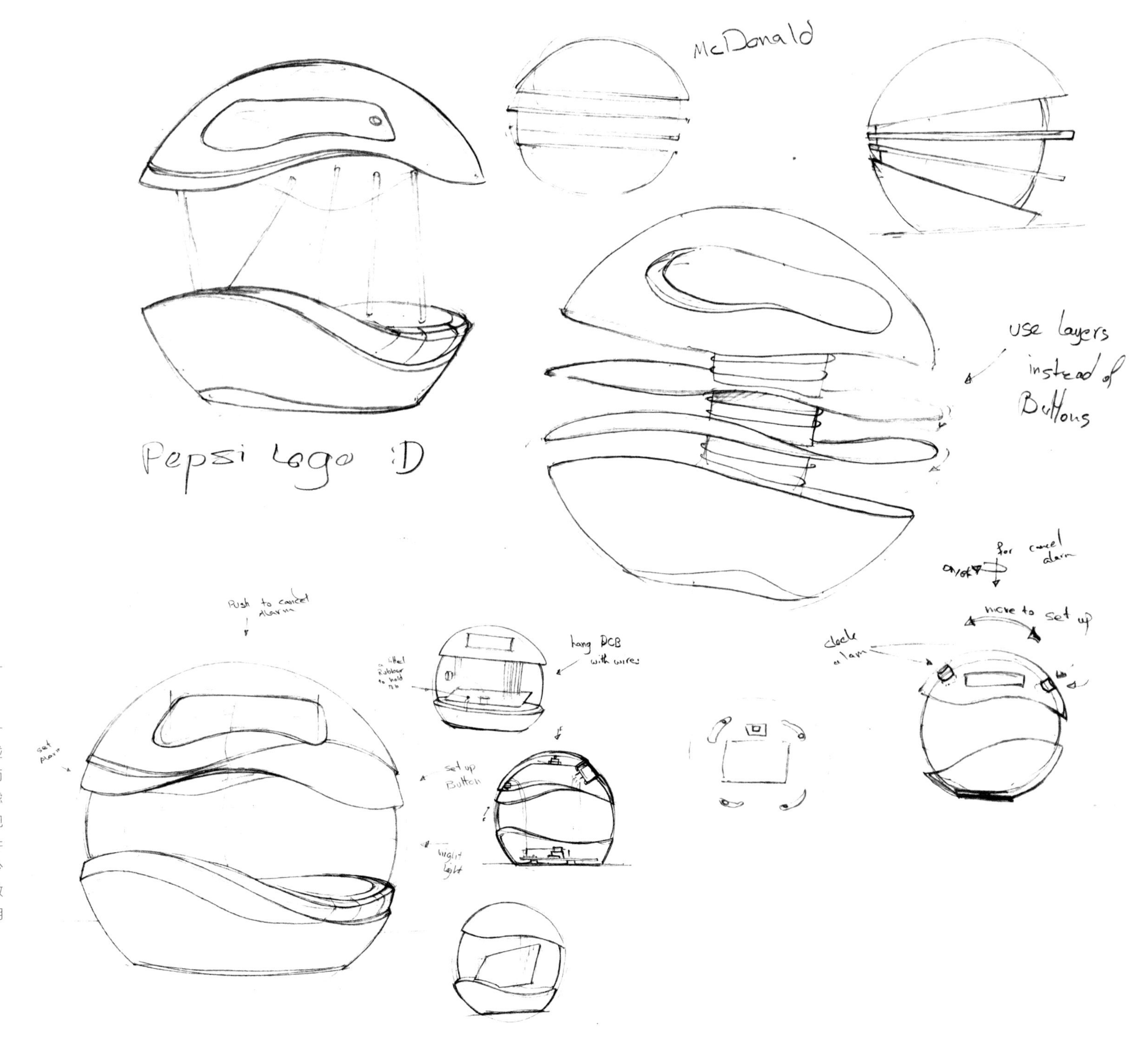

设计师：阿拉什 · 赛派（Arash Sepahi）
客户：学校项目，赫特福德大学

闹钟的创新解决方案——使用者可以触碰或拍打闹钟上部的任何区域使其关掉，而无需寻找开关按钮。该设计项目着重考虑产品的规格、构件的装配、用户交互、视觉设计以及相关的生产问题等。产品构件的装配非常重要，设计时应尽量使用较少的部件。简洁是该设计的核心理念，多数消费者会选择比例协调、材质精良、实用性强的产品。

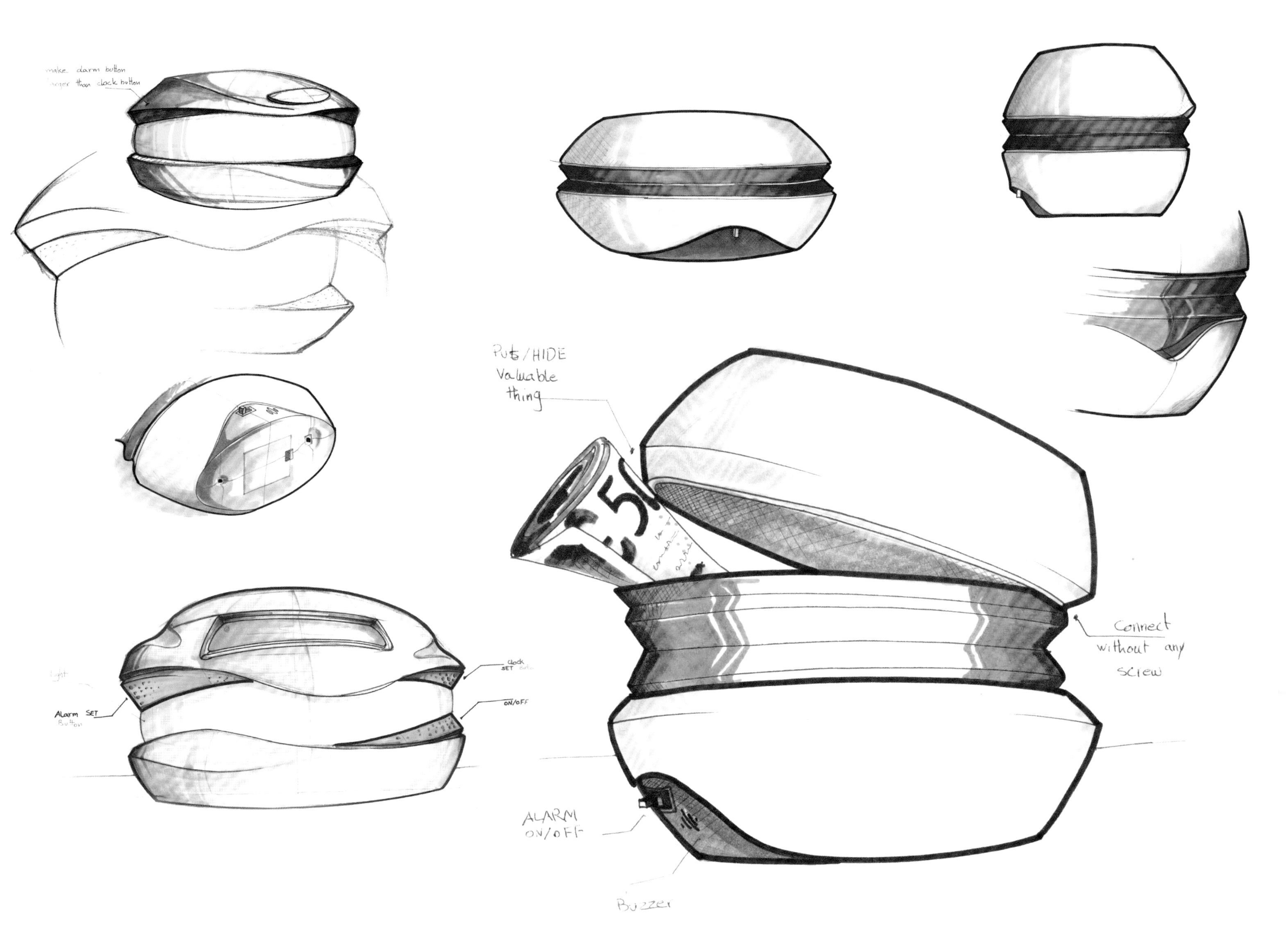

make alarm button
larger than clock button
light
Alarm SET
Button
Clock
SET
ON/OFF
Puts/HIDE
Valuable
thing
Connect
without any
screw
ALARM
ON/OFF
Buzzer

Automatic Toilet Flusher
全自动便器冲水器

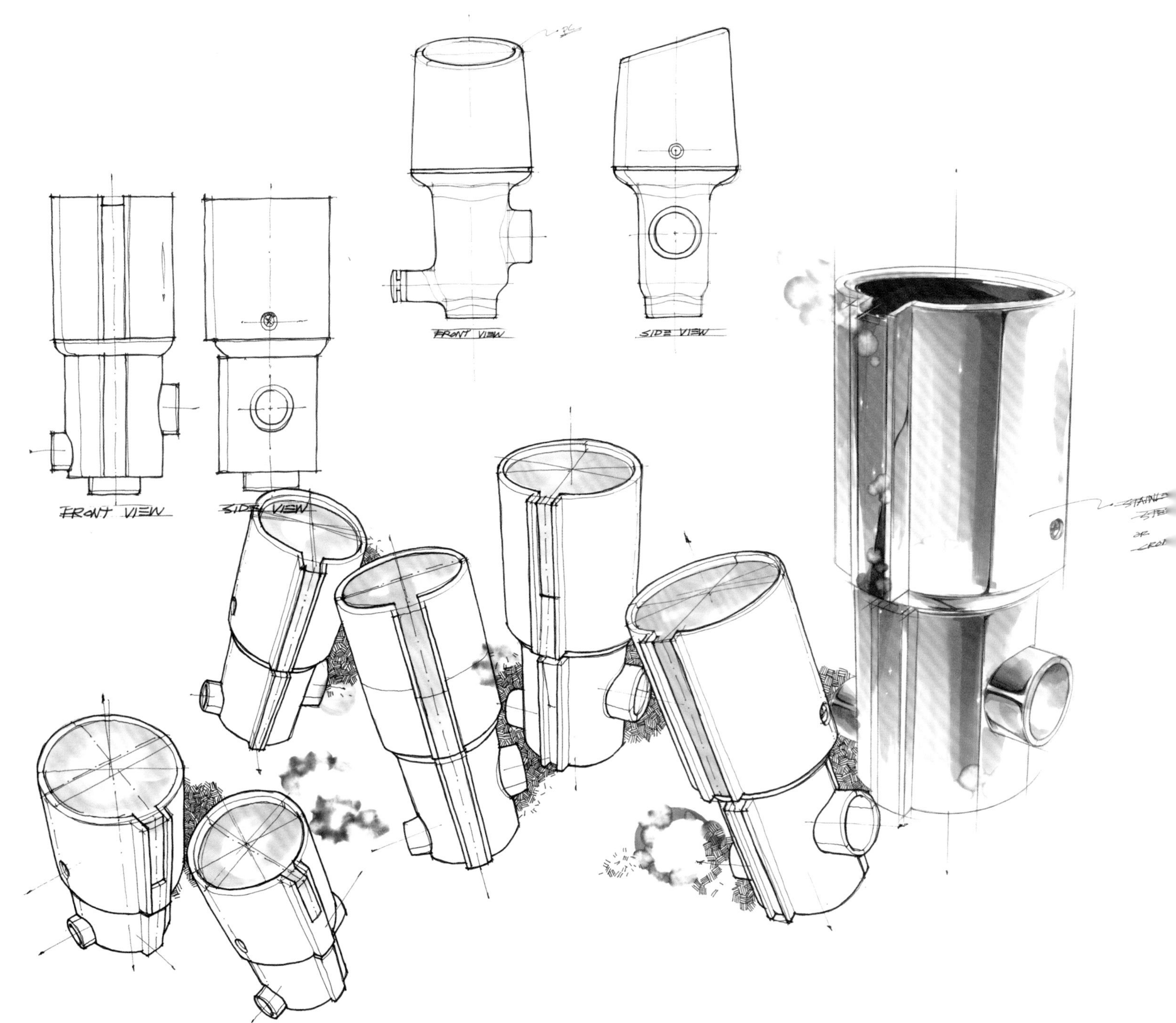

设计师：柳时彤
客户：TS JAVA公司

这款便器冲水器带有全自动感应装置。很多人认为产品设计即造型设计，其实设计的不仅是造型——设计不但赋予产品外壳的形式，而且赋予产品功能和结构。该产品便将感应技术与产品形态相融合。

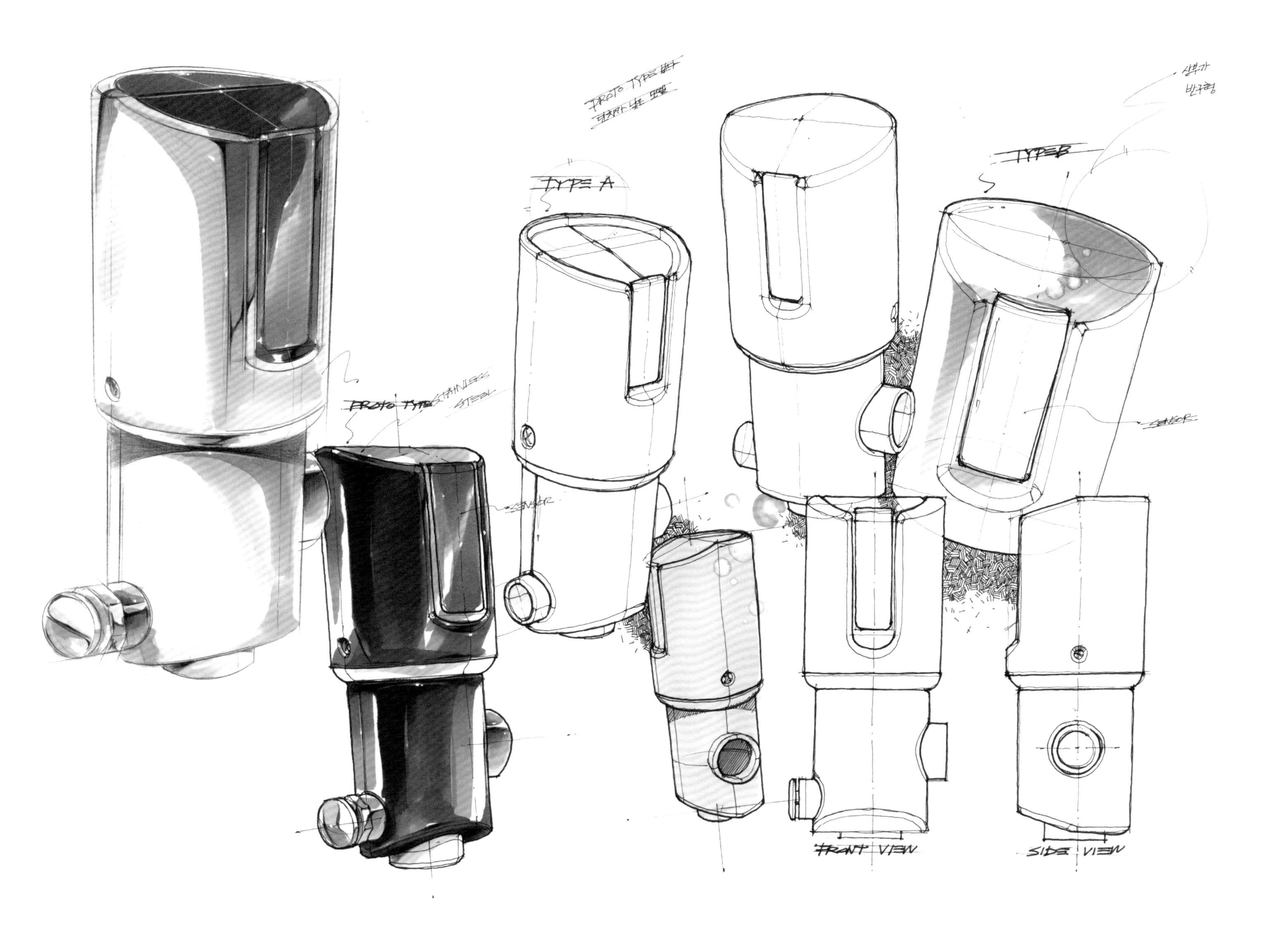
TYPE A
TYPE B
PROTO TYPE
STAINLESS STEEL
SENSOR
SENSOR
FRONT VIEW
SIDE VIEW

Automatic Toilet Flusher

全自动便器冲水器

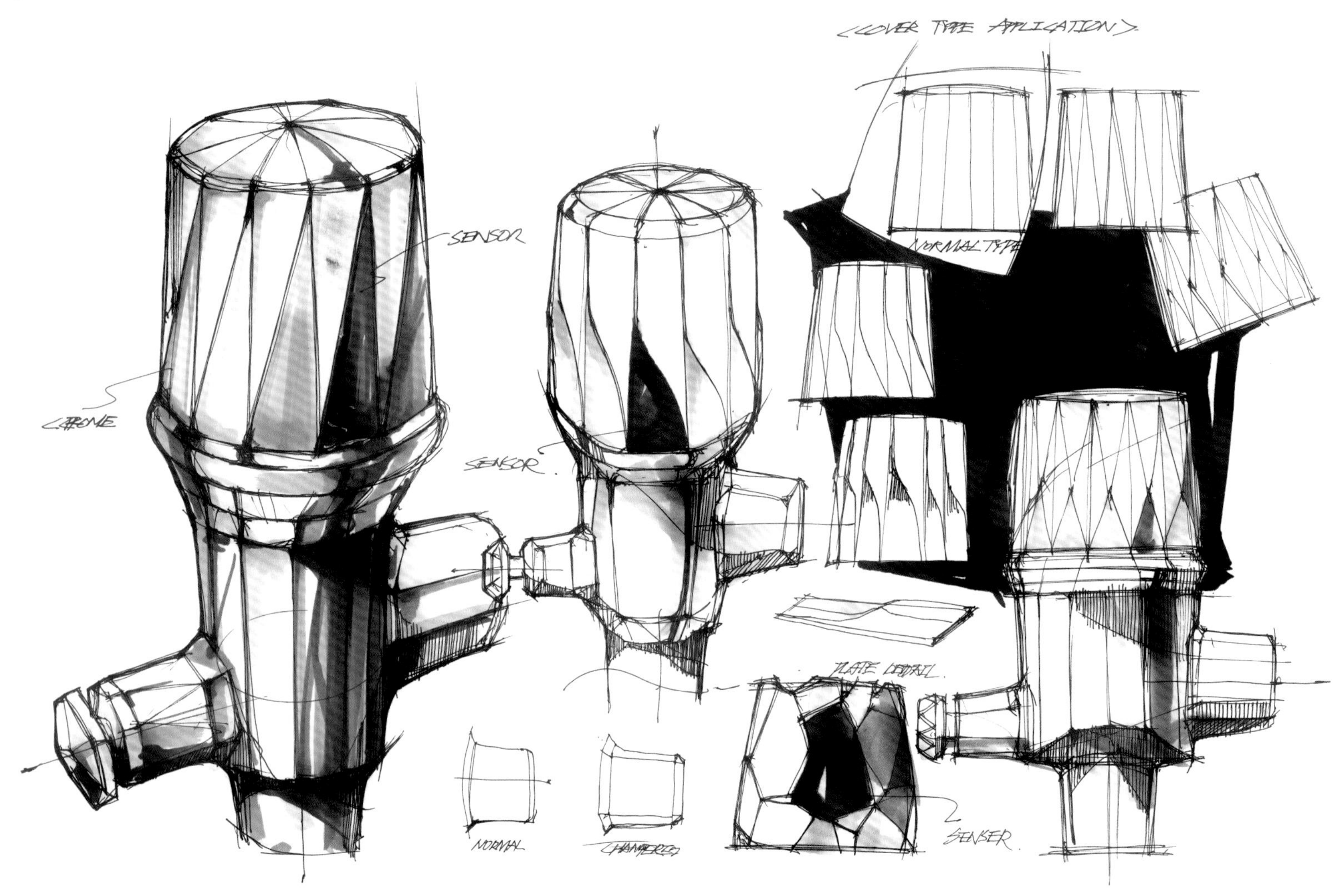

设计师：柳时形

客户：Hanseo公司

这是为Hanseo品牌创造的全自动便器冲水器的概念设计，旨在为全自动便器冲水器创造一种全新的设计语言。设计师所面临的最大挑战，是探讨如何在产品金属外壳上嵌入传感器。

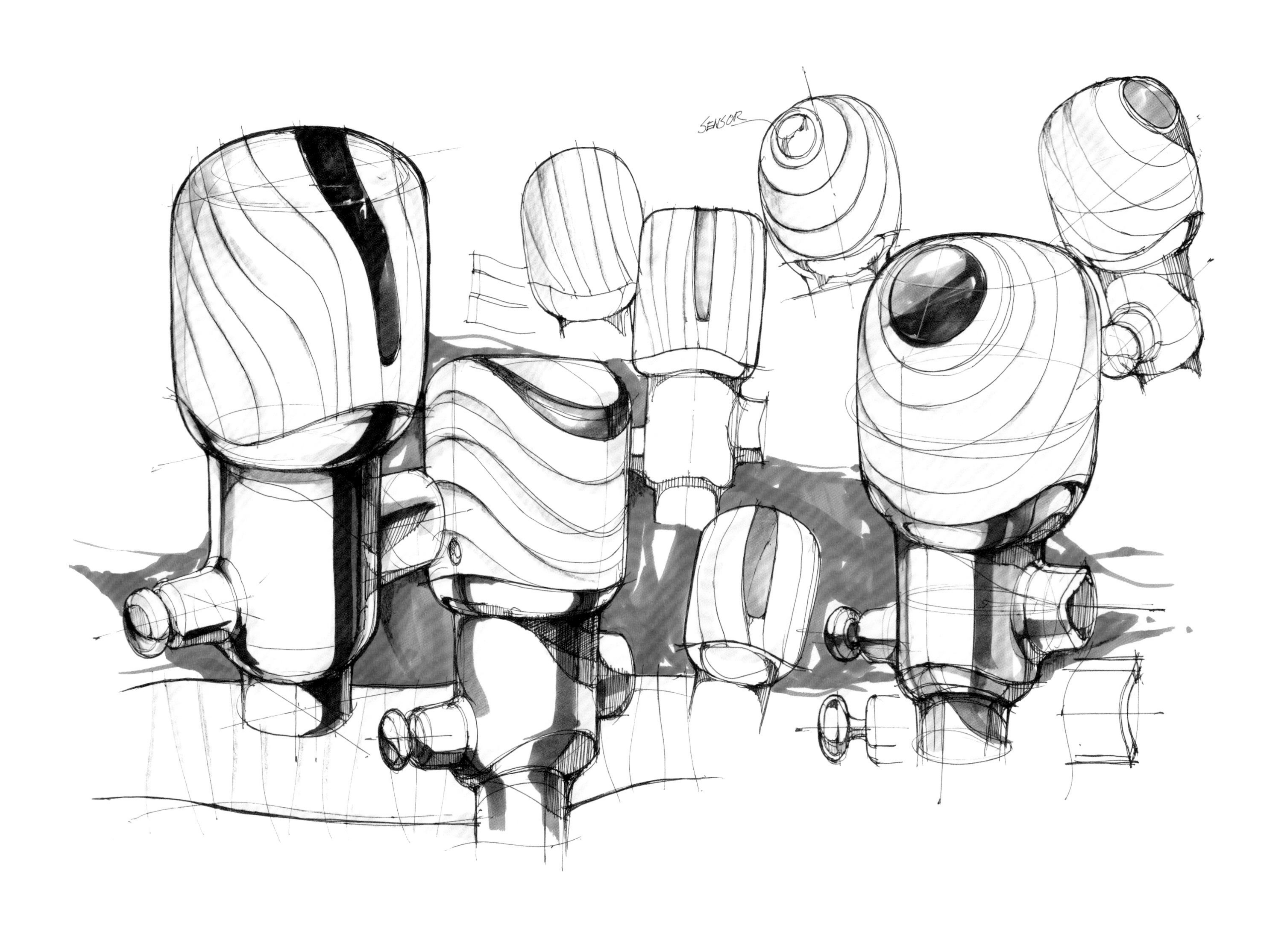
SENSOR

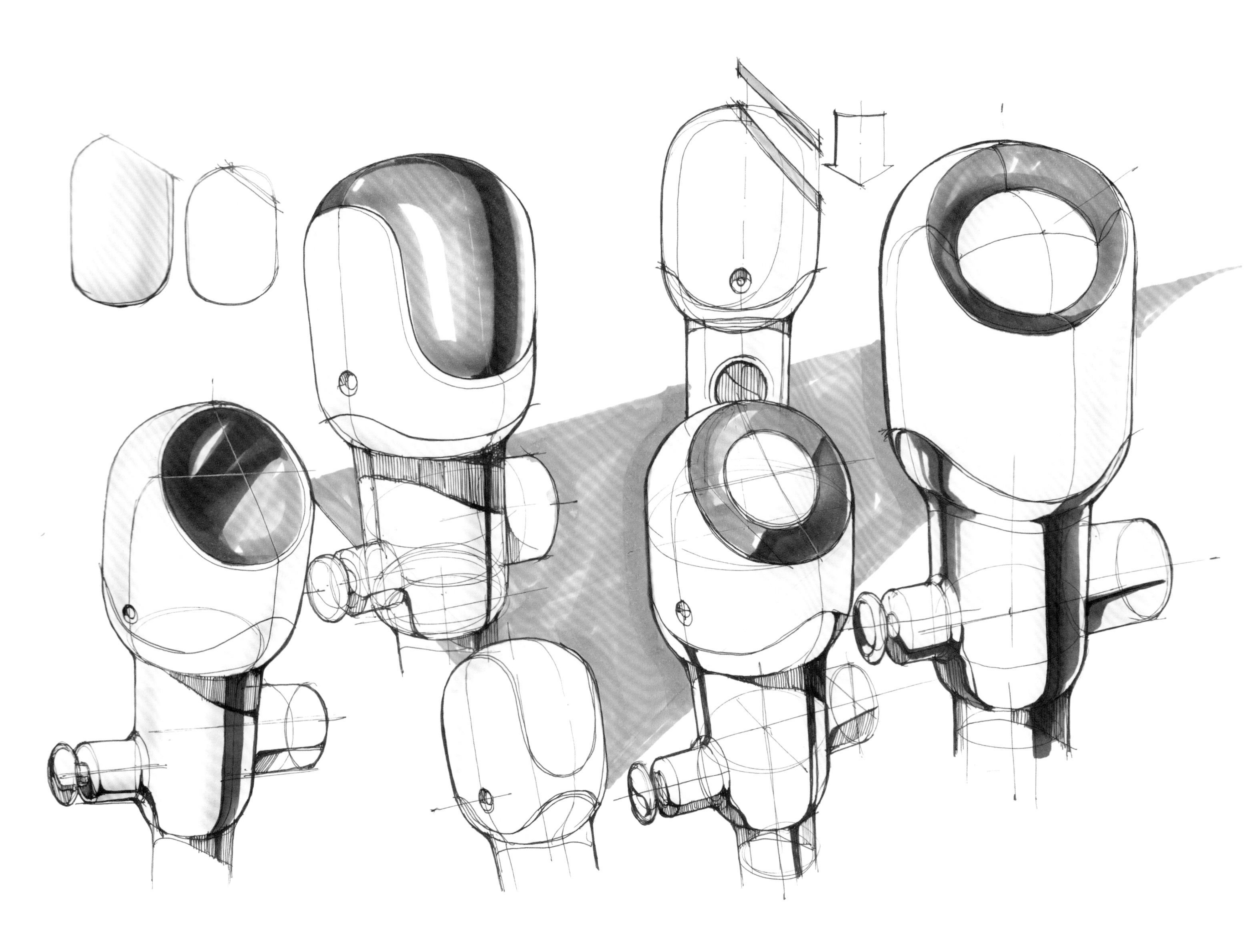

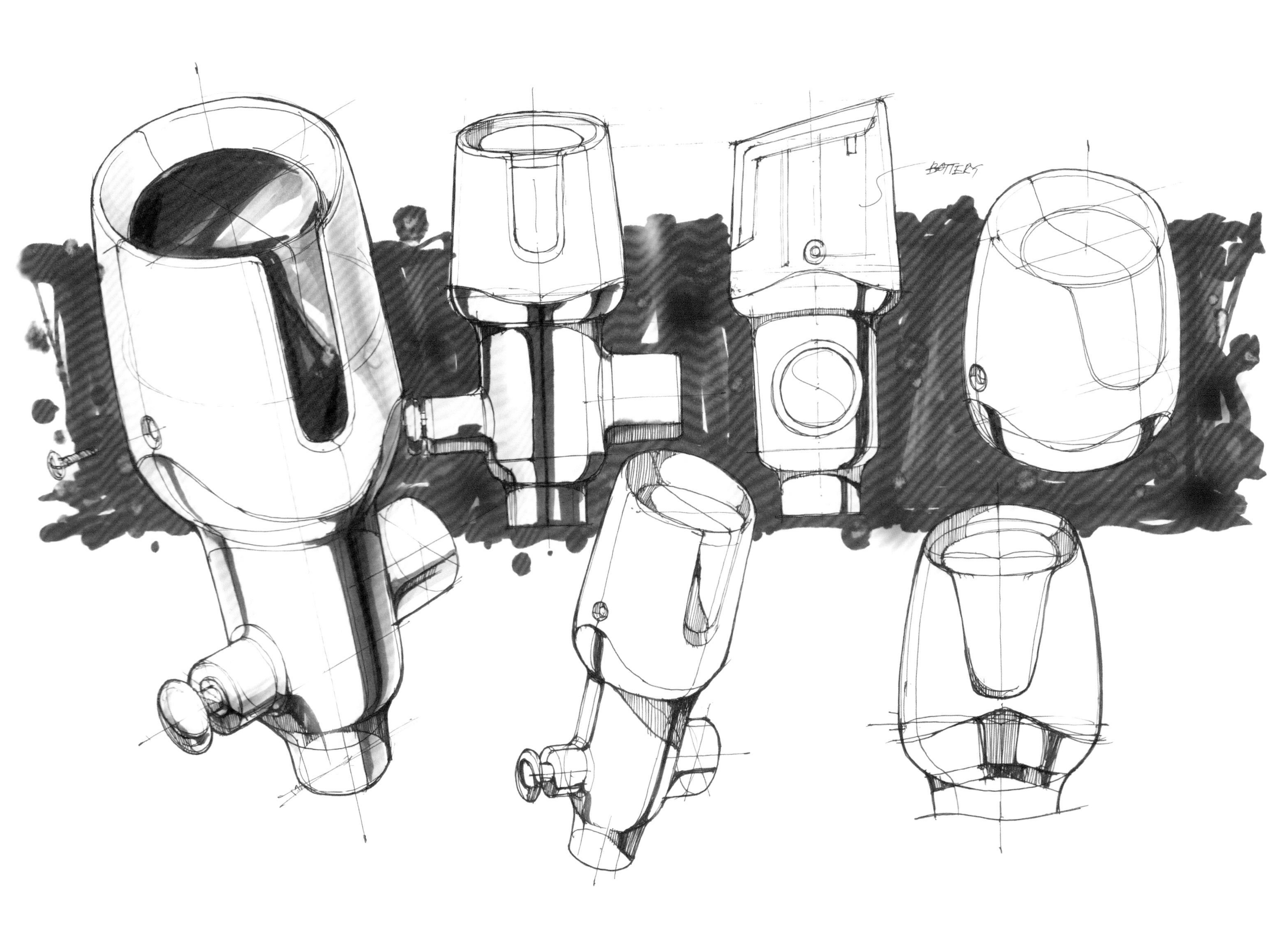
BATTERY

Body Shower System
淋浴装置

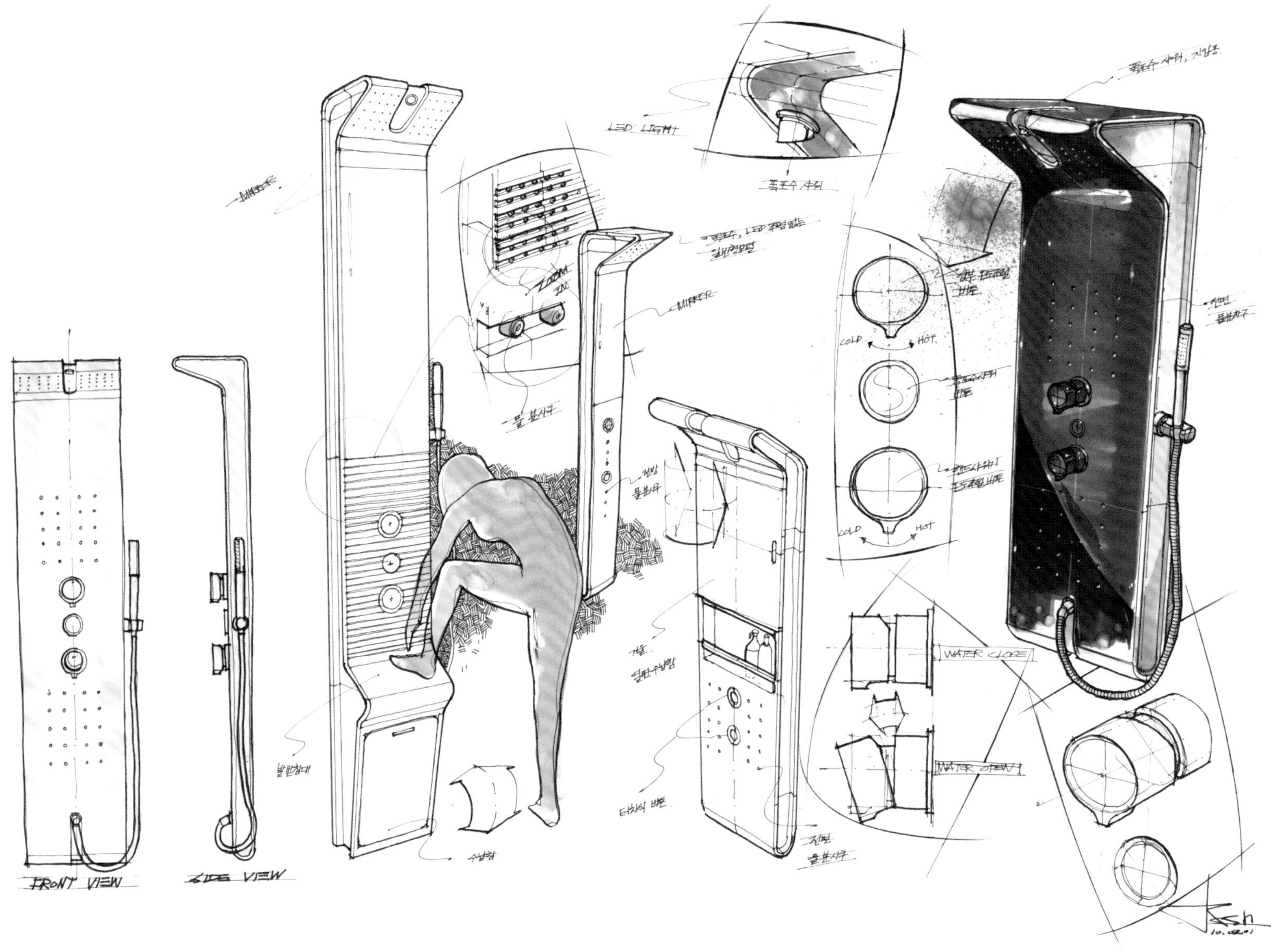

设计师：柳时彤

客户：TS JAVA公司

设计师运用探讨用户体验流程的方式寻找设计创新的可能性。通过一系列场景草图，诠释了使用者在洗澡时的行为以及这样做的原因。

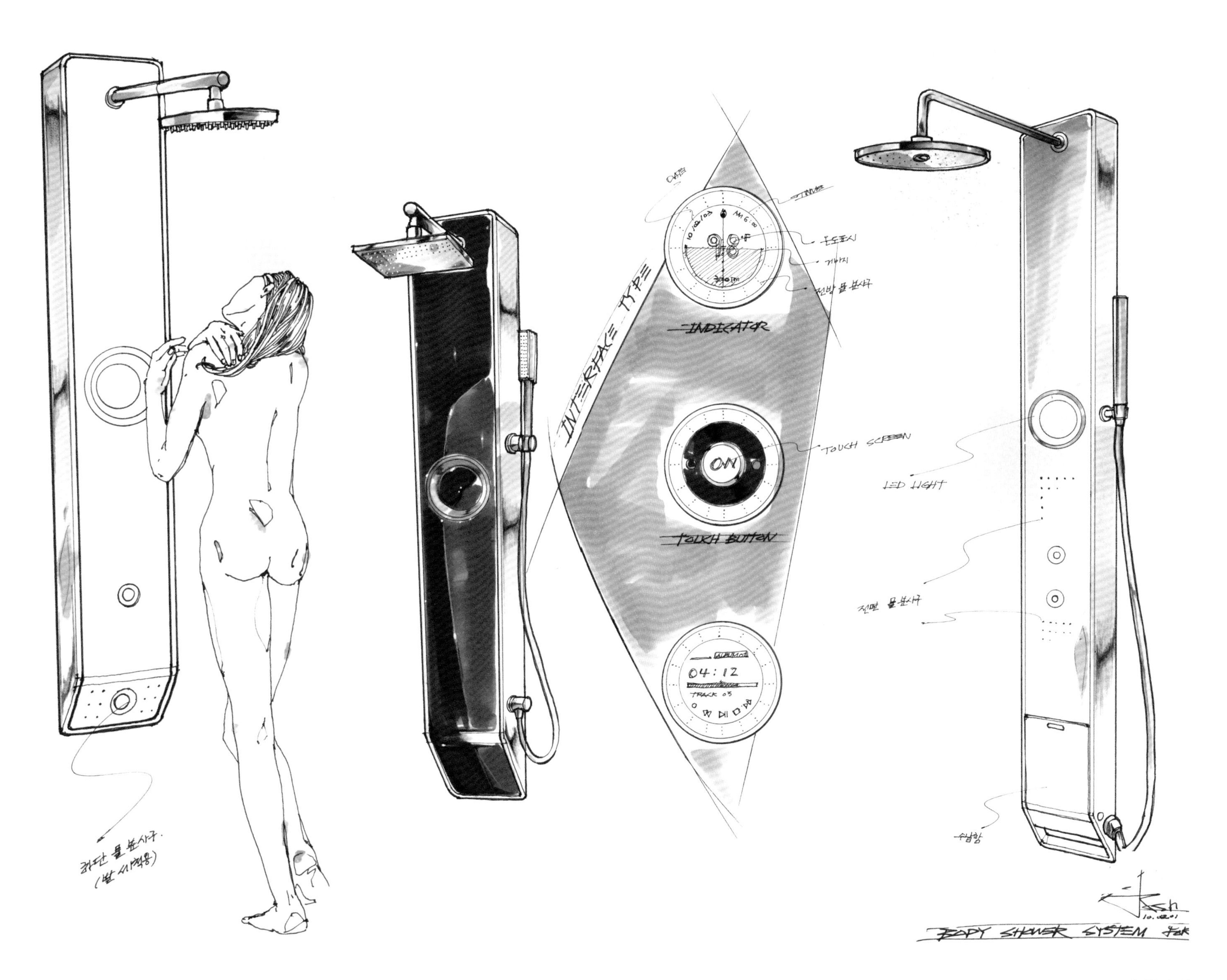
INTERFACE TYPE
DATE
TIME
10/02/03
AM 6:00
98°F
온도표시
거리
300FM
전방 물 분사구
INDICATOR
ON
TOUCH SCREEN
TOUCH BUTTON
04:12
TRACK 05
LED LIGHT
전면 물 분사구
수납함
BODY SHOWER SYSTEM FOR
10.02.01

Dustroyer: the Vacuum Duster

真空吸尘掸

设计师：柳时形

客户：个人作品

人们通常采用鸡毛掸子进行屋内清洁，然后使用扫帚或者真空吸尘器清理地面，这样往往花费很长的时间。设计师采用了这样的设计流程：定义—了解—观察—寻找契机—构思—概念深化—视觉化—最终方案。设计师以设计准则和趋势分析为基础明确了设计的方向，最后提出了真空吸尘掸的概念——一款能够进行真空吸附的除尘掸。

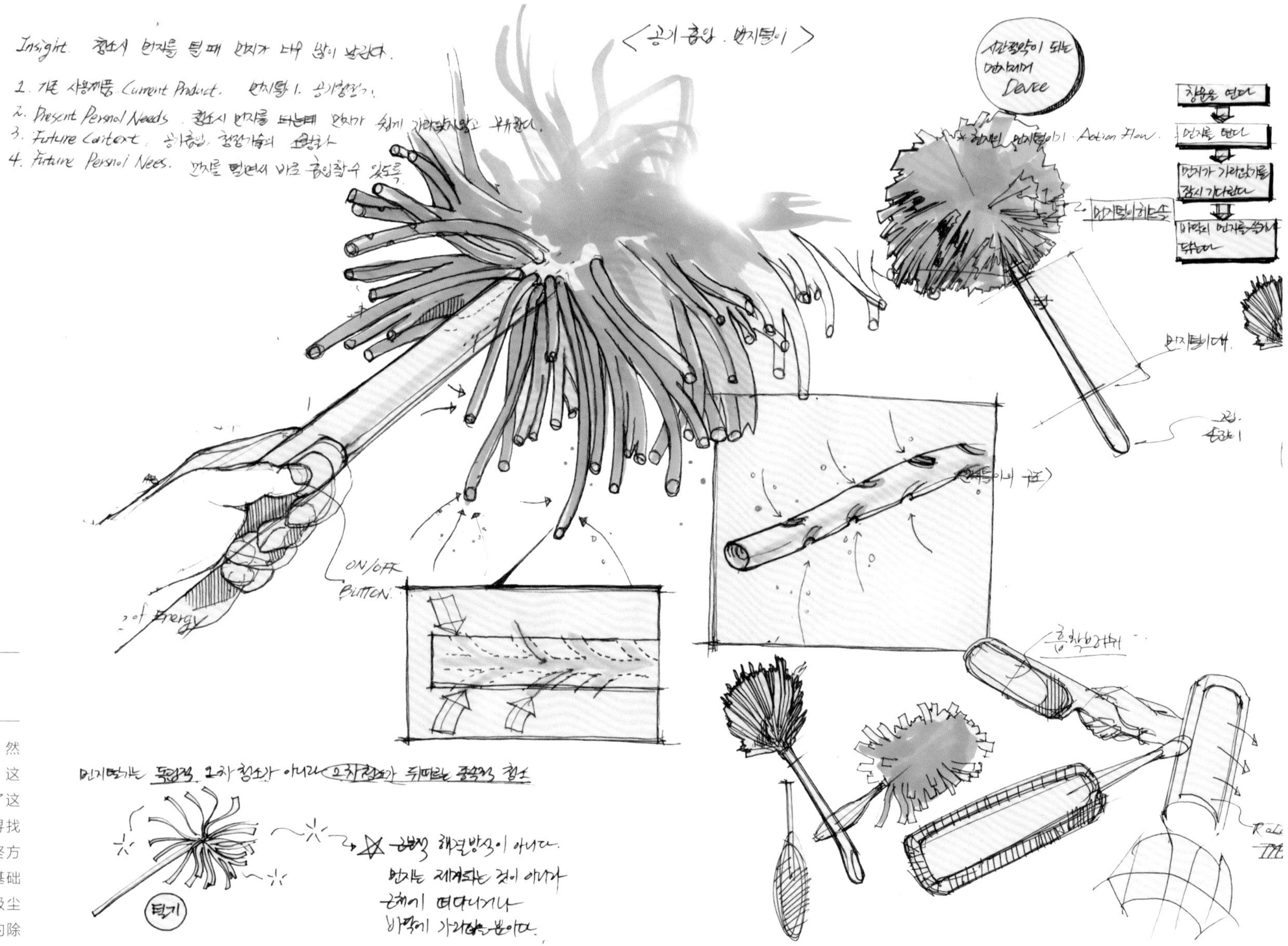

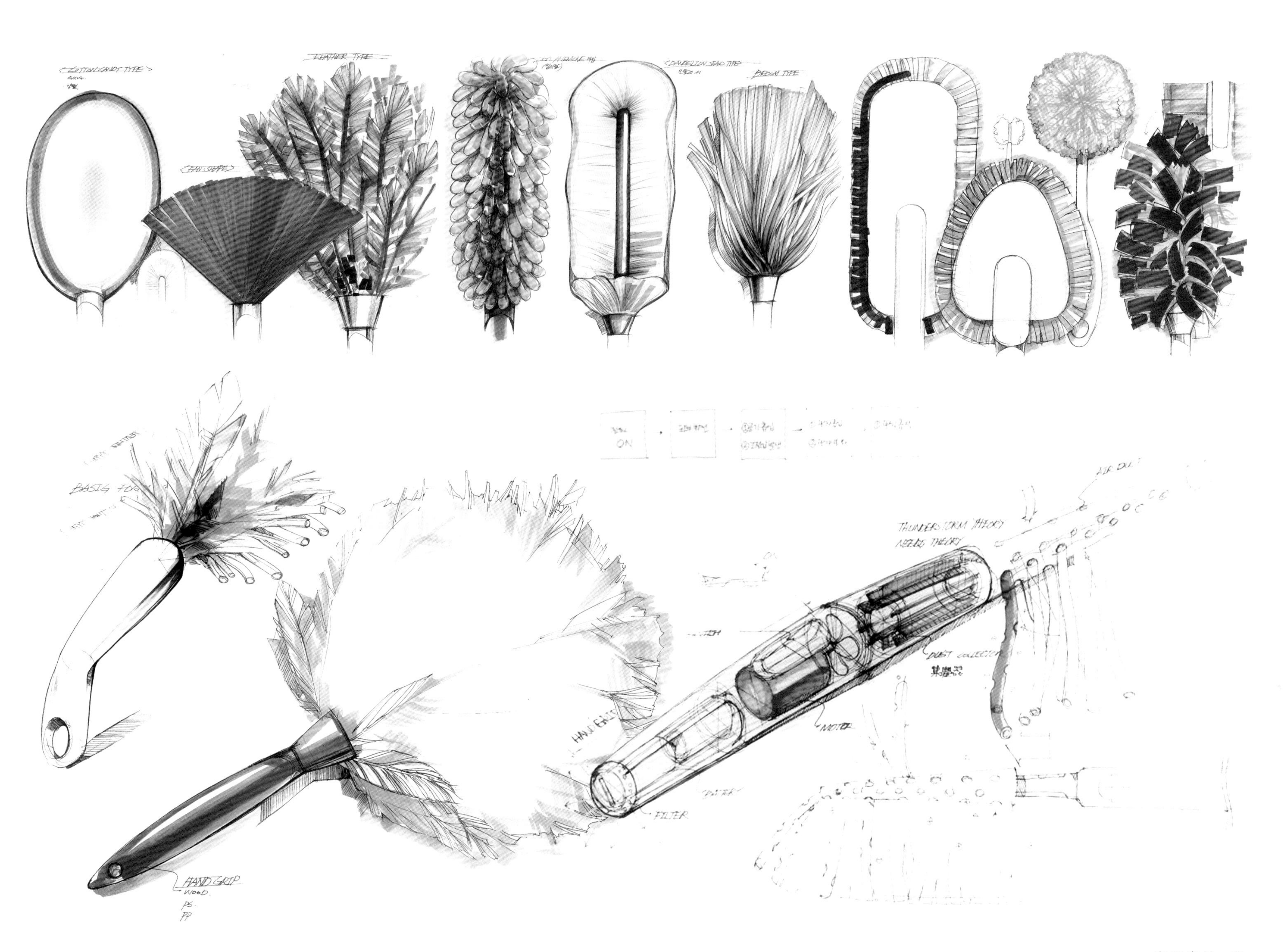
FEATHER TYPE
BROOM TYPE
ON
HAND GRIP
WOOD
PS
PP
MOTOR
FILTER

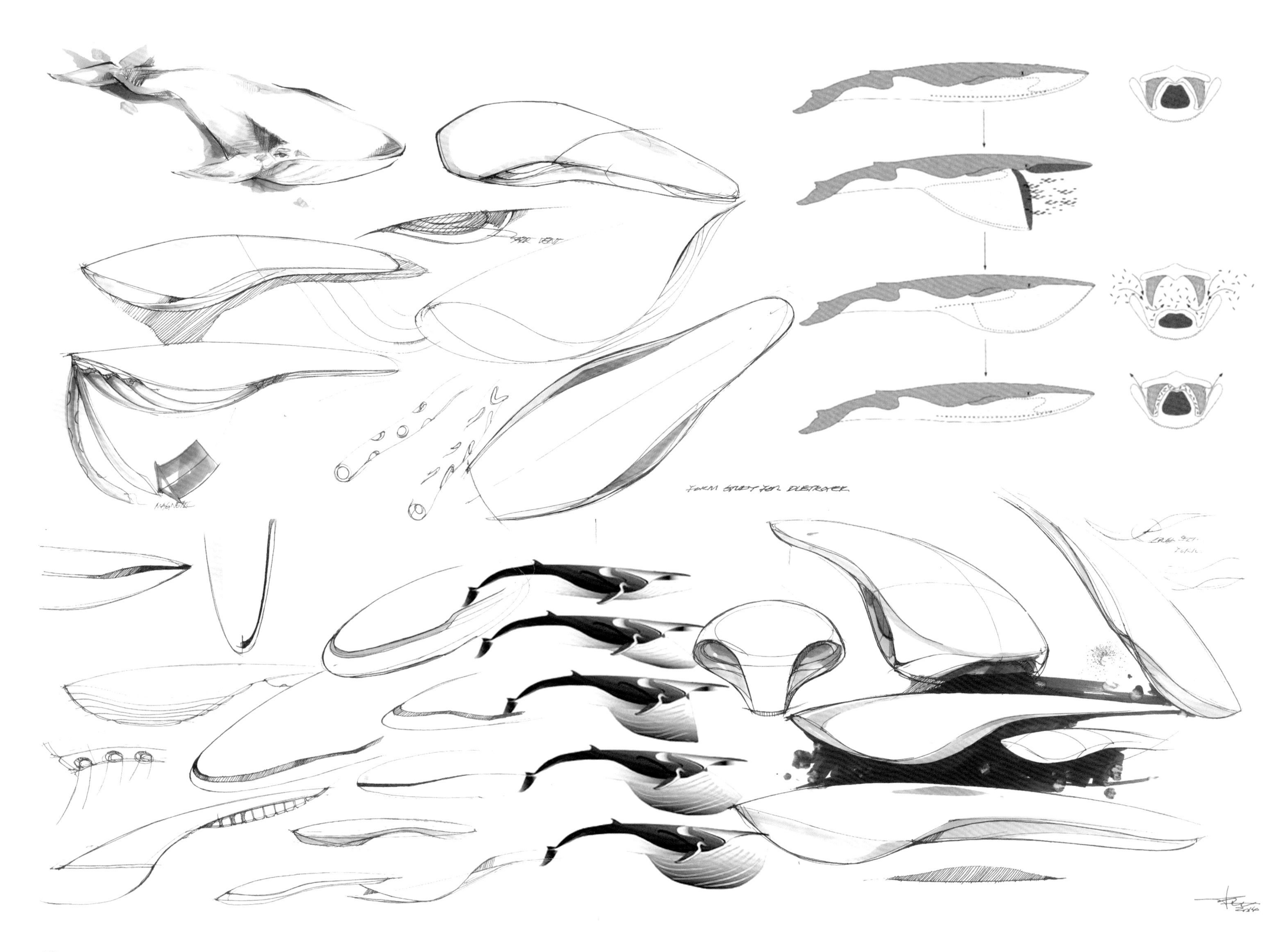
FORM STUDY FOR

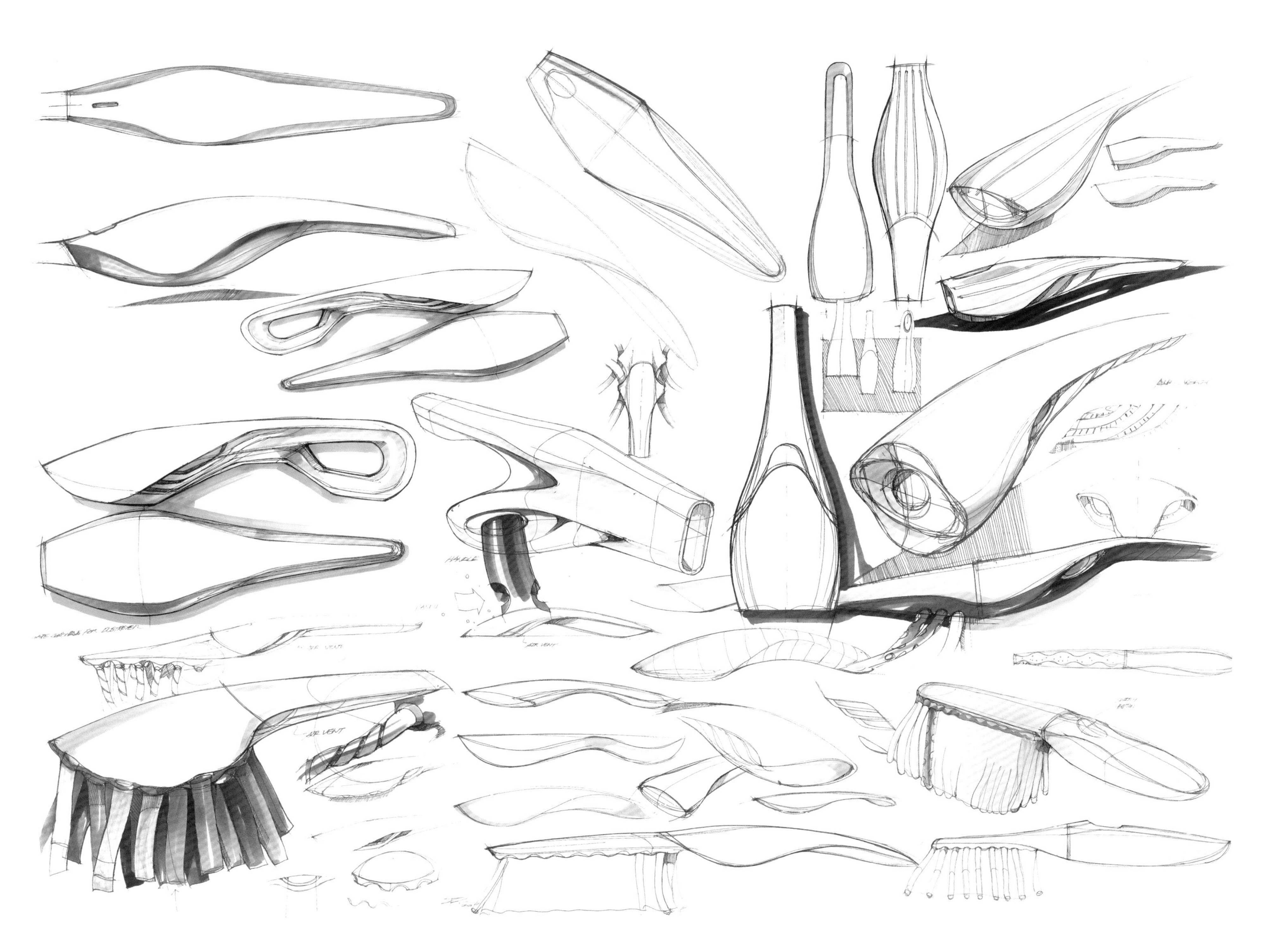

3

BAGS & SHOES
箱包鞋履

Cycling Bag

骑行背包

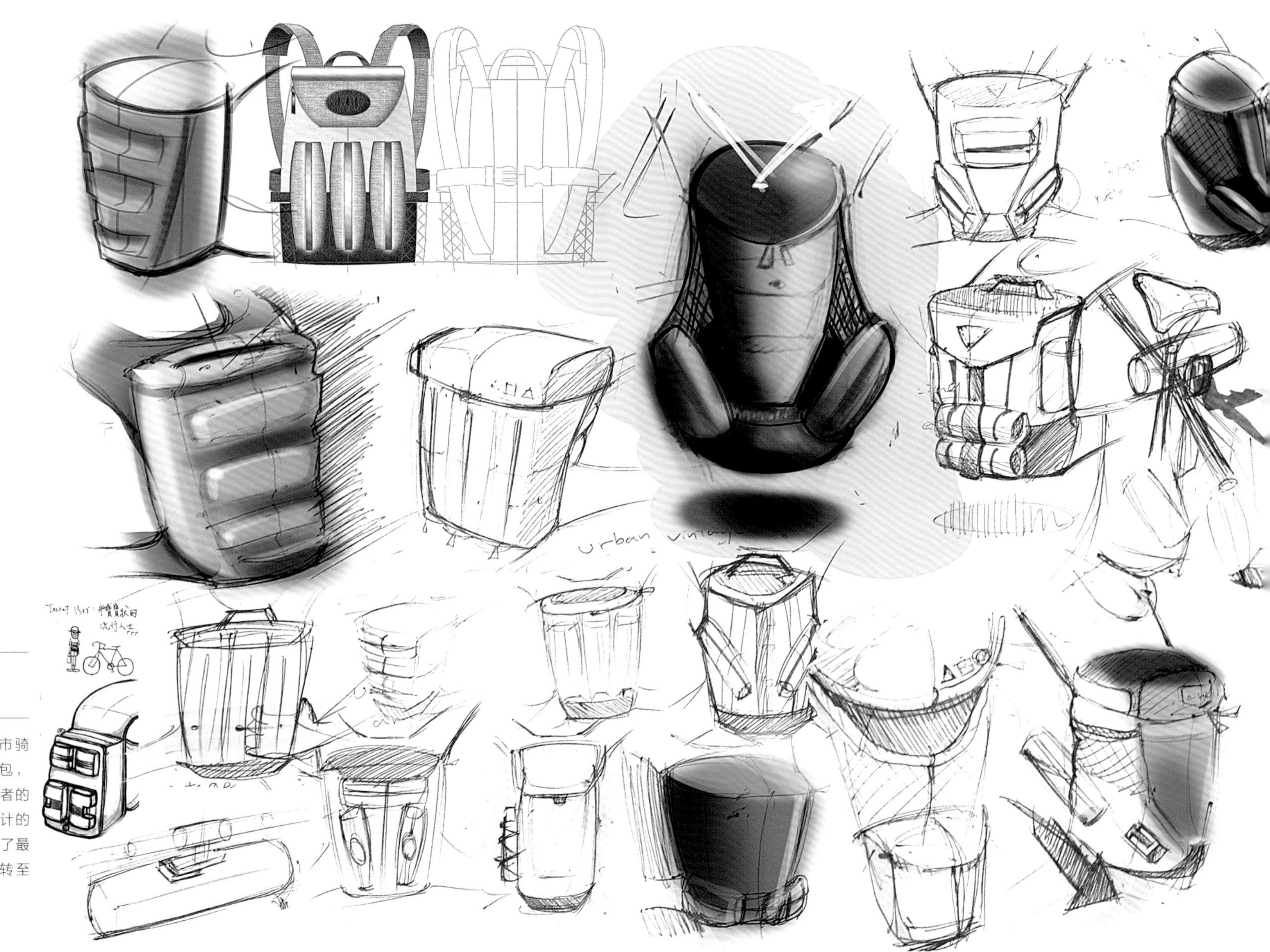

设计师：王子维（Tzu-Wei Wang）

客户：学校项目

这是一个学校设计项目，要求为城市骑行爱好者打造一款实用、时尚的背包，项目为期一个月。项目从对骑行爱好者的访谈调研开始，随后便进入到方案设计的草图绘制阶段。随后，经过讨论选定了最终的方案。最后，设计师再将草图转至Photoshop软件中进行完善和渲染。

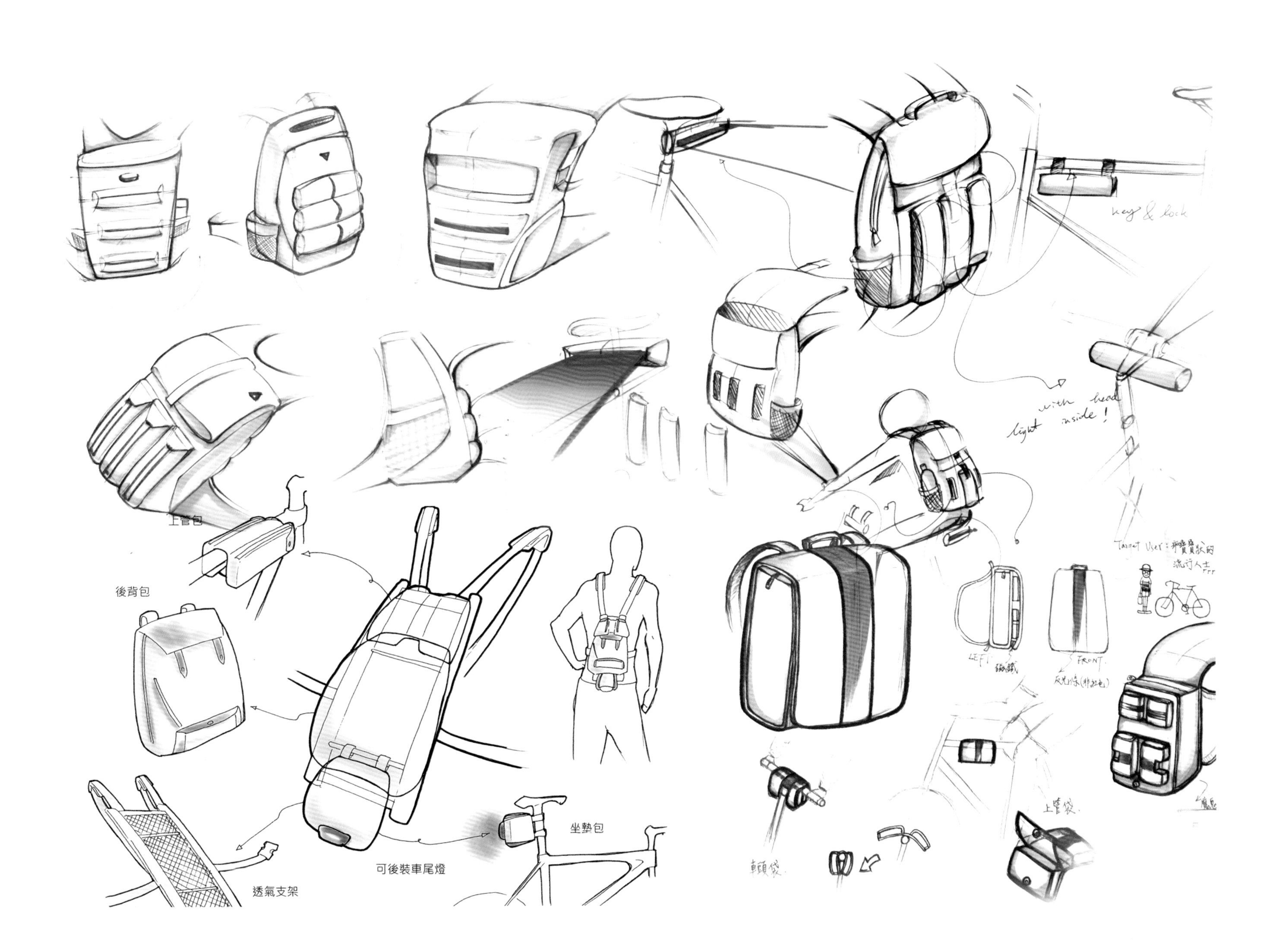
key & lock
with head light inside!
上管包
後背包
LEFT
FRONT
上管袋
透氣支架
可後裝車尾燈
坐墊包

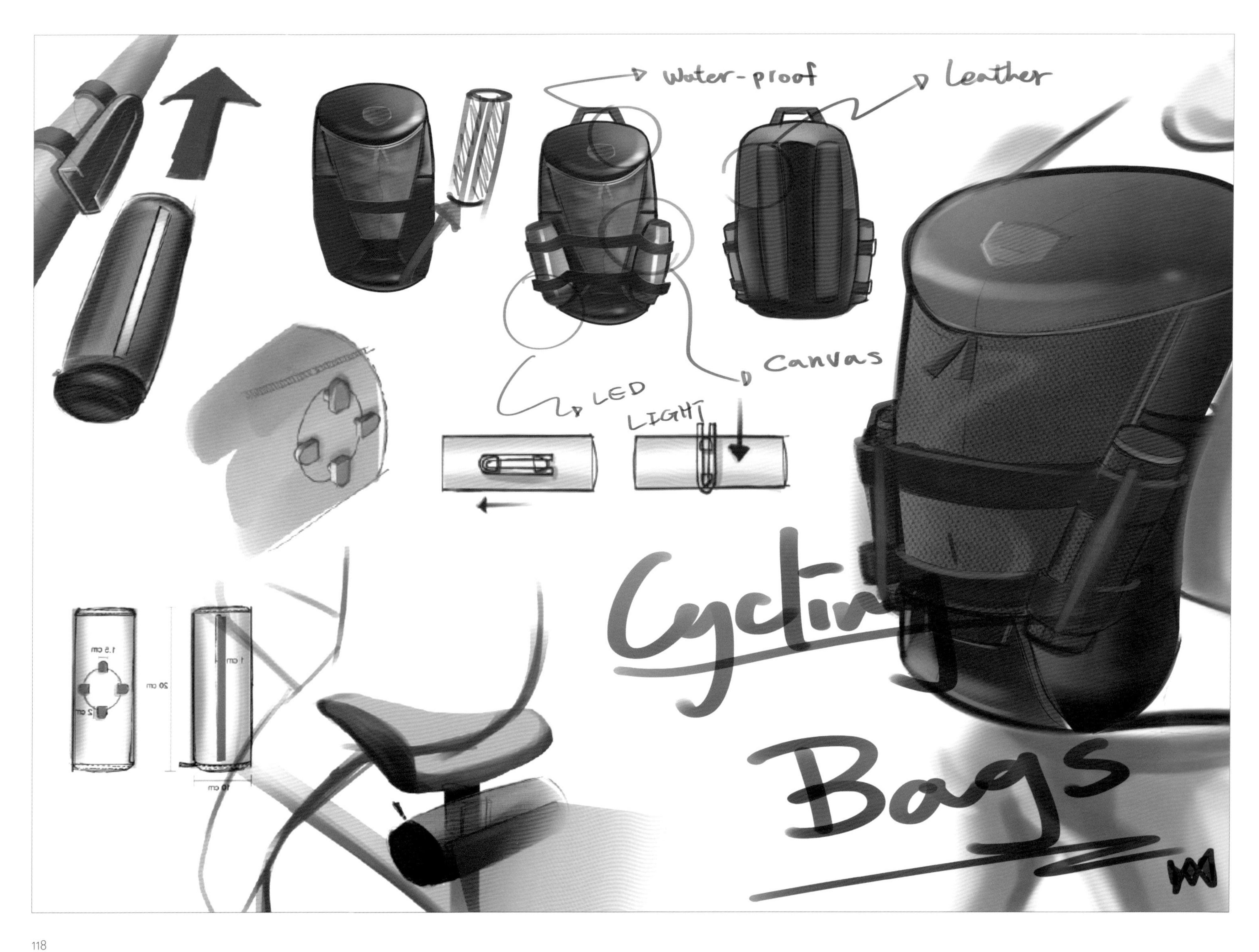
Water-proof
Leather
Canvas
LED
LIGHT
1.5 cm
2 cm
20 cm
1 cm
10 cm
Cyclin
Bags

Mesh Swim Bag
泳具包

设计师：马可·佩佐利（Marco Pezzoli）
客户：Arena公司

这款泳具包全部采用网状织物打造，无异味、易清洁、易携带、易折叠，可压缩折叠放到包顶端的小袋子里。由于泳具包的一大特点就是要保证透气，因此在进行产品设计时特别强调并表现了泳具包织物材料的科技。

BACK
hand strap
FABRIC 420 D
· COATED
· POLYESTER
· PVC
"plikè" FABRIC
fast clip
MESH S-M-L?
fast pocket X2
Shoulder strap
ZIP
ZIP
Adjustable shoulder strap
COATED FABRIC
WATERPROOF POCKET
arena
arena
arena
arena

X Fastpack
双肩泳具背包

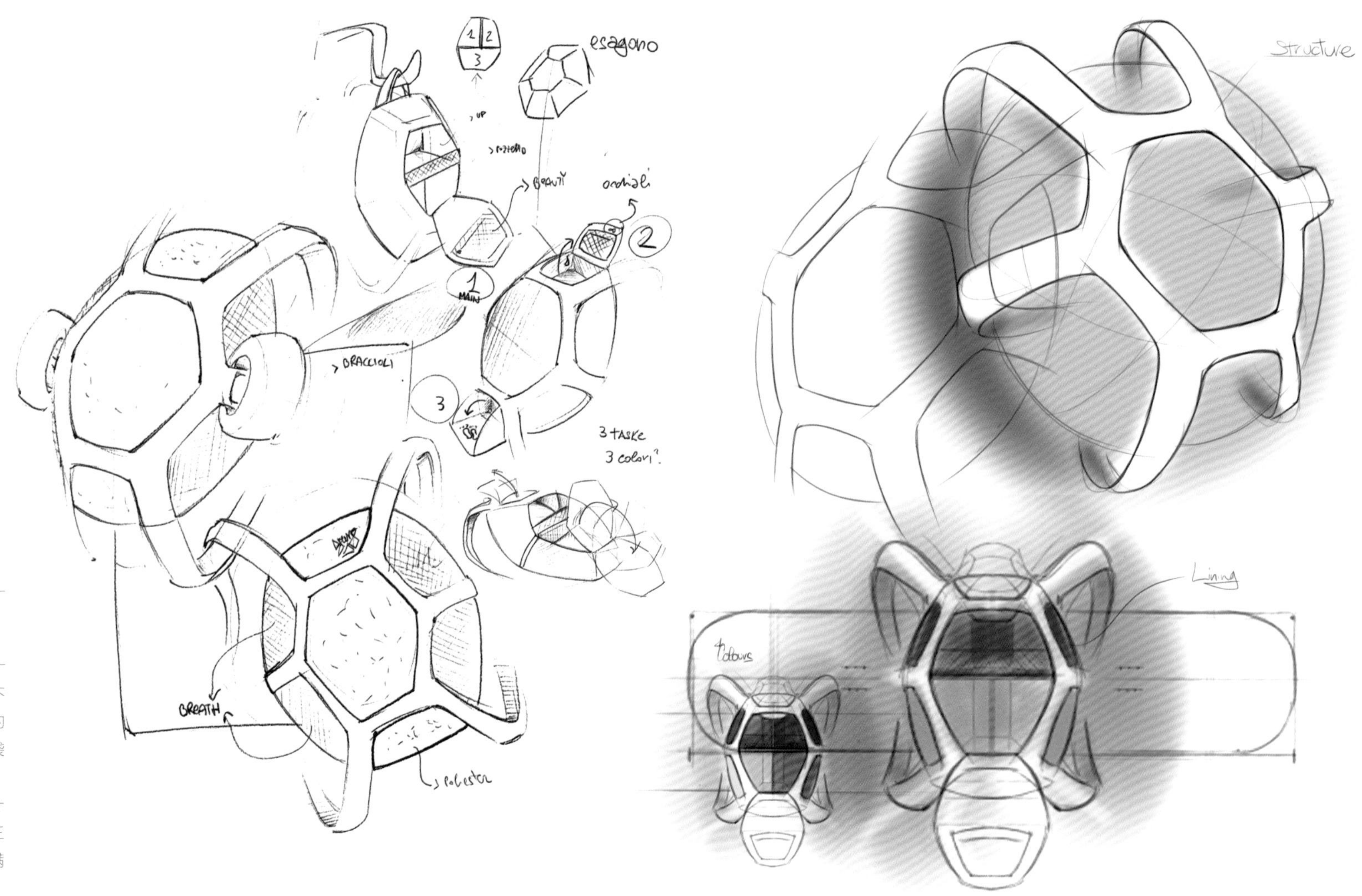

设计师：马可 · 佩佐利
客户：Arena公司

在这个设计中，书包中的每个口袋都有不同的颜色，这样可以通过色彩区分包内的物品。这款包适合盛装泳具，内部的包袋适合装湿的物品，前袋可分成两个区域，包可以被挂在挂钩上。而书包造型的设计灵感源自龟壳。其主要特点可总结为三点：仿生的造型设计、功能分区以及充满活力的色彩。该产品原本是为儿童设计，但上市之后却受到广大成年人的欢迎。

arena
X FASTPACK

The Freelancer: Casual Organizer

Freelancer
休闲包

设计师：萨拜因·安东尼娅·莱特纳
(Sabine Antonia Leitner)
客户：个人作品

这款包是系列产品中的一个，专为生活繁忙的人士打造。既可作为宽大而紧凑的手提包，也可作为挎包，有多个口袋，非常适合通勤族使用，可满足多种需求。设计师的初衷即希望将包中的随身物品收纳得井然有序。

LIGHTER SHADE
CONTRAST MATERIAL
CONTRAST
ZIP
BRUSHED SILVER
RIVET
POCKET WITH
CONTRAST ZIP & STITCHING
TWIN NEEDLE
TOP STITCH
BRUSHED SILVER
BUCKLES
CONTRAST
TOP STITCH
COIN/KEY
PURSE
DECORATIVE
LOGOTYPE
FRONT POCKET
WITH MESH FLAP POCKET
CONTRAST LINING
REMOVABLE
i-PHONE COVER &
CARD STORAGE

Umbro Field Hockey

曲棍球鞋

FOREFOOT SUPPORT

设计师：里奥 · 奥乔亚（Leo Ochoa）

客户：Nike公司、Pensole创意鞋履设计学院、艺术中心设计学院

这款鞋是钉鞋和跑鞋的混合体。Nike公司收购了Umbro品牌，这一设计要求打造一款全新运动装备——曲棍球男子运动鞋。设计师之前从未打过曲棍球，因此在设计之前首先展开了全面的调研。之后便明确了设计理念，即着重提升鞋及运动员的表现力。在构思的过程中，设计师选择了最佳的设计方案，并为此找到了最适合这一运动环境的材质，从而以2012年伦敦奥运会为主题设计了这一曲棍球鞋。

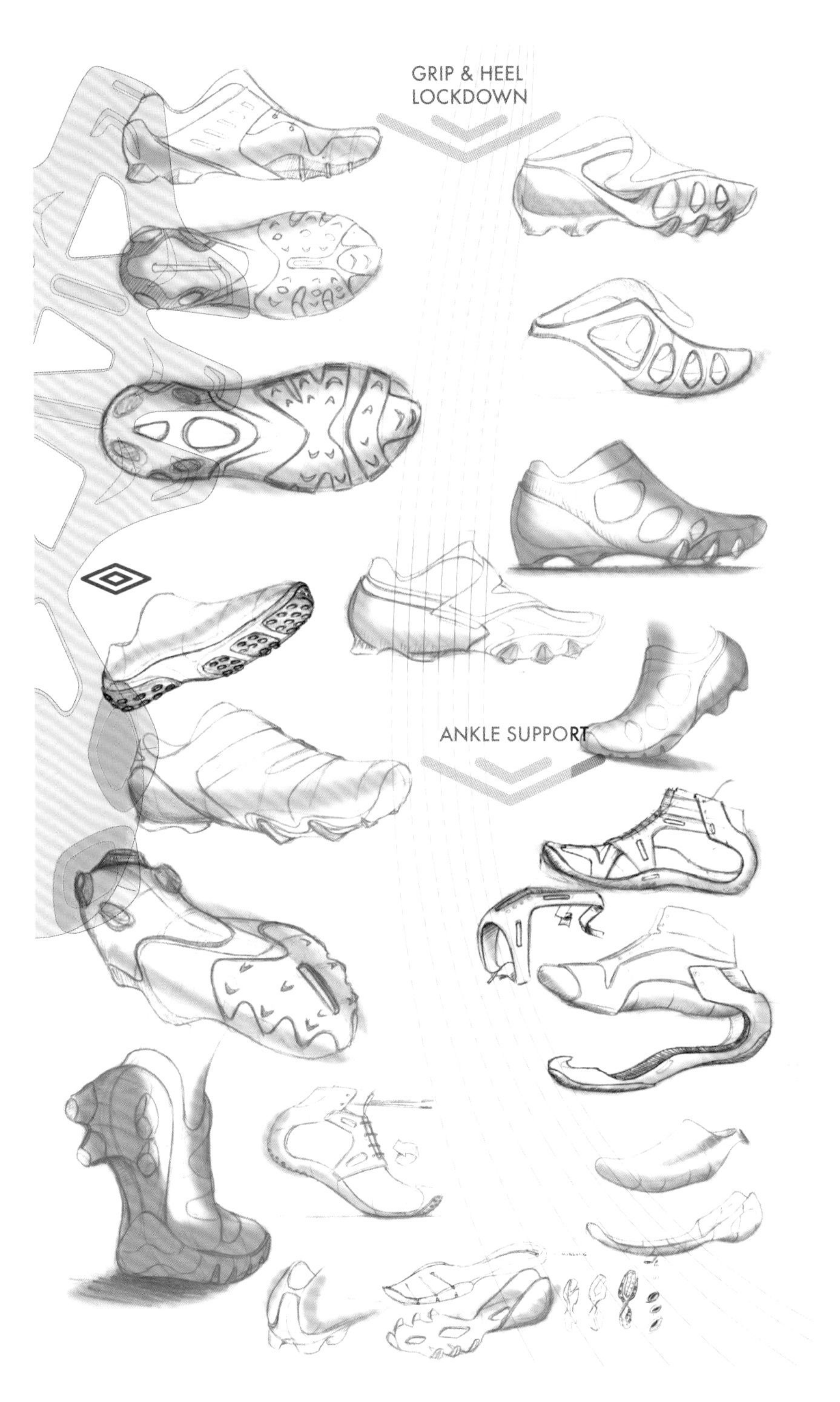
GRIP & HEEL
LOCKDOWN
ANKLE SUPPORT

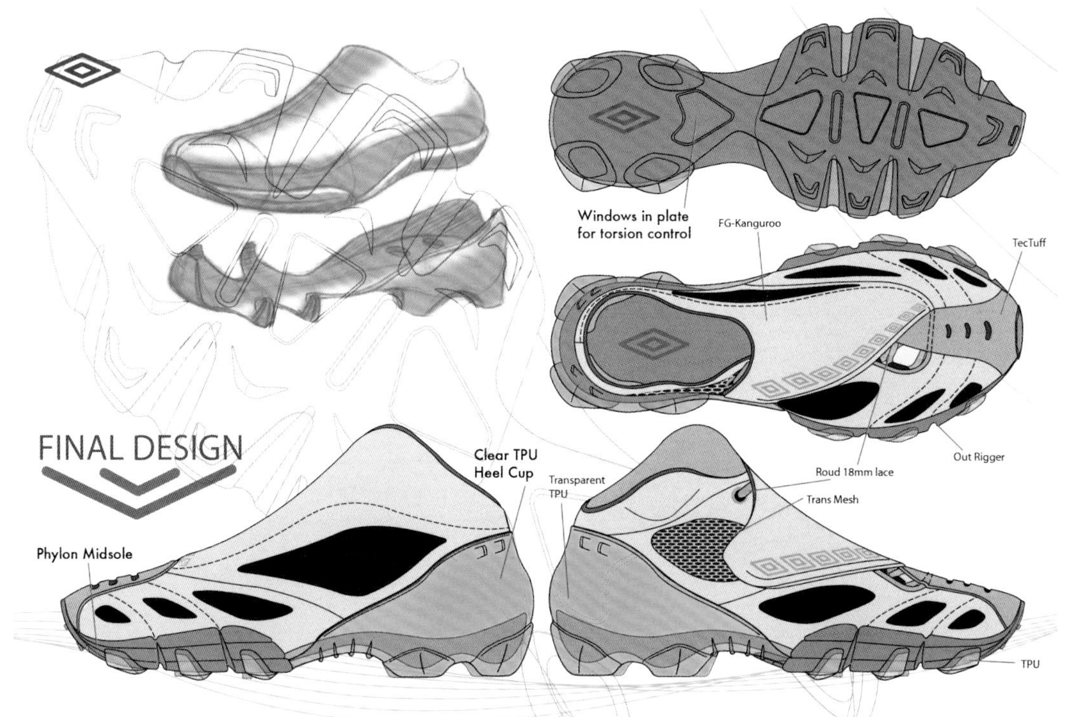
Windows in plate
for torsion control
FG-Kanguroo
TecTuff
FINAL DESIGN
Clear TPU
Heel Cup
Transparent
TPU
Roud 18mm lace
Out Rigger
Trans Mesh
Phylon Midsole
TPU

X-Country
滑雪靴

设计师：里奥·奥乔亚
客户：Adidas公司、Pensole创意鞋履设计学院

这一概念设计是Adidas公司Adipower系列产品之一，汲取了1960年代的设计风格，旨在专为17-18岁的运动健将打造一款滑雪靴。

设计师首先研究了Adidas公司1960年代的一系列滑雪靴及冬季运动鞋的设计，并与滑雪用具商店的人士深入交流，从而了解这一项运动。随后又从Adidas跑鞋技术中获得了灵感，设计了这款带有弹簧系统的滑雪鞋，使滑雪者每一步都可以获得强大的助力。

IDEATION performance

PROCESS

adiBOUNCE

"S" shape Spring Bounce

Helps lock heel for stability

adidas SKIINNOVATION

LEO OCHOA | LEOOCHOADESIGN.COM | leoochoadesign@gmail.com

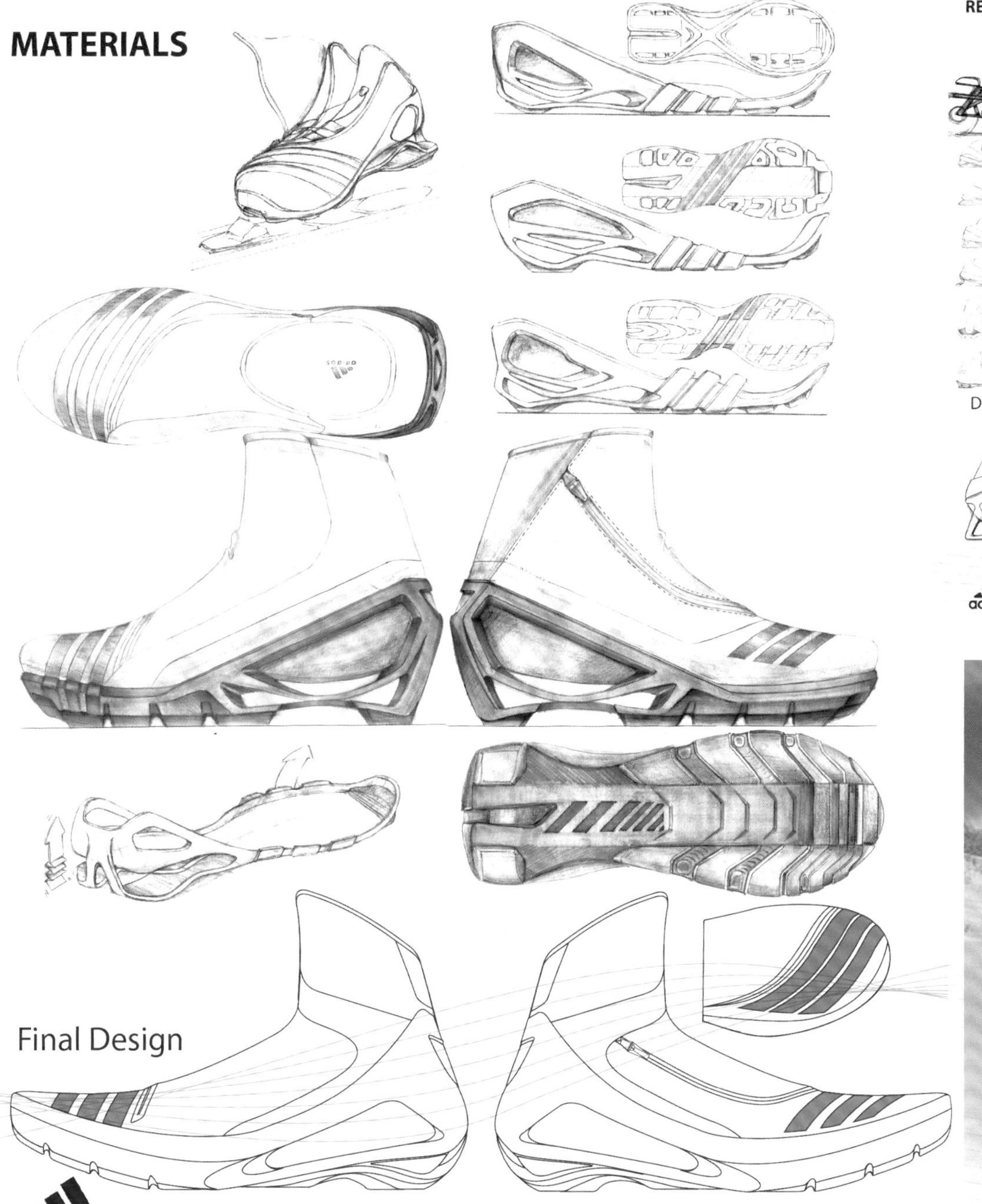
MATERIALS
Final Design
adidas SKIINNOVATION

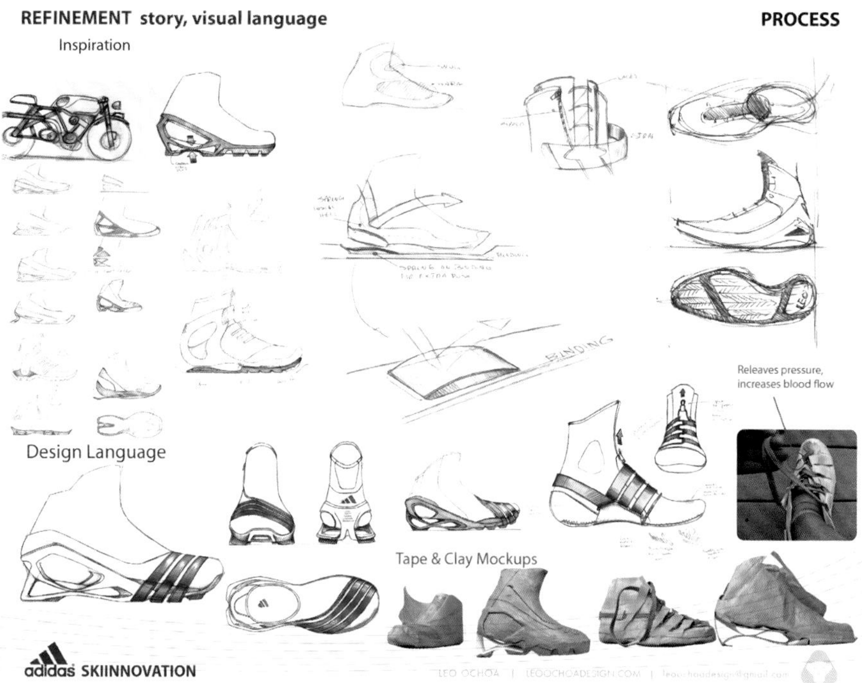
REFINEMENT story, visual language
PROCESS
Inspiration
Design Language
Releaves pressure, increases blood flow
Tape & Clay Mockups
adidas SKIINNOVATION

ADIPOWER X COUNTRY SKI BOOT
2014
adidas SKIINNOVATION PENSOLE

Nike Basketball Ideation

篮球鞋

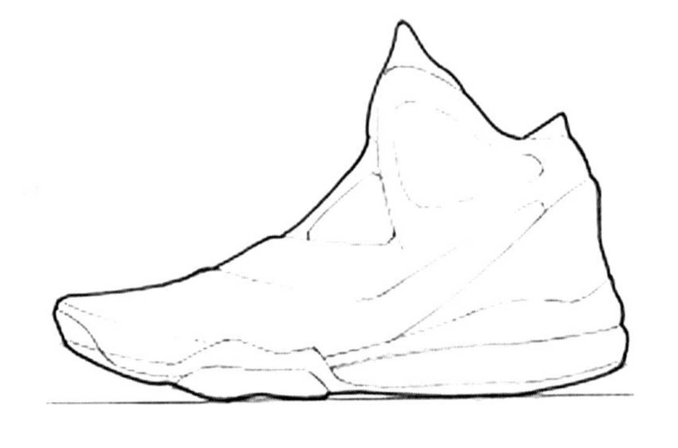

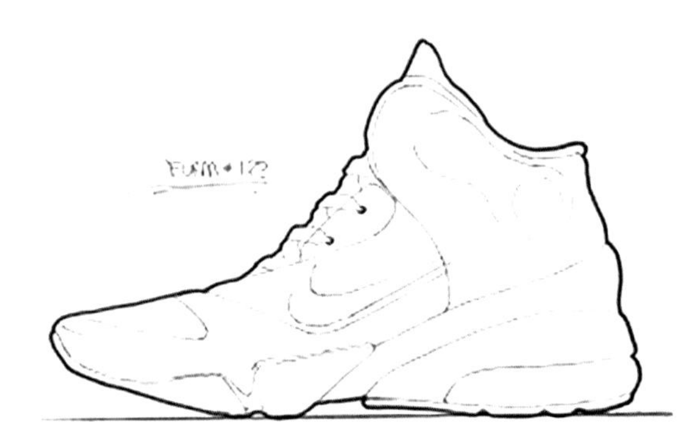

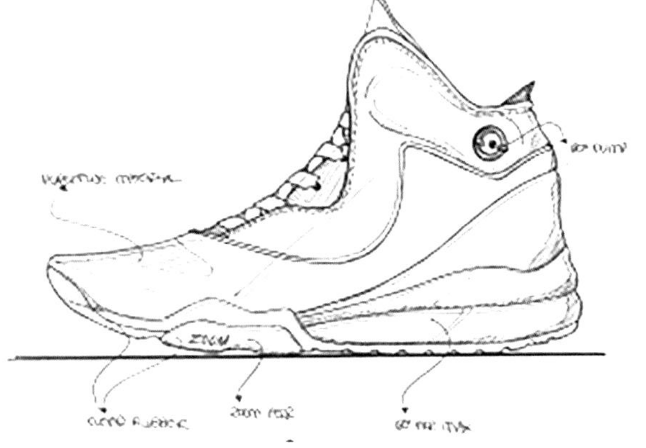

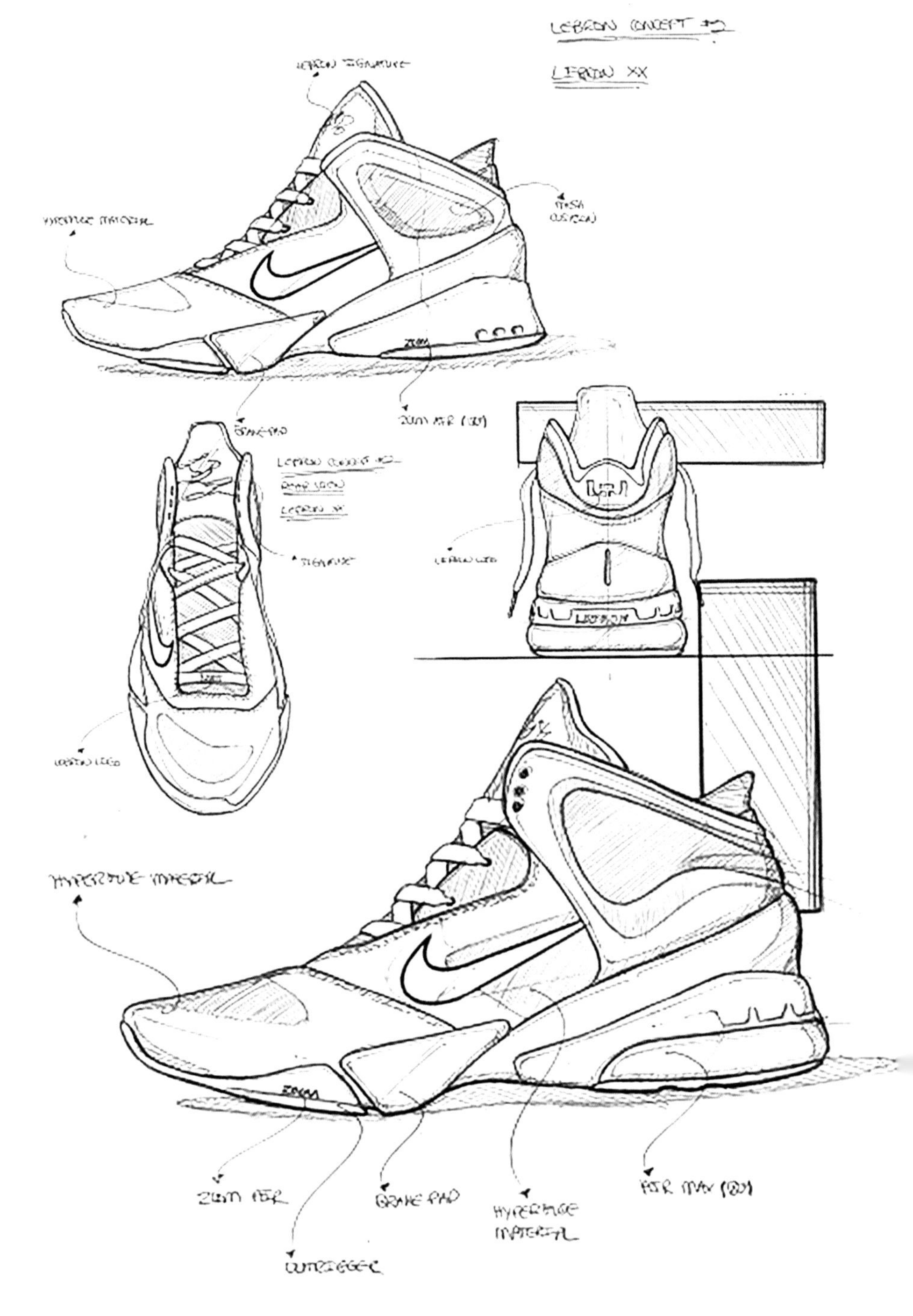

设计师：昆廷 · 威廉姆斯（Quintin Williams）
客户：Nike公司

Nike独家发布的LeBron9代是备受欢迎的鞋款。设计师将经典的“Nike Air Force 180Pump”技术同LeBron9代的时尚风格结合在一起，设计了这款Nike Hyperdunk。该款篮球鞋还加入了Flywire高强度细线支撑技术和Hyperfuse新型鞋面材质，以确保球鞋的轻量化及高透气性。

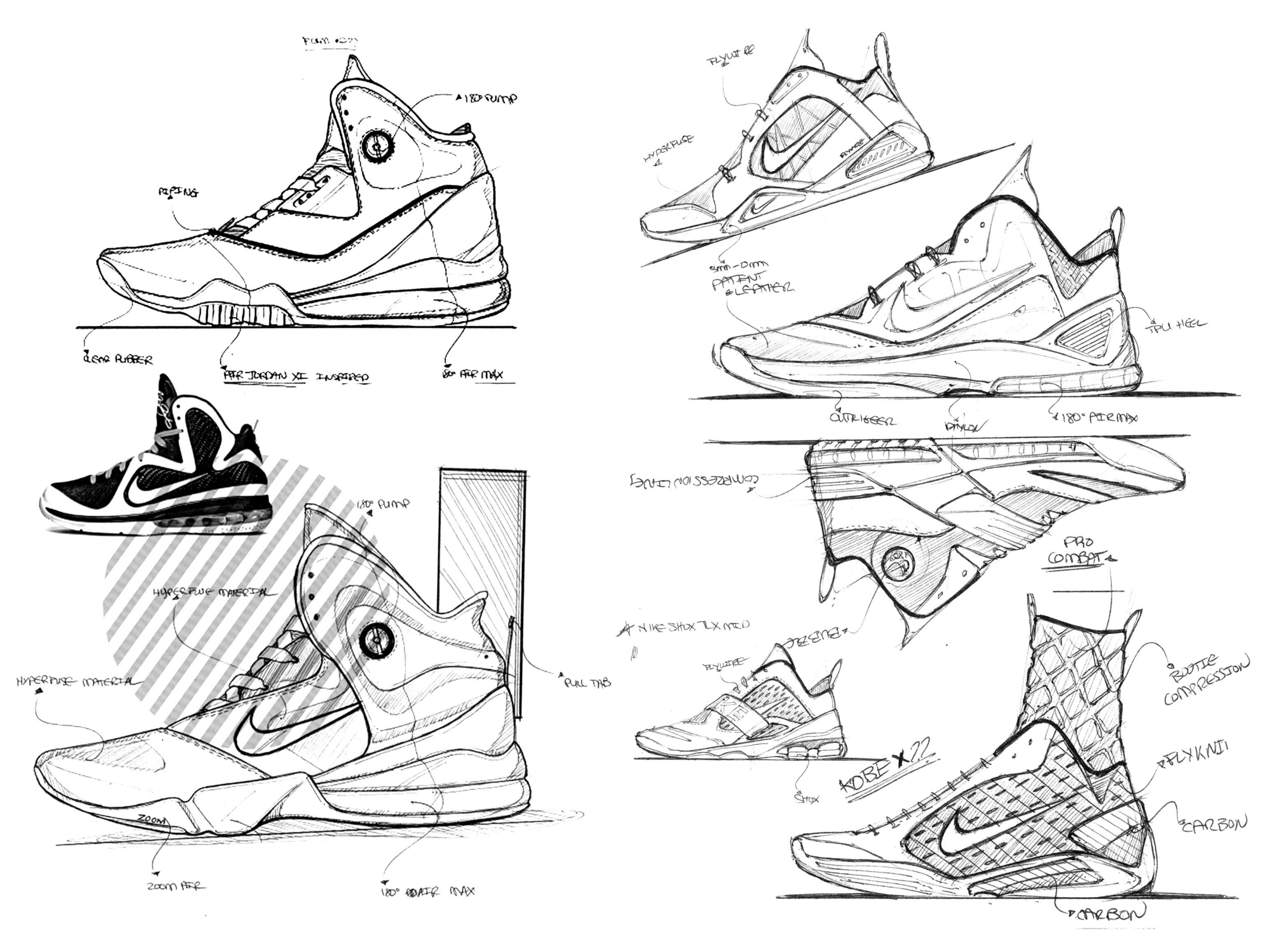
180 PUMP
PIPING
CLEAR RUBBER
AIR JORDAN XI INSPIRED
180 AIR MAX
FLYWIRE
HYPERFUSE
PATENT LEATHER
TPU HEEL
OUTRIGGER
180° AIR MAX
180° PUMP
HYPERFUSE MATERIAL
HYPERFUSE MATERIAL
PULL TAB
ZOOM
ZOOM AIR
180° AIR MAX
NIKE SHOX TLX MID
FLYWIRE
SHOX
PRO COMBAT
BOOTIE COMPRESSION
FLYKNIT
CARBON
CARBON

Basketball Shoes
篮球鞋

设计师：昆廷·威廉姆斯
客户：Q.Designs公司

这一组草图旨在探索全新的篮球鞋样式。设计过程始于搜集儿时的老鞋款，并将其重新设计以满足如今的市场需求。左侧的草图以Nike Flightposite系列为基础，鞋面的绑带融入了TPU弹力材质。而设计师希望在上面加入反光条，并更新一个大底及鞋面图案，使其更加新潮。右侧的设计草图则是从Nike Huarache Trainer系列鞋款中获得灵感。

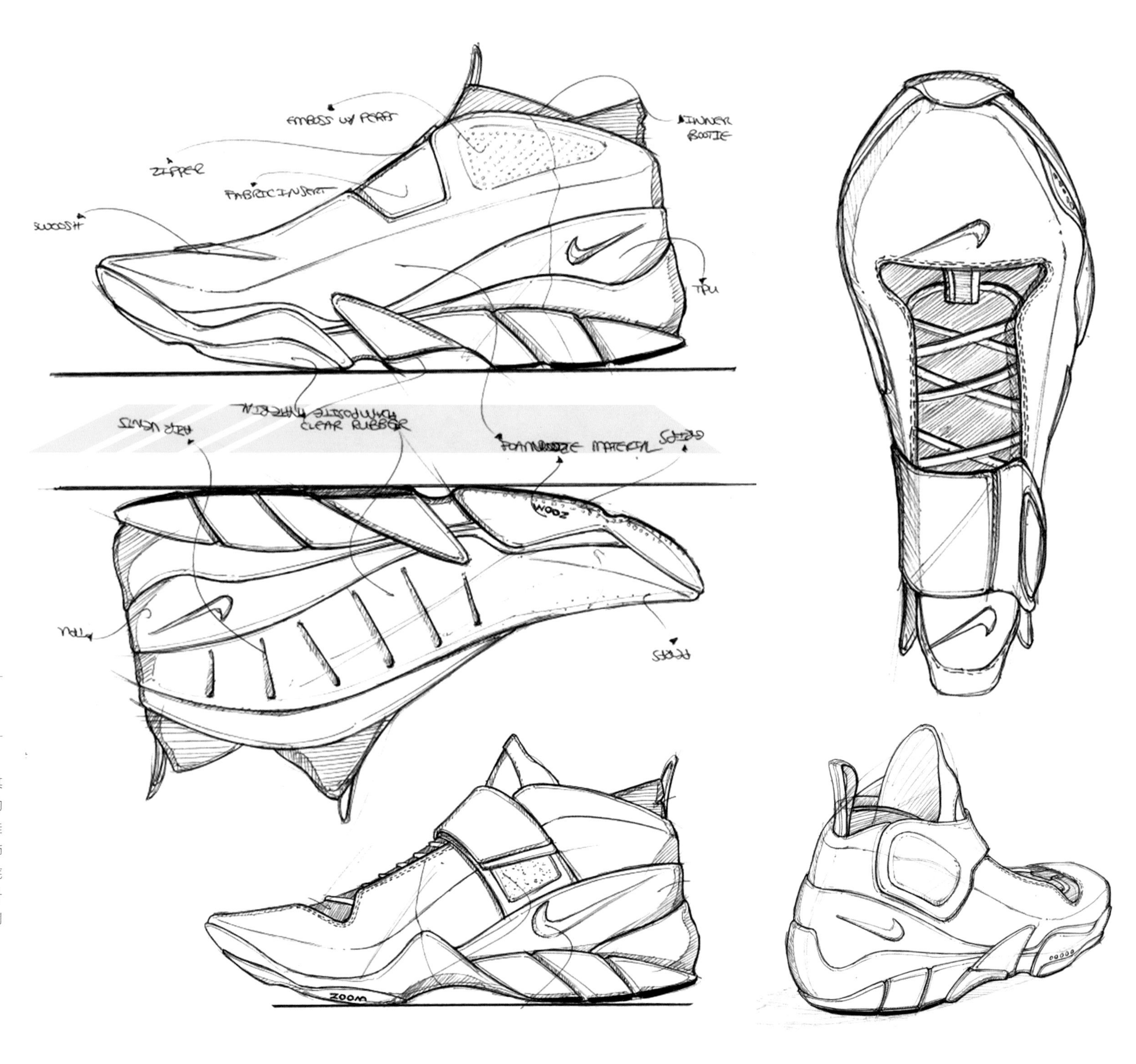

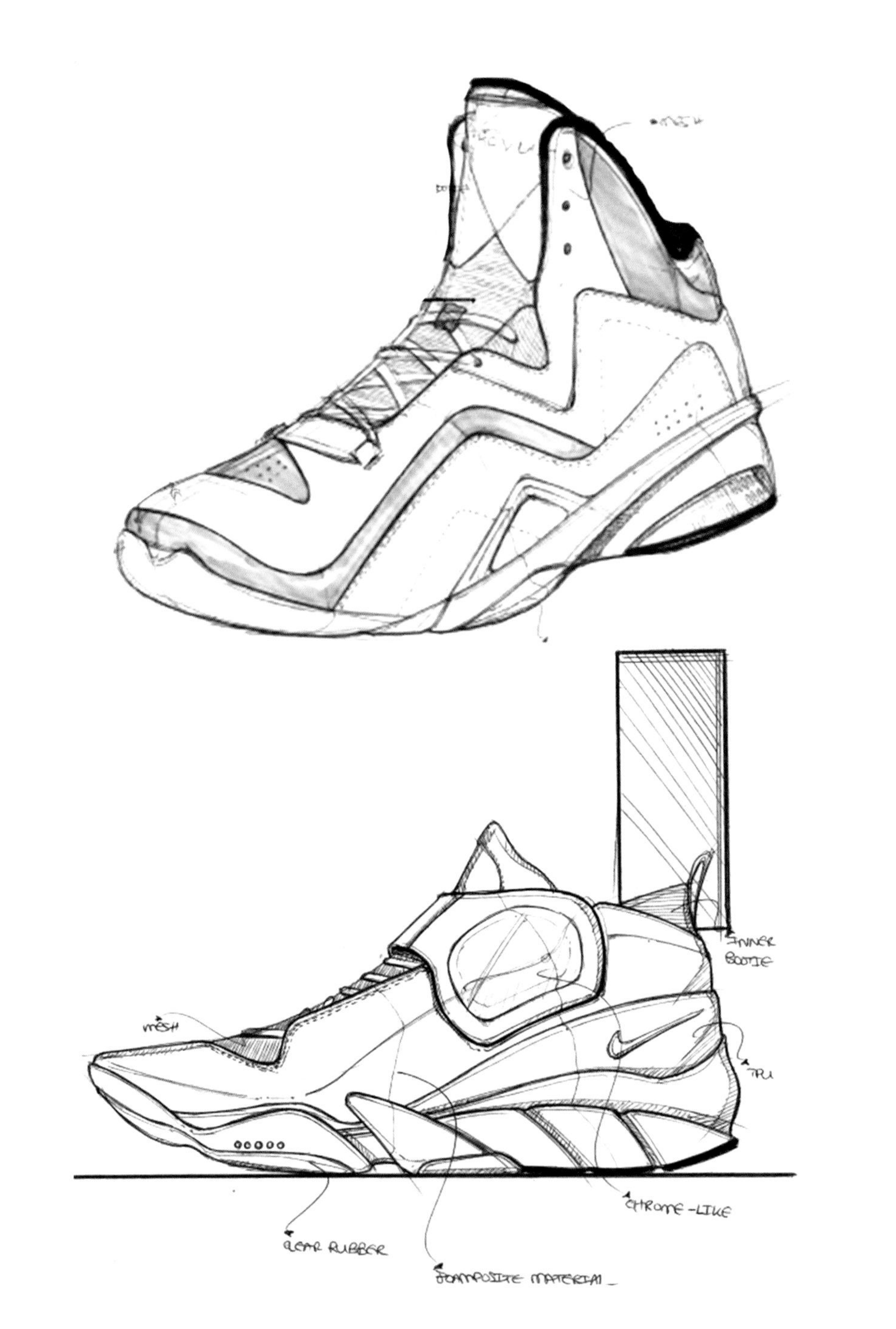
MESH
INNER
BOOTIE
MESH
TPU
CHROME-LIKE
CLEAR RUBBER
FOAMPOSITE MATERIAL

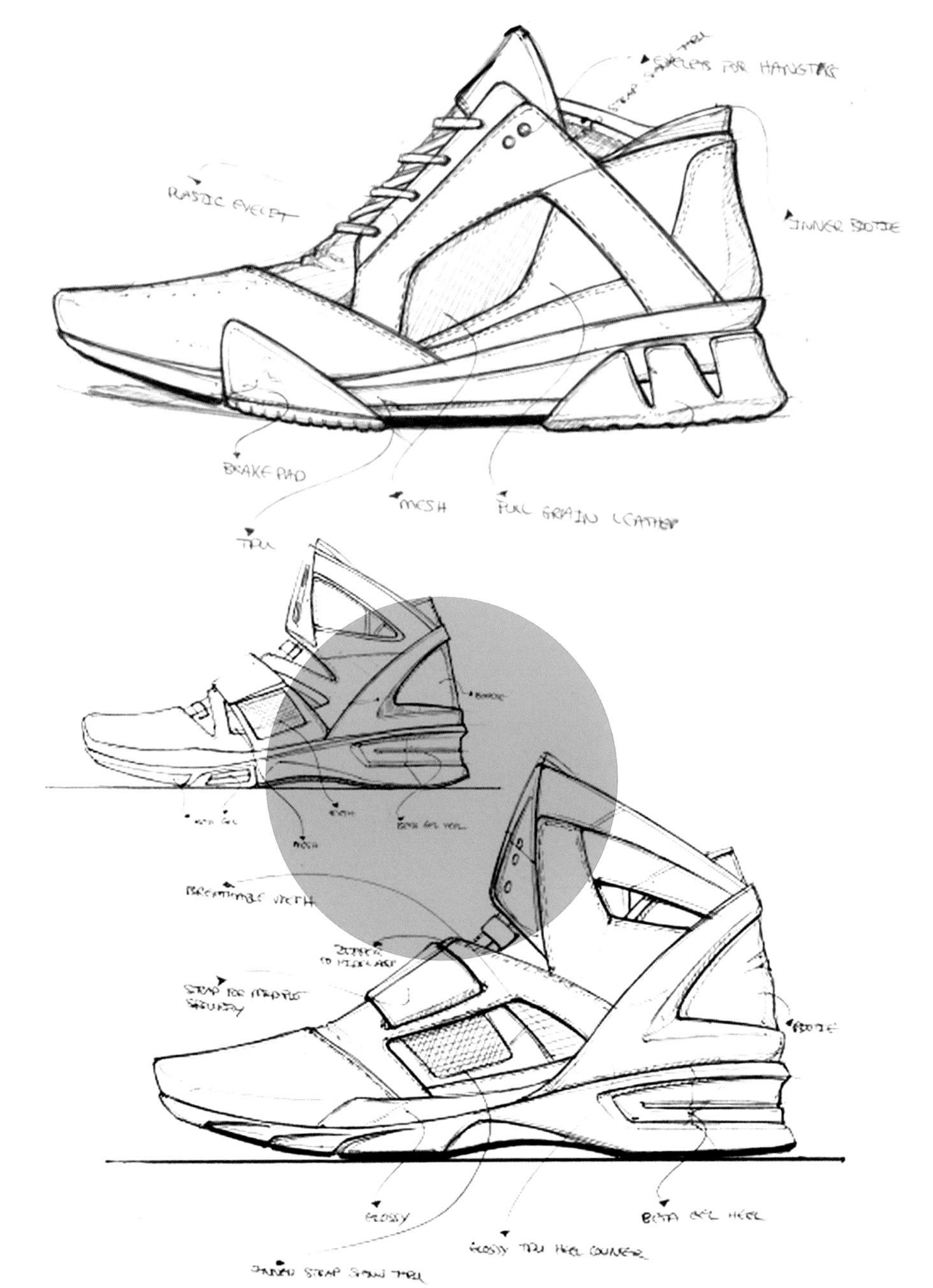
PLASTIC EYELET
INNER BOOTIE
BRAKE PAD
MESH
FULL GRAIN LEATHER
TPU
BOOTIE
BREATHABLE MESH
STRAP FOR MIDFOOT
SECURITY
BOOTIE
GLOSSY
BETA GEL HEEL
GLOSSY TPU HEEL COUNTER
INNER STRAP SHOW TPU

Skechers Runner/Jogger
慢跑鞋

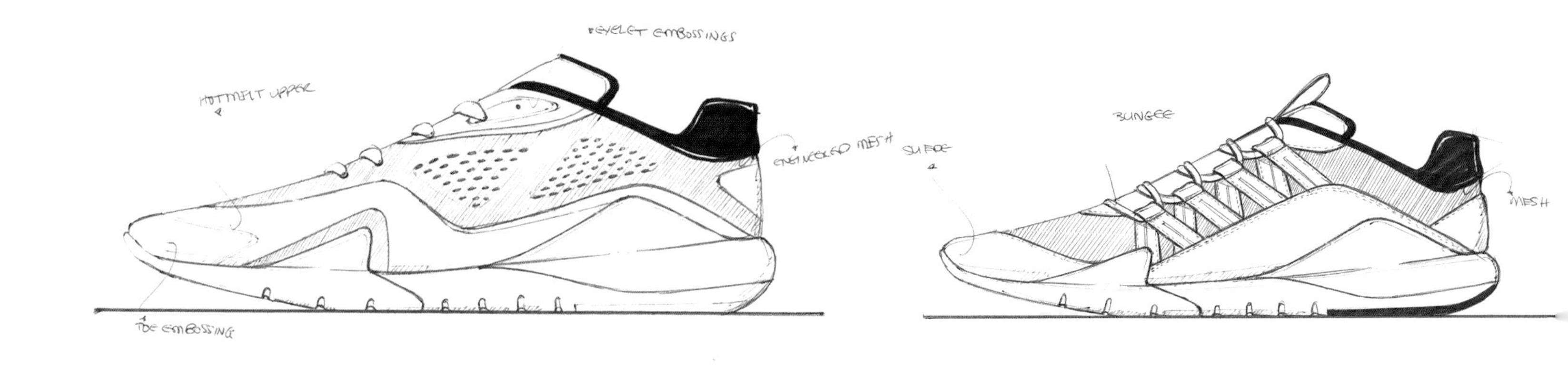

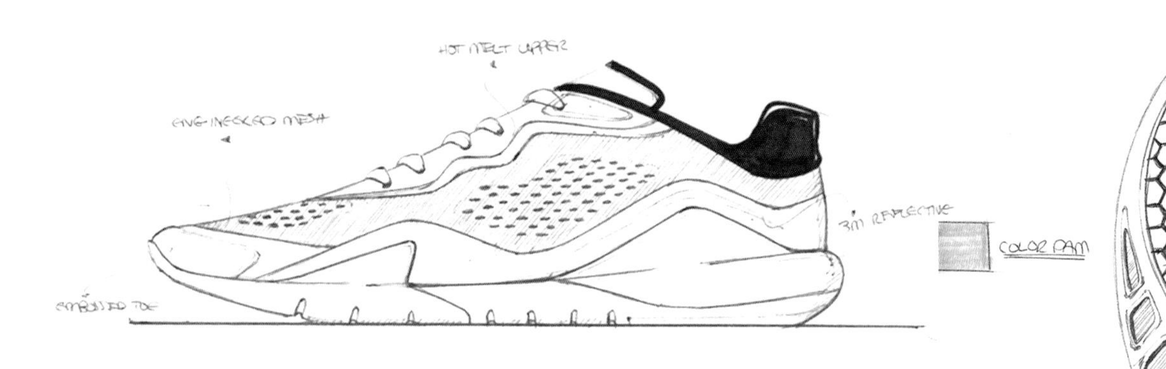

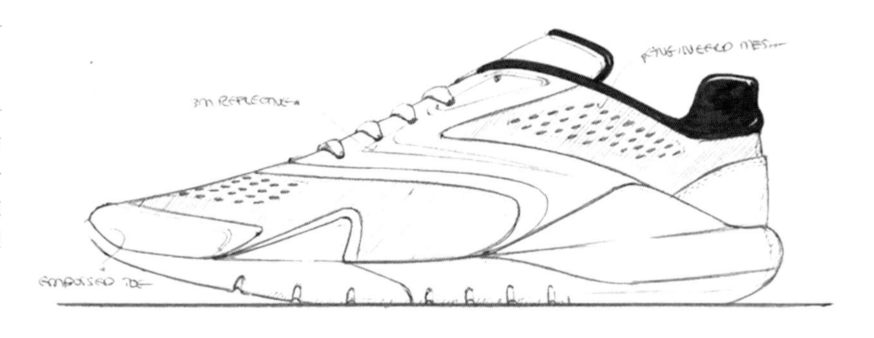

设计师：昆廷·威廉姆斯
客户：Skechers公司

这一项目最初目标是设计一款慢跑鞋，与市场上其他大品牌的同类产品竞争。设计师牢记这一点，在设计中增添了最新的技术，如工程网眼面料、热熔嵌边及轻便灵活的中底等。

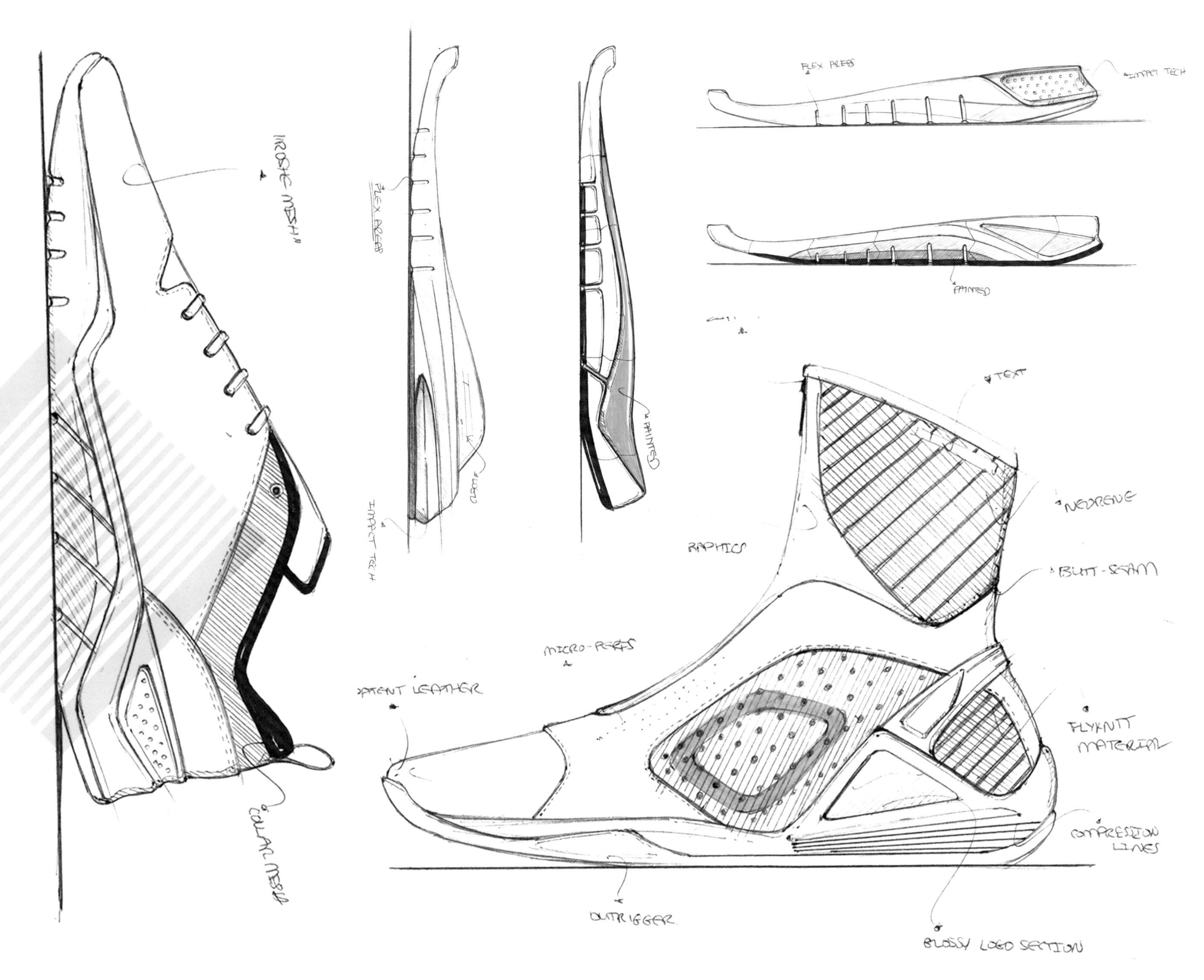

FLEX PERFS
IMPACT TECH
PAINTED
TEXT
NEOPRENE
RAPHICS
BUTT-SEAM
MICRO-PERFS
PATENT LEATHER
FLYKNIT MATERIAL
COMPRESSION LINES
OUTRIGGER
GLOSSY LOGO SECTION

Styling of Shoes

鞋履习作

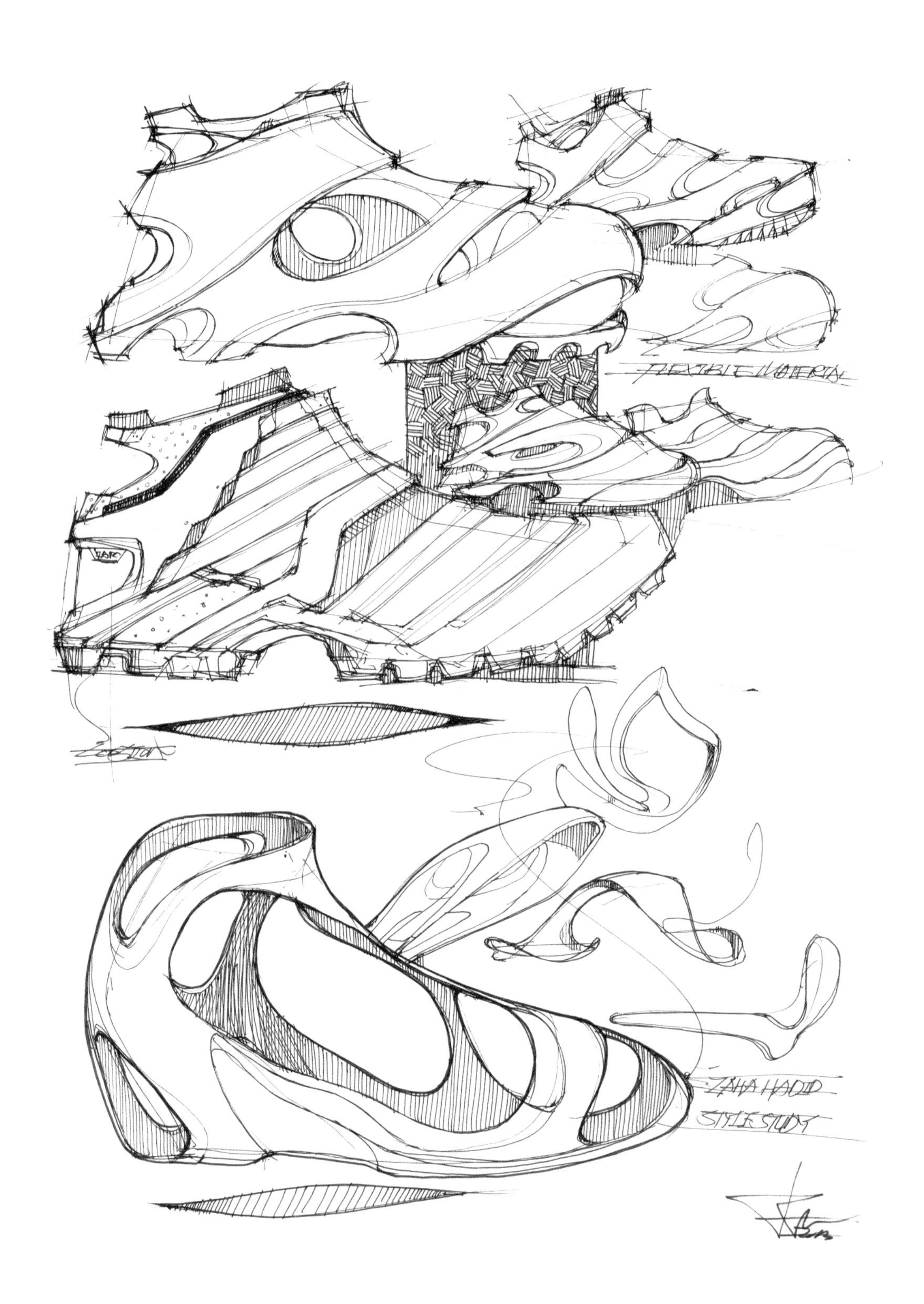

设计师：柳时彤
客户：个人作品

设计师喜欢搜集一些鞋履款式，仔细观察后把它们描绘下来。因为鞋子的款式通常与新材料与时尚趋势紧密相关，设计师可以通过手绘鞋履而学到很多，并在原有基础上进行再设计。

Bull 0.1 Original Boots

Bull0.1牛仔靴

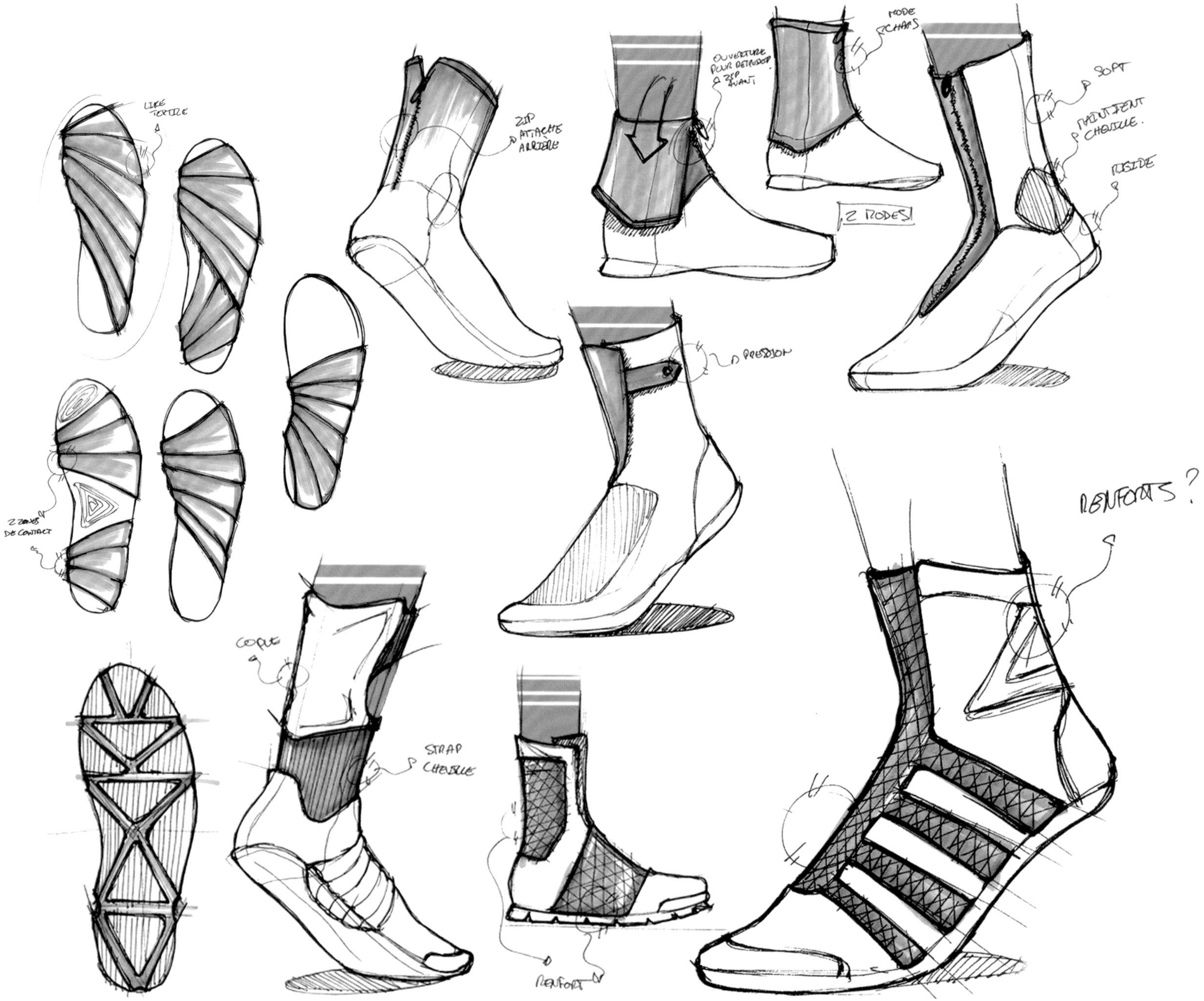

设计师：费奥娜·勒塞克（Fiona Lesecq）
客户：学校项目

Bull0.1是一款专门为牛仔打造的靴子。其分为两部分：前半部分的设计参照传统的牛仔靴，其他部分则采用现代材料。这款靴子的设计初衷即将传统和现代的设计元素相互融合。

该设计的目的是提升传统牛仔靴的性能，使其更加舒适。设计师尝试了多种方式，旨在以保留传统牛仔靴的造型为基础，最大限度地满足靴子穿着的平衡性、灵活性和稳固性的要求。

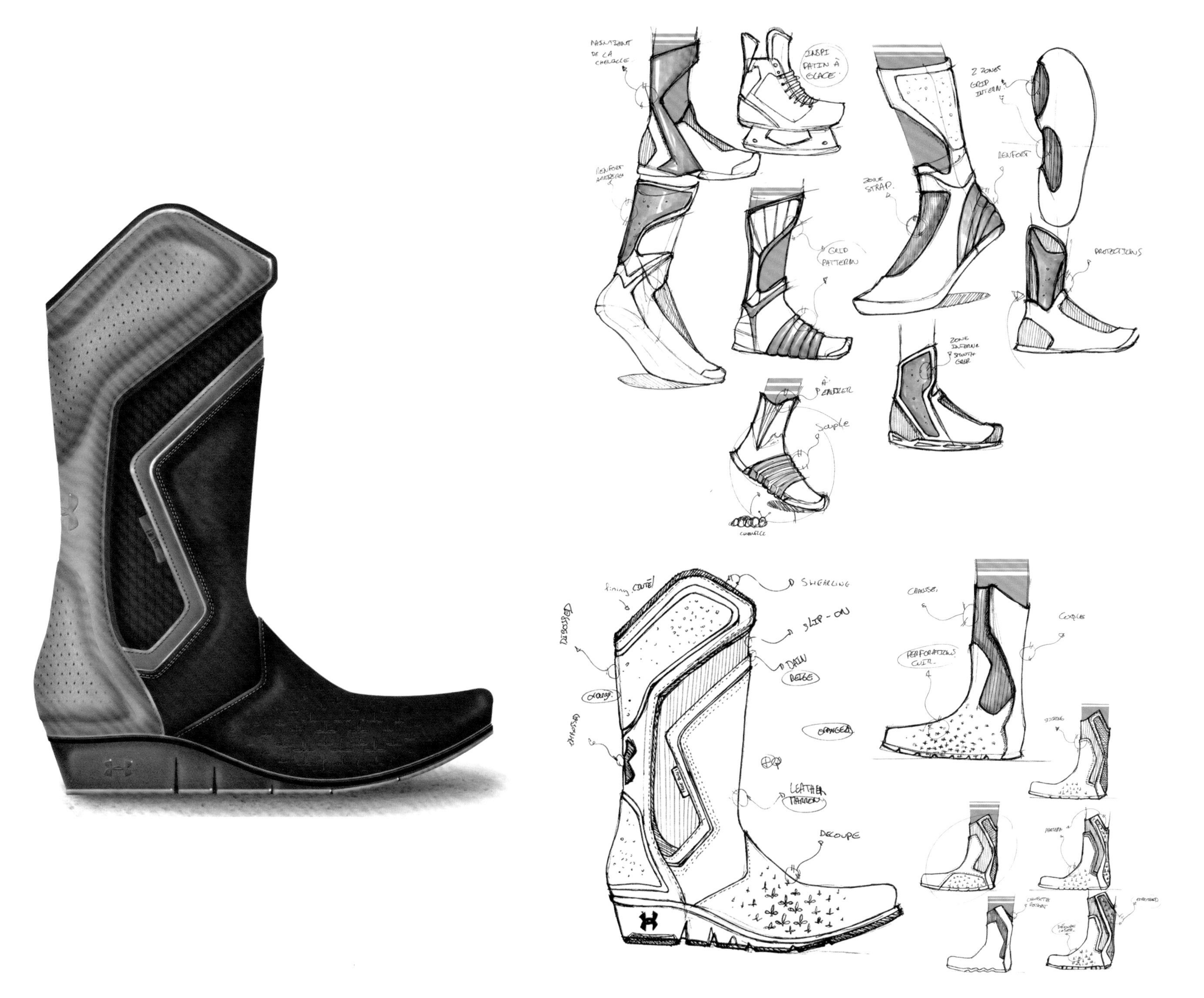
INSPI PATIN À GLACE
RENFORT
ZONE STRAP
GRID PATTERN
PROTECTIONS
Souple
SHEARLING
SLIP-ON
LEATHER
DECOUPE
PERFORATIONS CUIR

Cam, Desert Running Boots

Cam沙漠跑鞋

设计师：费奥娜 · 勒塞克

客户：个人作品

这款沙漠跑鞋的设计理念是其形态可在跑步的过程中发生改变：开始时，鞋面是闭合的；随着鞋内温度的升高，外侧的鞋面将逐渐打开，进而调节温度。

设计之初，设计师研究了足部解剖学，旨在了解热源的位置。随后，使用模拟肌肉、静脉和骨头来确定鞋子的造型。鞋底的设计上，设计师借鉴了骆驼的足部结构，从而便于获得更好的支撑和平衡。

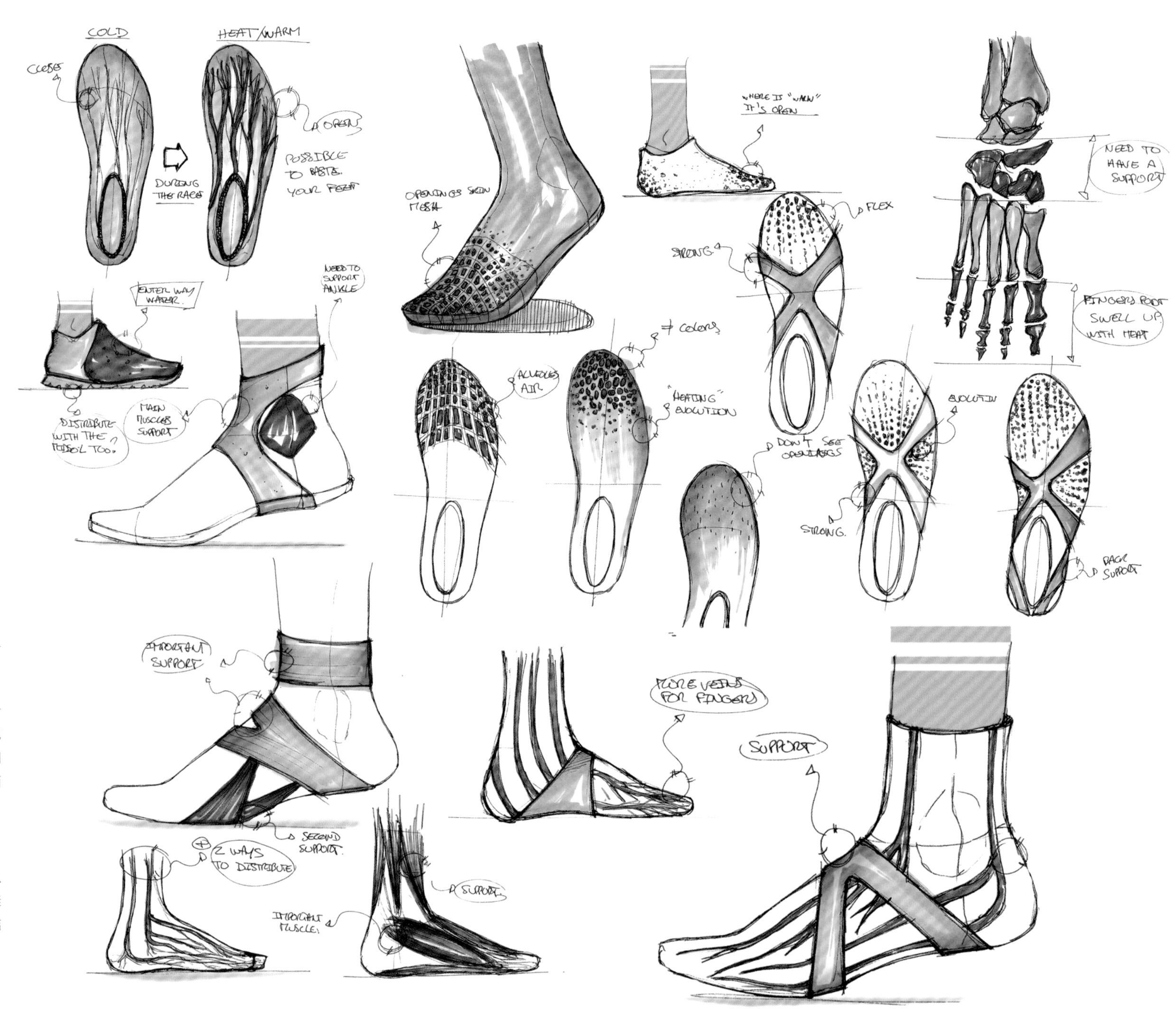

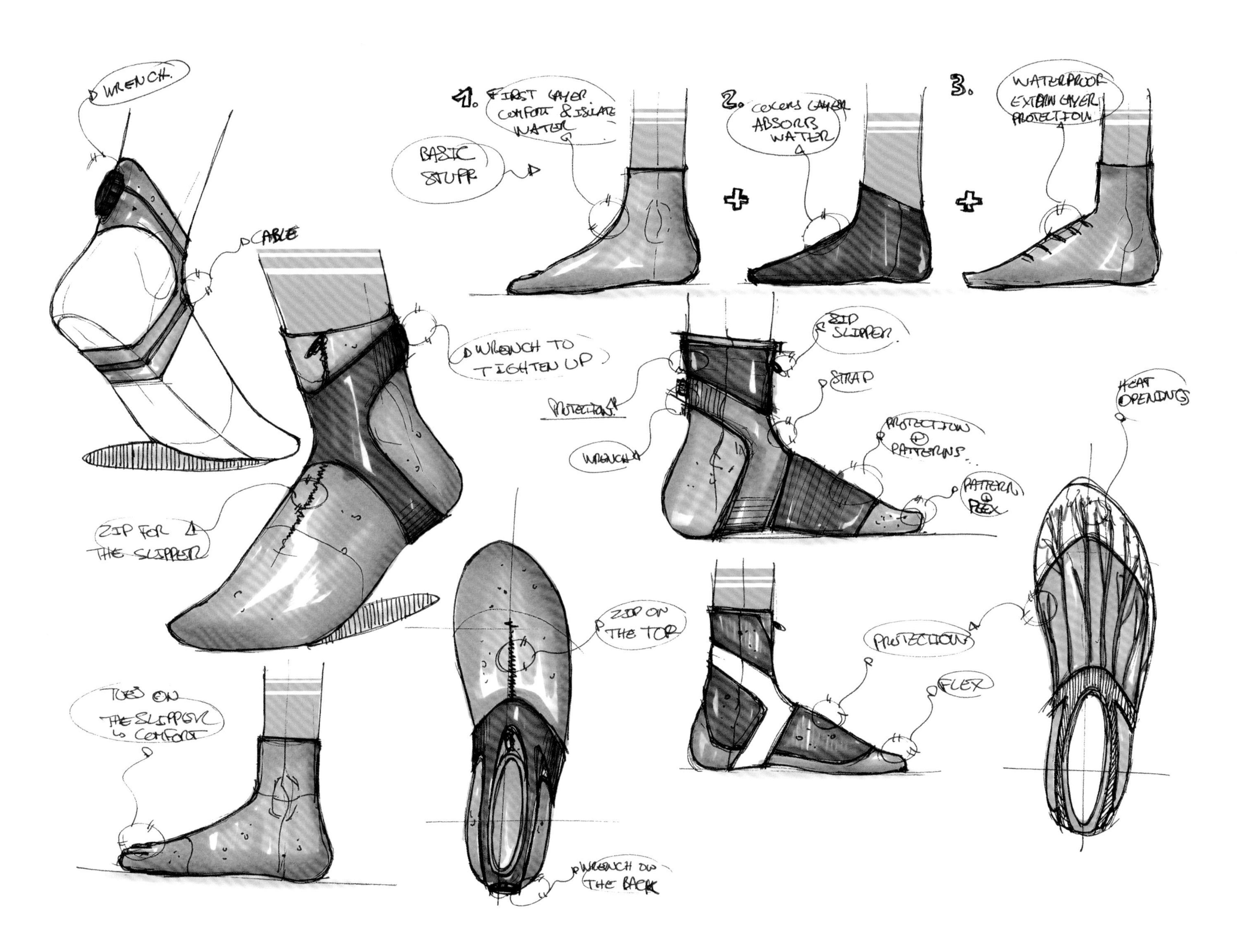
WRENCH.
CABLE
WRENCH TO TIGHTEN UP
ZIP FOR THE SLIPPER
TOES ON THE SLIPPER
↳ COMFORT
1.
FIRST LAYER COMFORT & ISOLATE WATER
BASIC STUFF
2.
LAYER ABSORB WATER
3.
WATERPROOF EXTERN LAYER PROTECTION
ZIP SLIPPER
STRAP
PROTECTION
WRENCH
PROTECTION + PATTERNS...
PATTERN + FLEX
HEAT OPENINGS
ZIP ON THE TOP
WRENCH ON THE BACK
PROTECTION
FLEX

MAIN VEINS
VESSEL
WATERPROOF
ZIP.
PADDING
PADDING
FLEX
PADDING TO PROTECT
LAYER CLOSE
COLORS ON TECHNICAL PARTS
START OF THE RACE
NO PATTERN

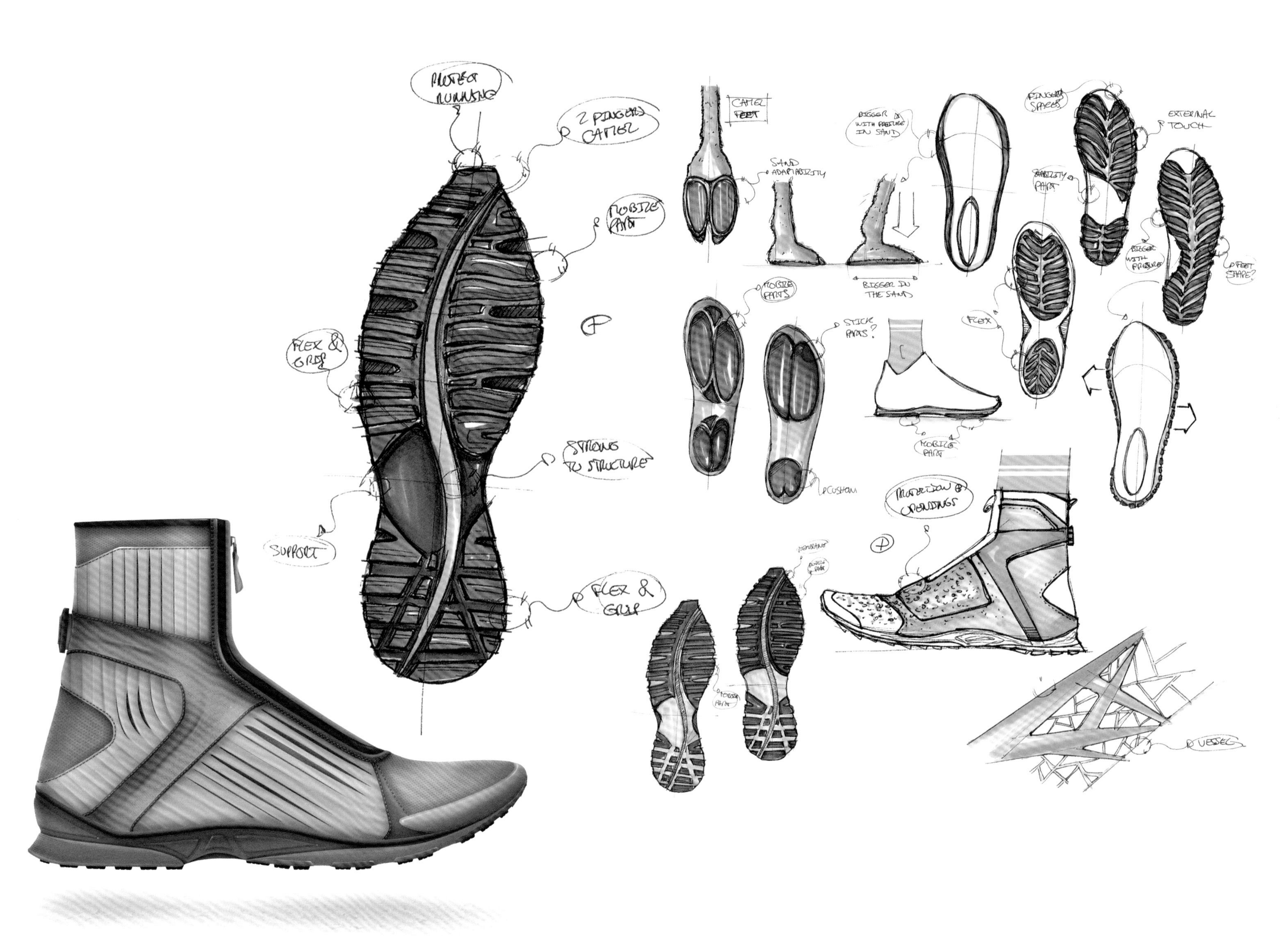
FLEX & GRIP
SUPPORT
EXTERNAL TOUCH
BIGGER IN THE SAND
CUSHION
FLEX

VEHICLES
交通工具

James Dean Porsche
詹姆斯·迪恩的Porsche跑车

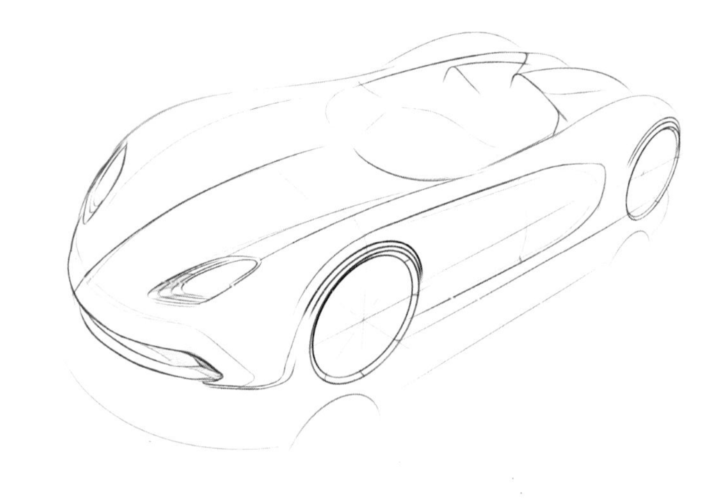

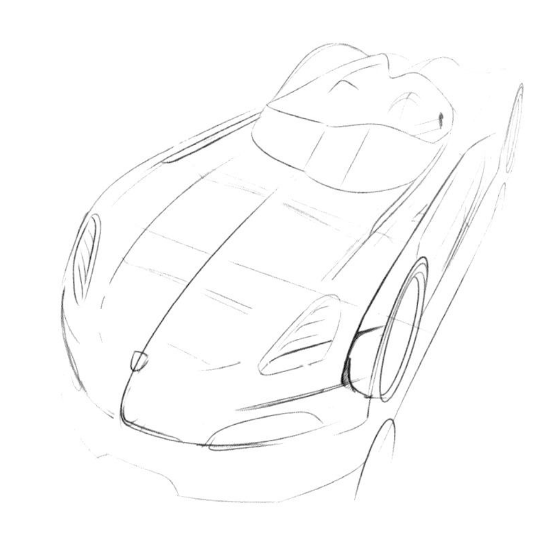

设计师：耶勒·特百思（Jelle Tjebbes）
客户：个人作品

在许多影片中，詹姆斯·迪恩（James Dean）似乎和Porsche跑车有着某种特殊的联系。但是目前的一些标志性Porsche跑车却没有体现出詹姆斯·迪恩的魅力，这一作品便是设计师打造的一辆专属詹姆斯·迪恩的Porsche概念跑车。

在草图的绘制过程中，设计师首先在Photoshop中用黑色笔刷画几个方案。虽然这种方法不能精细地表现出平滑的过渡，但却可以通过快速表现去推敲汽车的体量。在这个阶段，可以不断地复制图层，并通过添加白色高光或黑色阴影进行调整。将这样的草图再深化成精确的效果图大概需要一小时，可以用线条或喷笔工具完善跑车表面的柔和过渡。

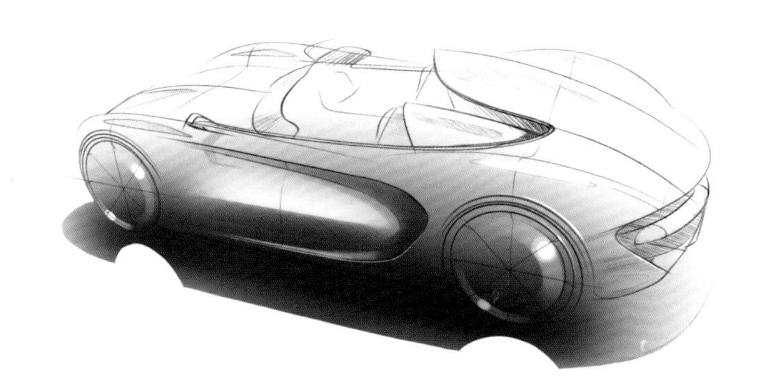

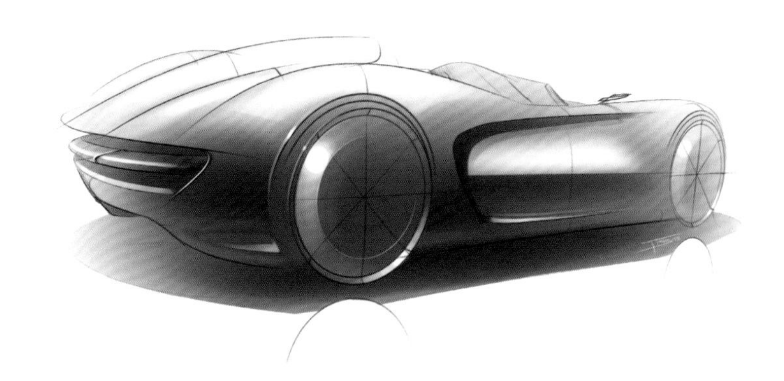

Premium
豪华版汽车

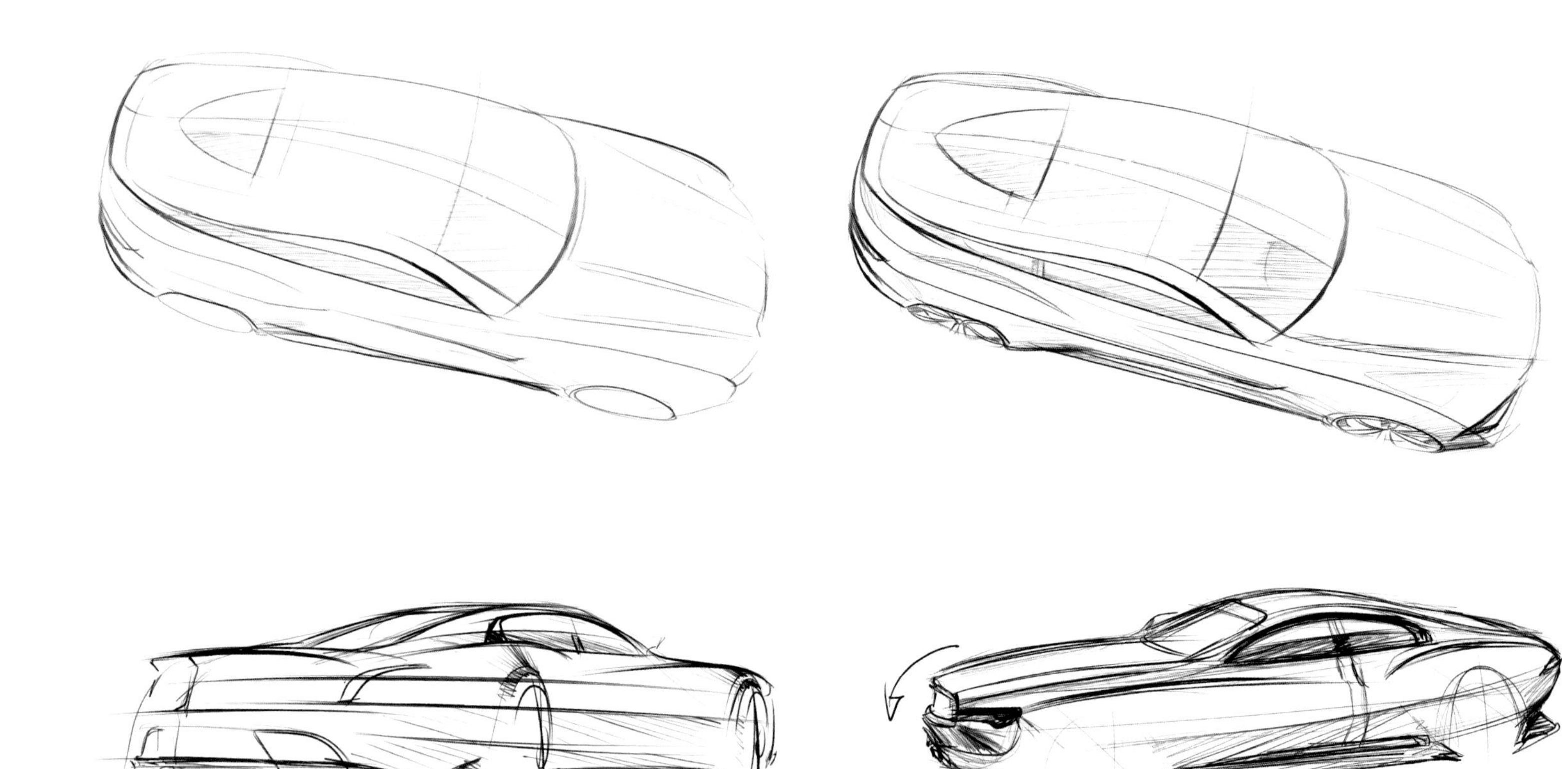

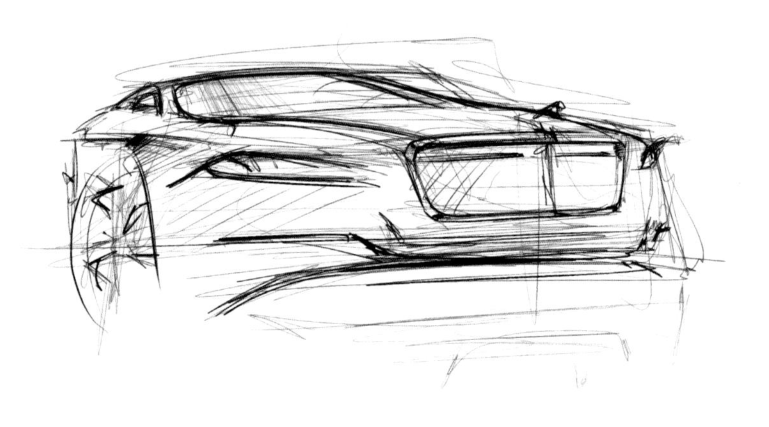

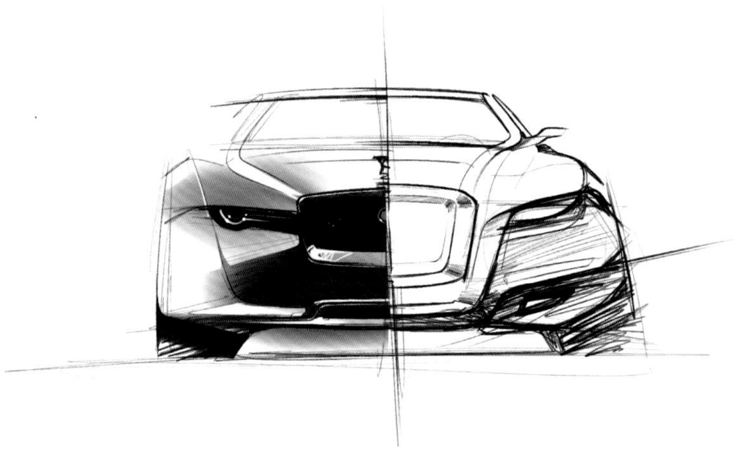

设计师：耶勒·特百思
客户：个人作品

在该设计中，设计师尝试从野生动物的形态中获得灵感，提炼出造型元素并运用到豪华版汽车的设计中。草图是在Photoshop软件中深化完成的。

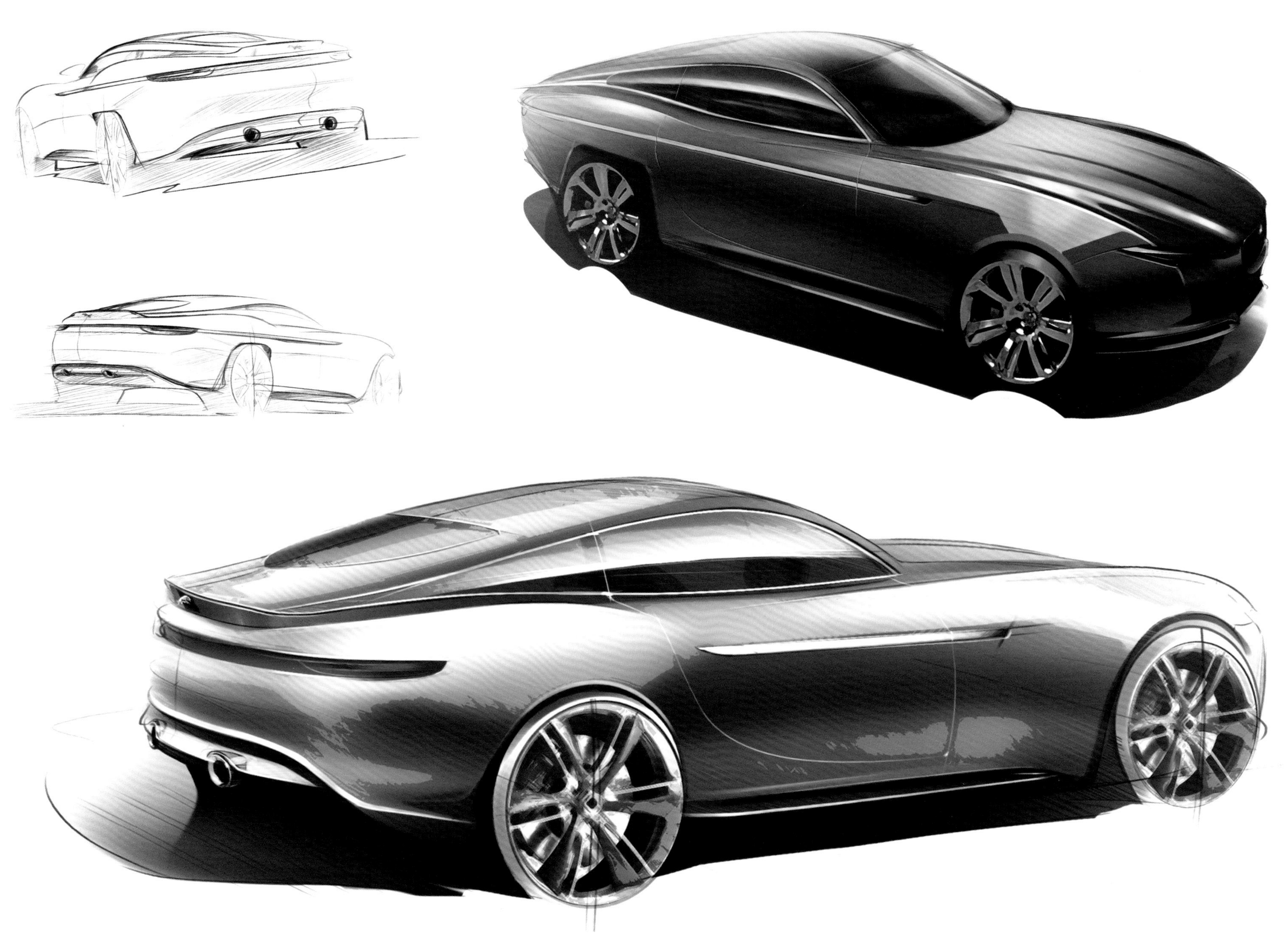

2030 Nissan Titan ET

Titan ET越野车

设计师：大卫·奥尔森（David Olsen）
客户：学校项目，艺术中心设计学院

这是设计师在位于美国加州帕萨迪纳市的艺术中心设计学院入学时的作品。这一设计将Nissan的电动车技术运用到下一代越野车上，独特的引擎布局方式增添了车厢的空间。汽车轮毂的电机提供了较低的重心并加大了扭矩。

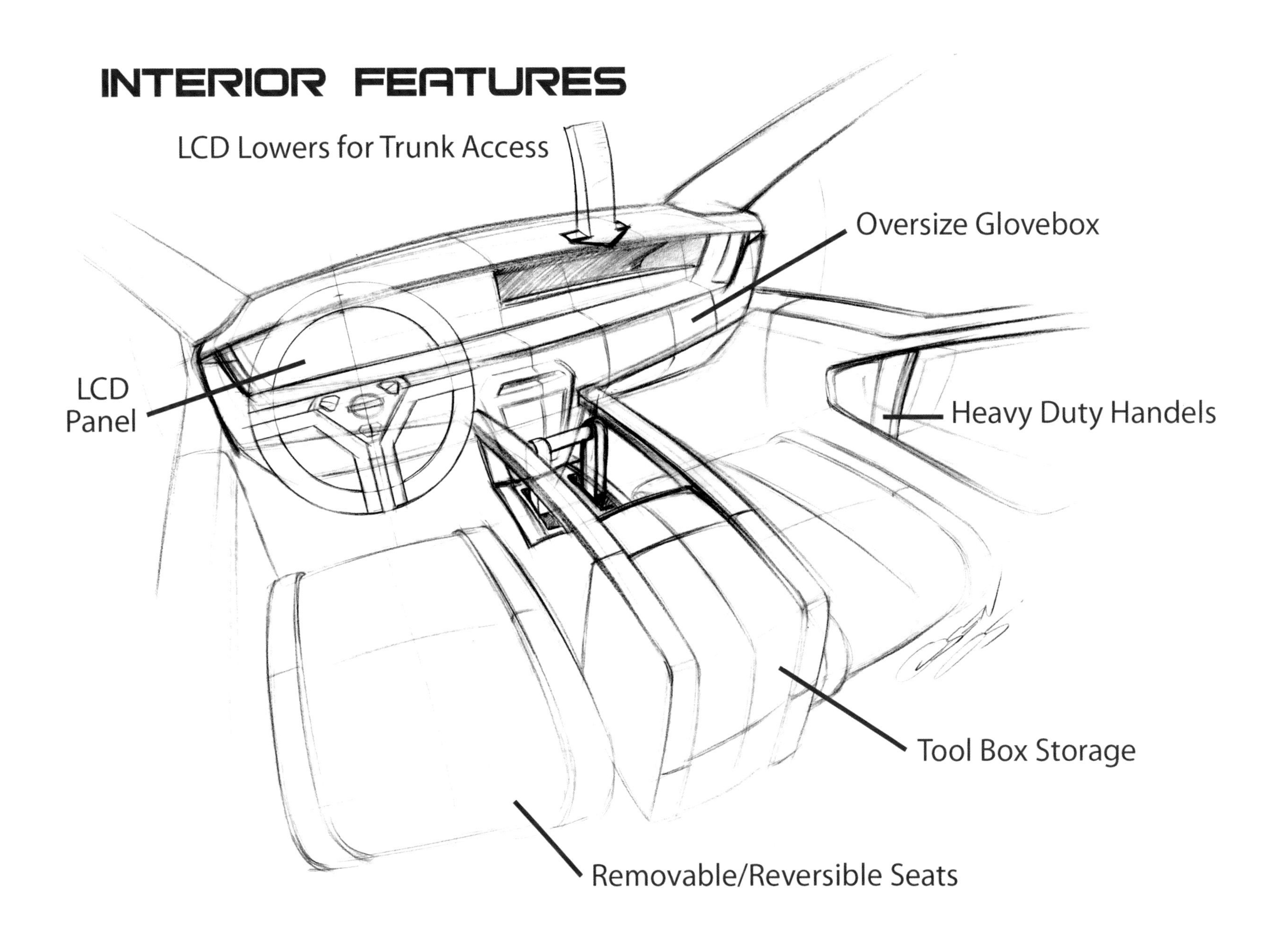
INTERIOR FEATURES
LCD Lowers for Trunk Access
Oversize Glovebox
LCD
Panel
Heavy Duty Handels
Tool Box Storage
Removable/Reversible Seats

NISSAN

NISSAN

A City Car Volkswagen
城市轿车

设计师：阿提姆 · 斯米尔诺夫（Artem Smirnov）
客户：个人作品

设计师在进行草图绘制之前，在脑海中已经勾勒出了这一城市轿车的造型轮廓。他独爱Volkswagen的Beetle型小轿车那种简洁纯粹的造型，并试图将这一经典车型的造型元素融入到自己的设计中。

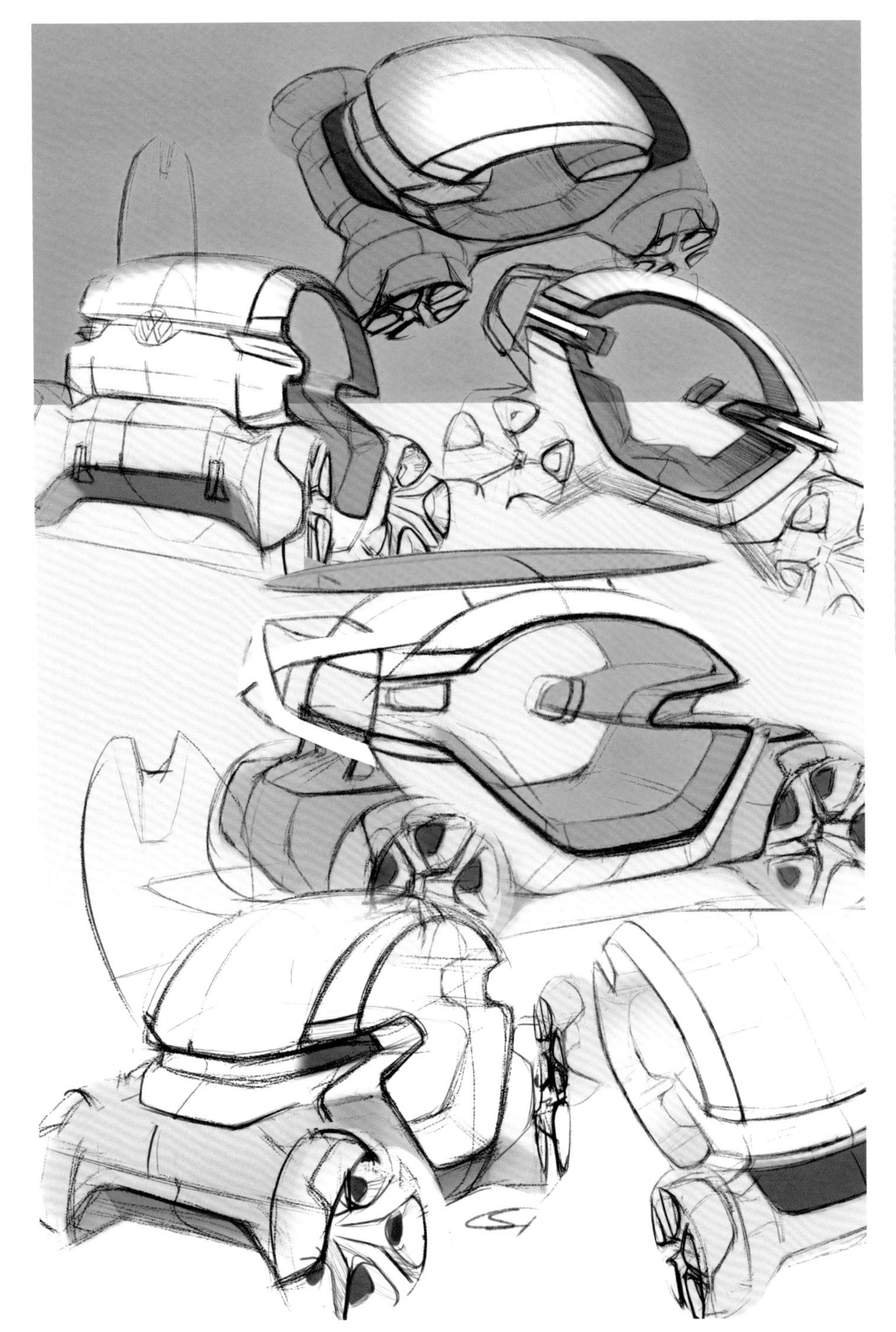

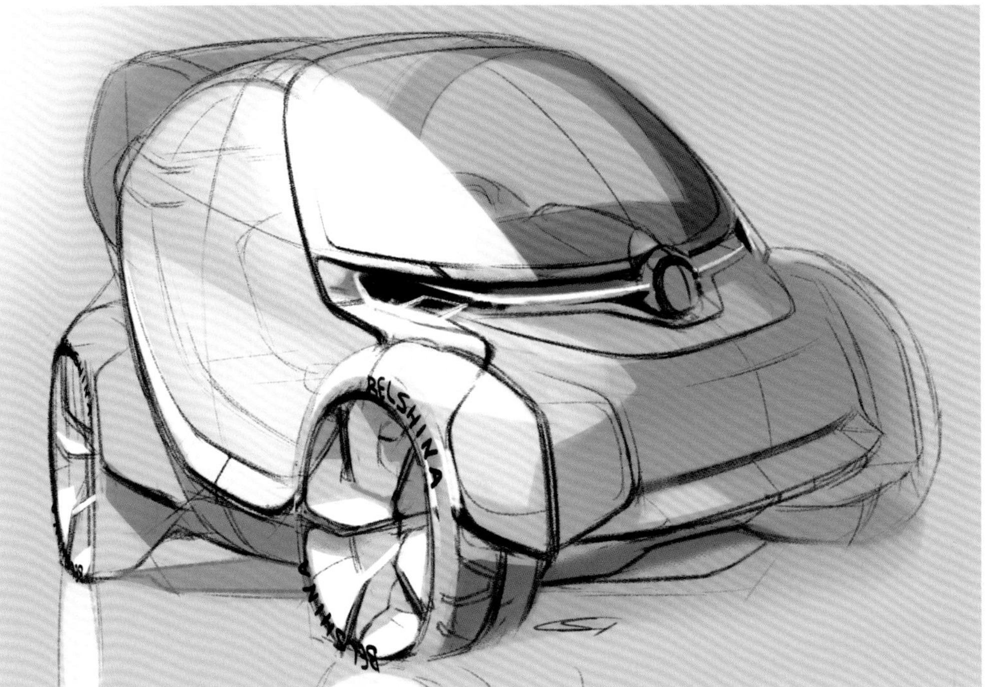
BELSHINA

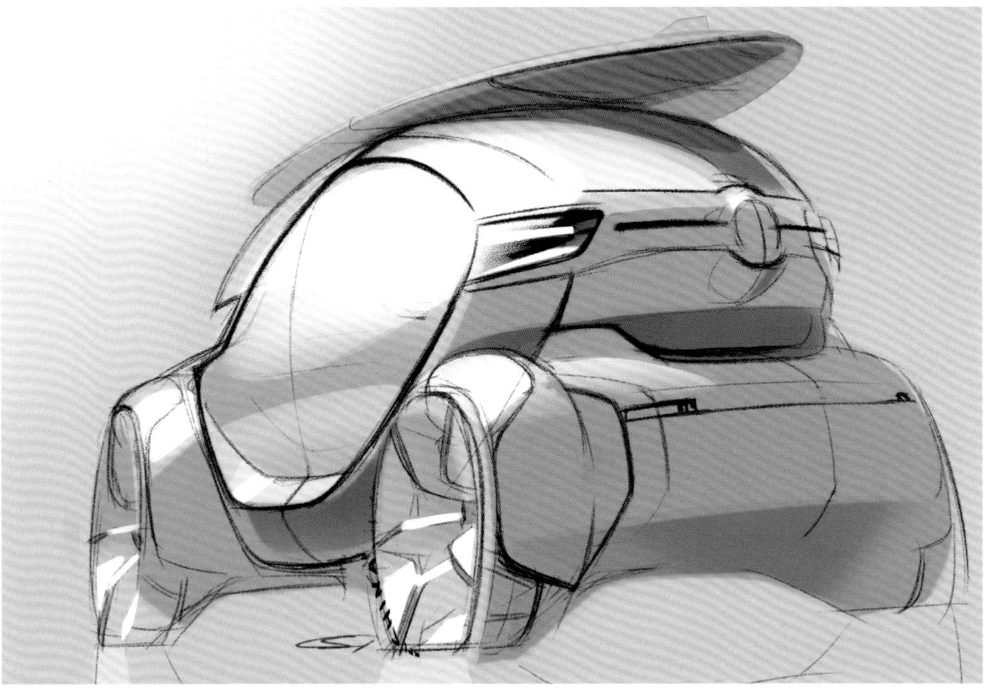

Bolwell Nagari

Nagari跑车

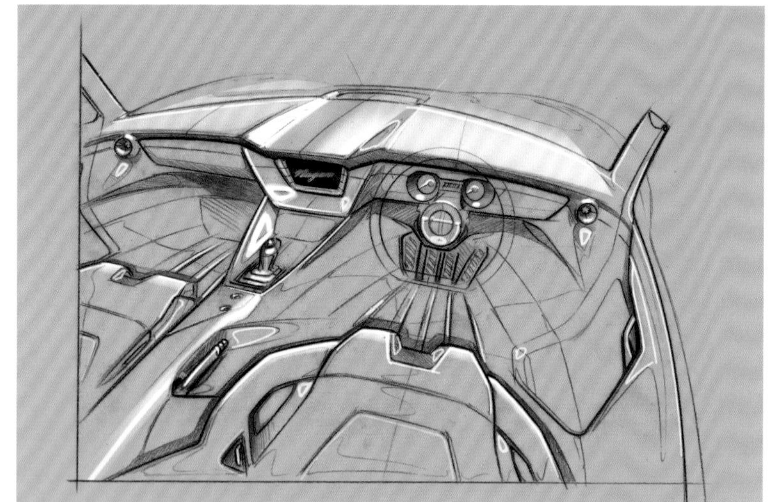

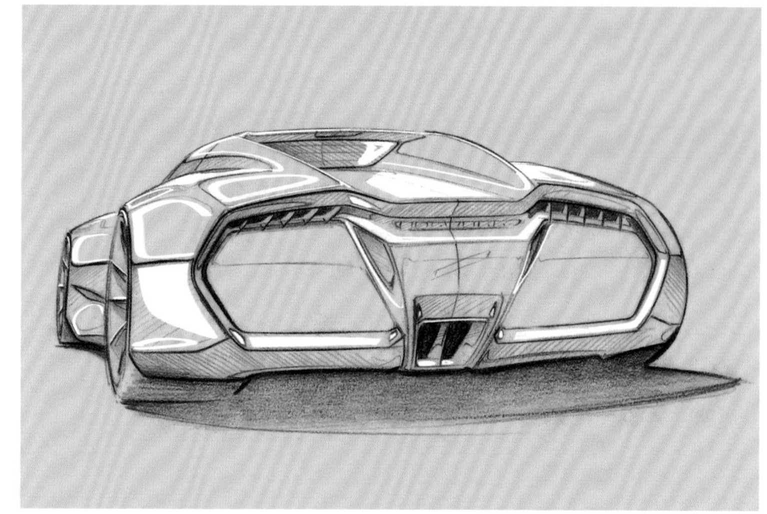

设计师：迈克尔·格雷（Michael Gray）
客户：个人作品

设计师从1970年代的Nagari跑车造型中获得灵感，在保留原有特色元素的基础上，依据当今的设计风格趋势，创造了全新的GT跑车造型。在绘制汽车草图时，应尝试从不同角度去描绘，这样可以全面地把控造型。

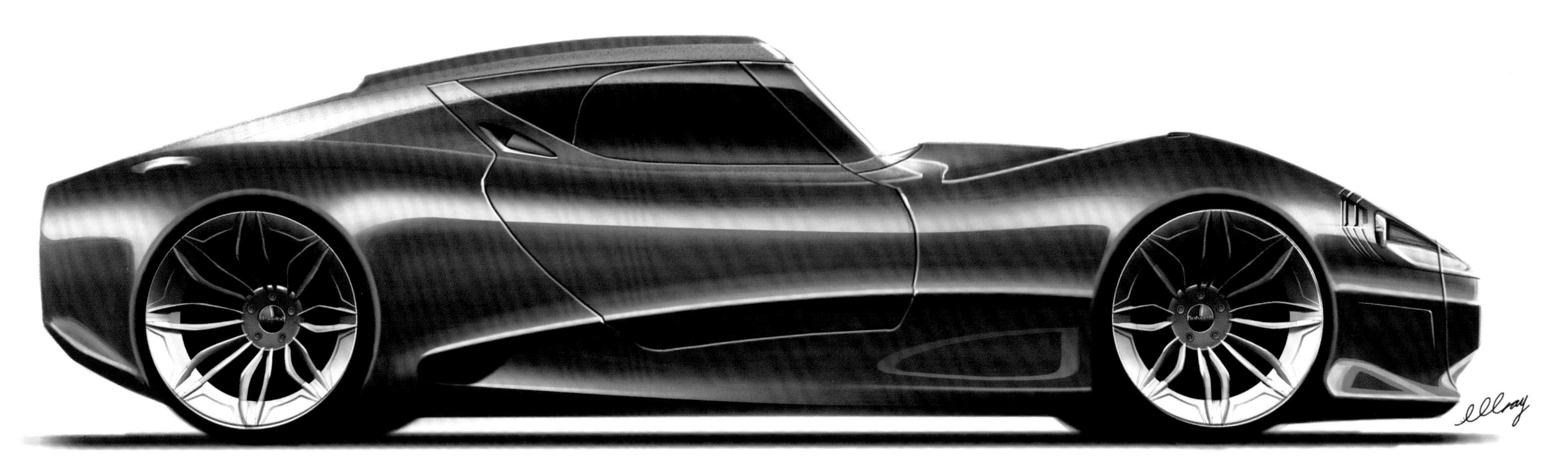

NAGARI

Sketches for VW Design Contest 2014
汽车设计大赛草图

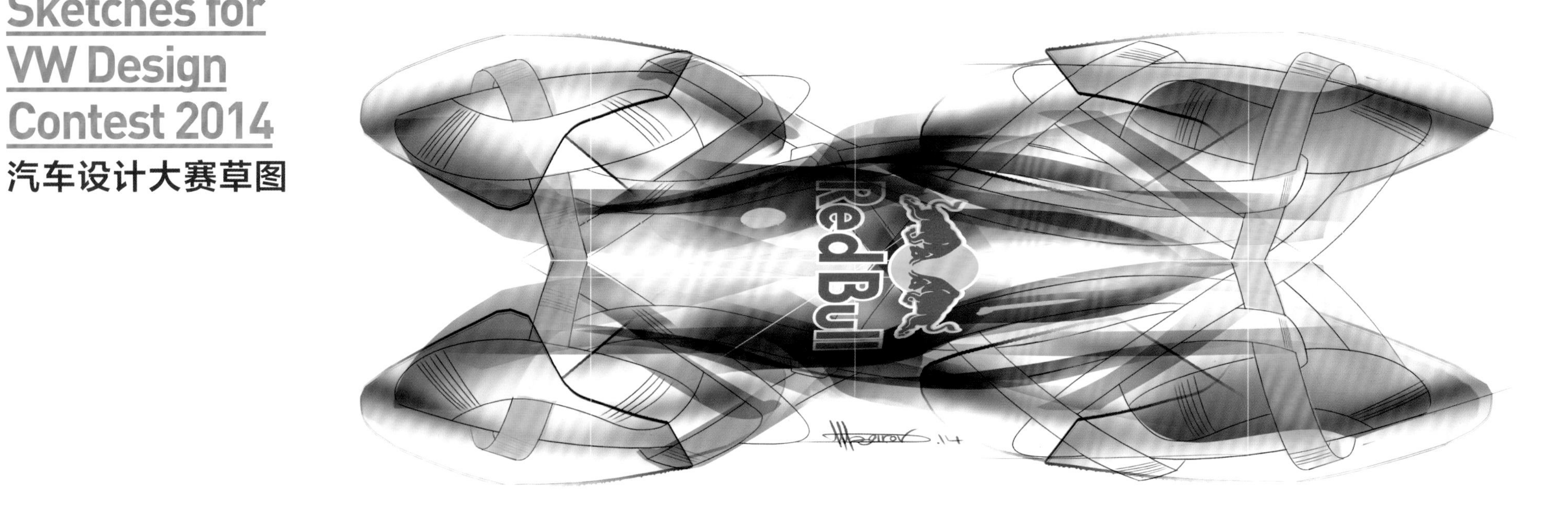

设计师：拉希德·塔吉罗夫（Rashid Tagirov）
客户：2014Volkswagen汽车设计大赛

设计师旨在打造一辆环保的汽车：其既可以在轨道上行驶，通过枢纽实现地点间的传输；又可以被司机自由驾驶。

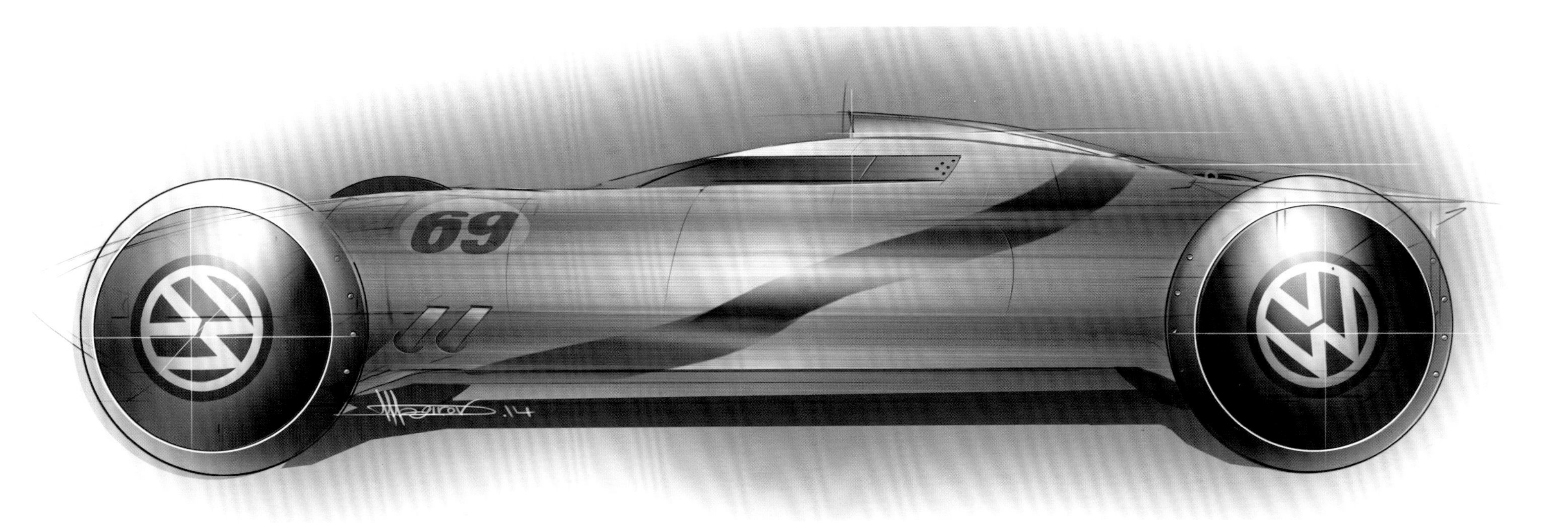
69

DUNLOP
DUNLOP
DUNLOP
DUNLOP

Mini Goo

MiniGoo
城市微型车

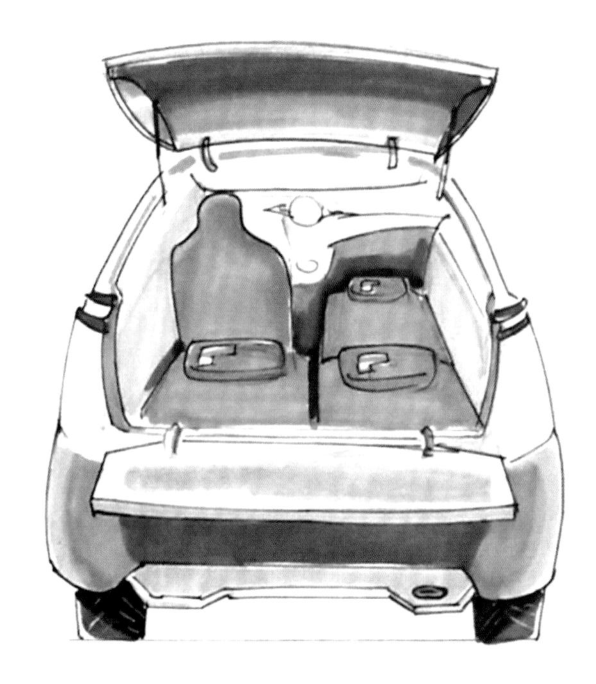

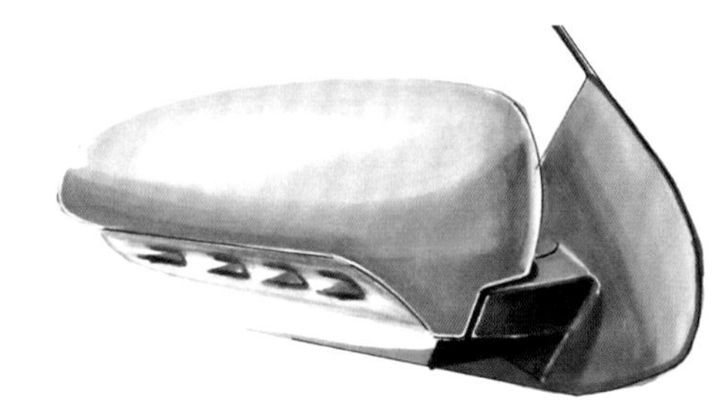

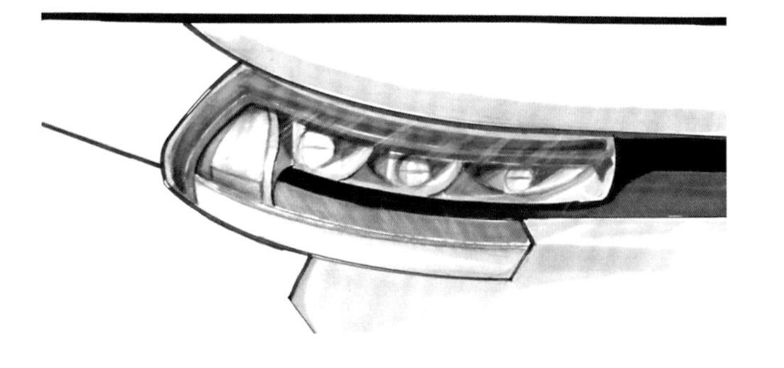

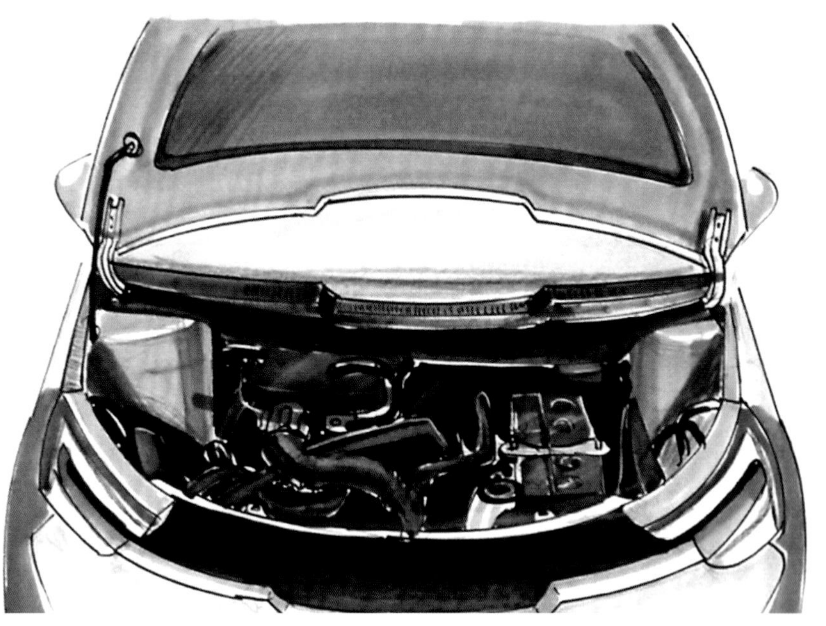

设计师：塔梅尔·尤塞克（Tamer Yüksek）
客户：学校项目

该设计旨在满足大城市居民的需求。通勤时，MiniGoo便于在拥挤的交通环境中驾驶；周末，便于居民外出郊游。因此，这款小车内部空间设计得紧凑而灵活。

HIZ GÖSTERGESİ SUTUNU
ARACIN HIZI DÜSÜKKEN
YESILDEDIR. HIZ ARTTIKCA
RENK KIRMIZIYA CIKAR.
ARKADAKI YOLCULARIN
INMESİ ICIN KOLTUKLAR
KAPIYA DOĞRU ACILIR.

Jeep Wrangler Interior

Jeep Wrangler 汽车内饰

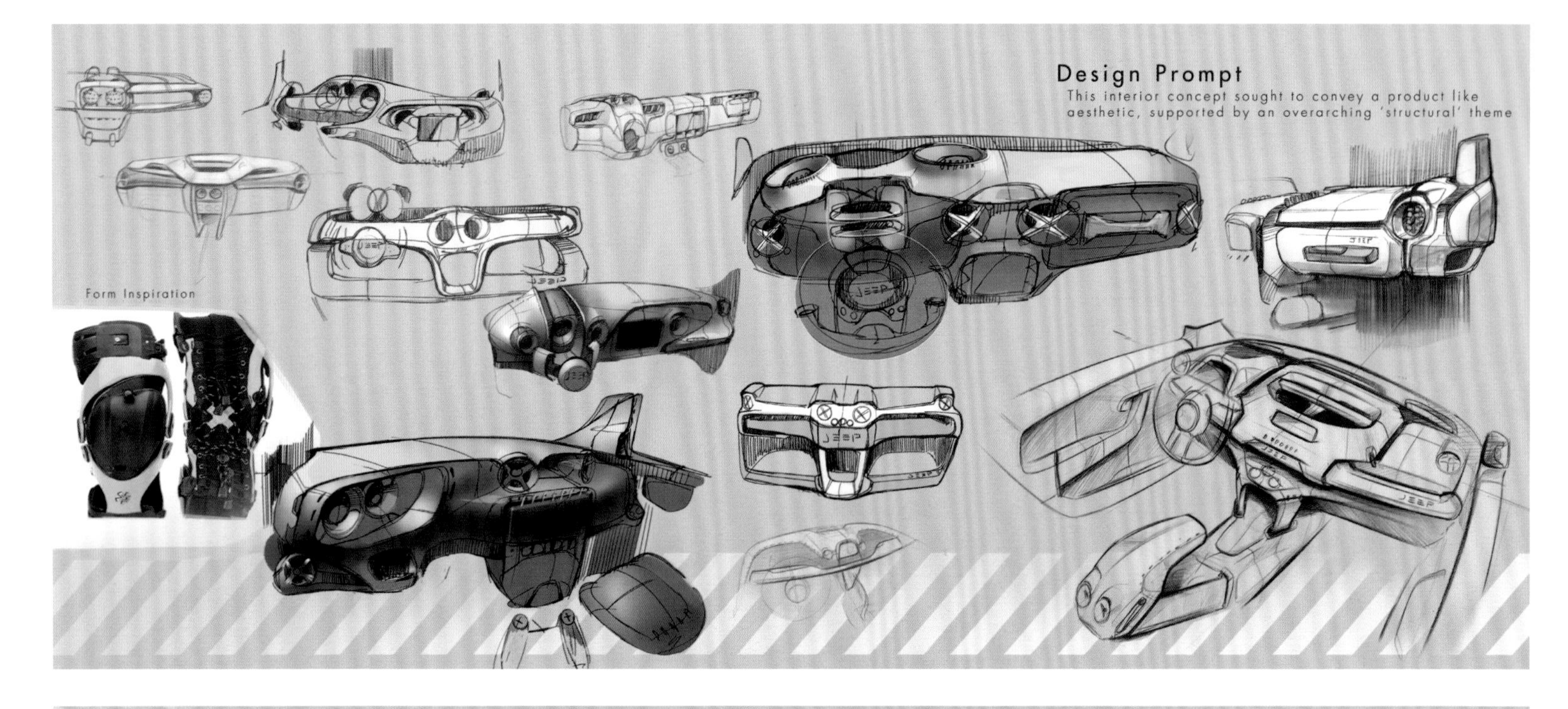

Design Prompt
This interior concept sought to convey a product like aesthetic, supported by an overarching 'structural' theme
Form Inspiration

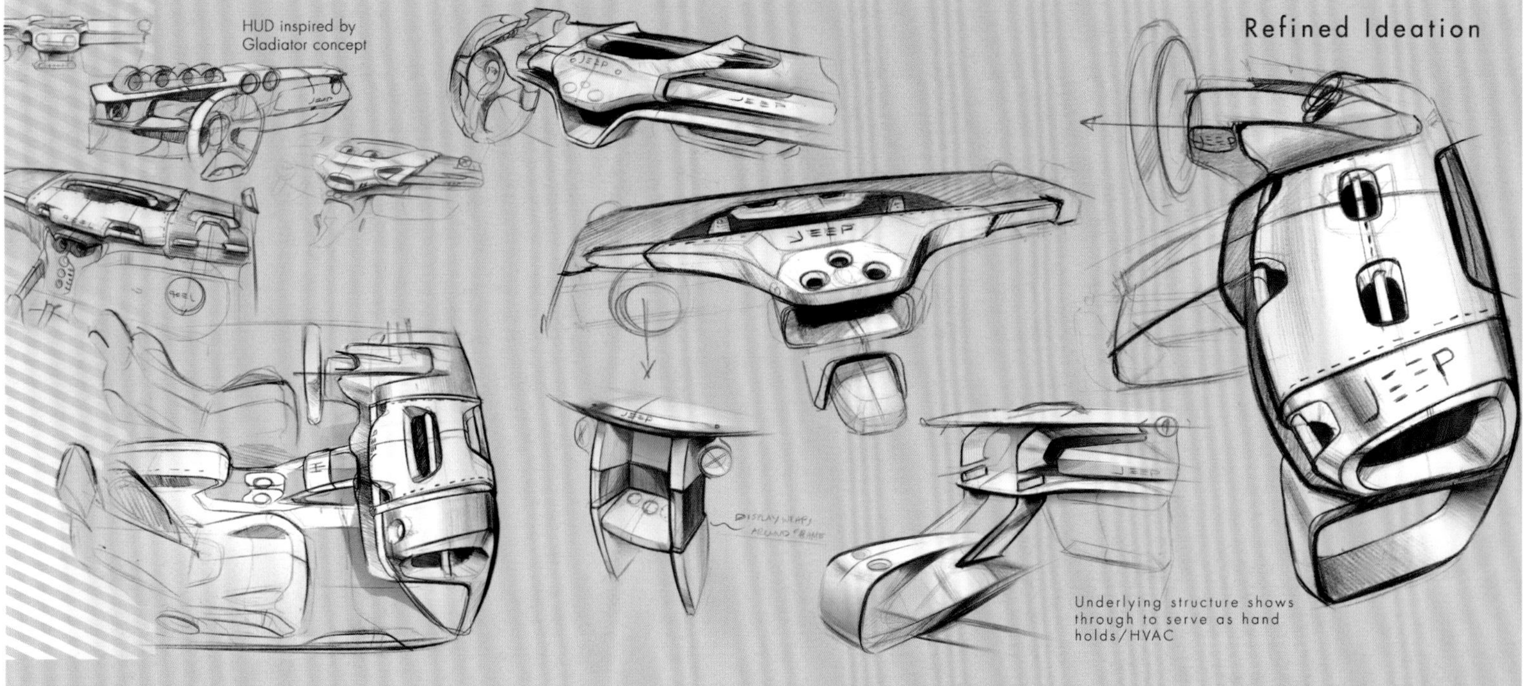

设计师：西蒙·韦尔斯（Simon Wells）
客户：Chrysler公司

这是设计师在Chrysler公司实习期间完成的作品，旨在以“骨架”为主题打造全新的车内空间。在草图绘制阶段，设计师对主题的烘托、车内空间的布局以及内饰材质的运用进行了深入探讨。

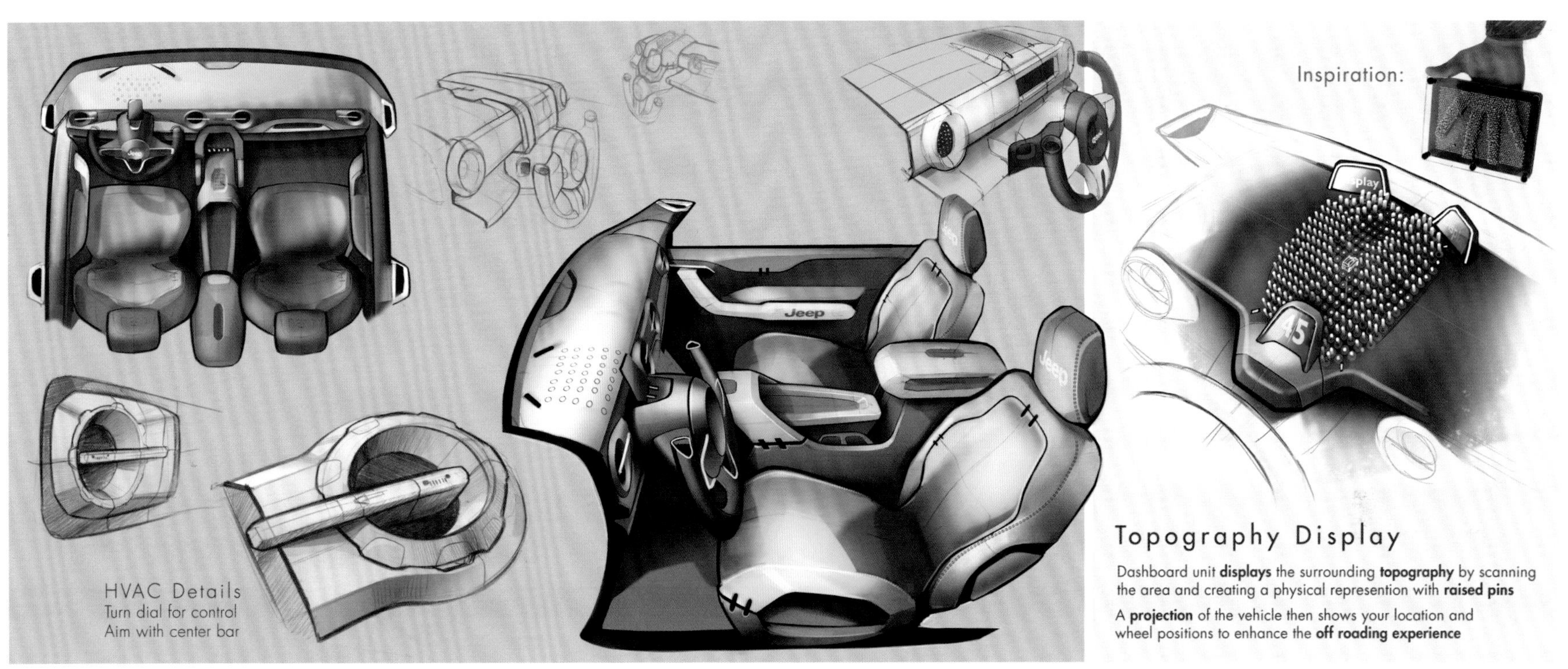
Inspiration:
play
45
Jeep
HVAC Details
Turn dial for control
Aim with center bar
Topography Display
Dashboard unit **displays** the surrounding **topography** by scanning the area and creating a physical represention with **raised pins**
A **projection** of the vehicle then shows your location and wheel positions to enhance the **off roading experience**

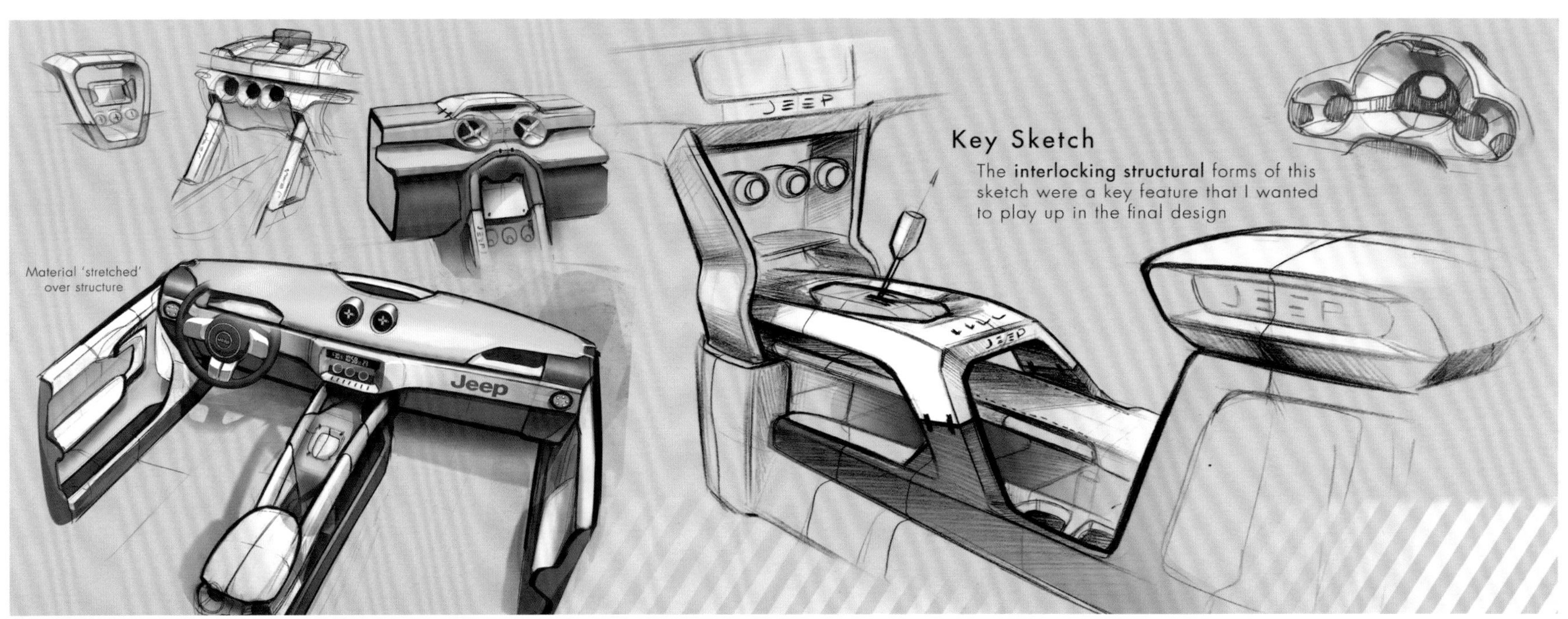
Key Sketch
The **interlocking structural** forms of this sketch were a key feature that I wanted to play up in the final design
Material 'stretched' over structure
Jeep
JEEP

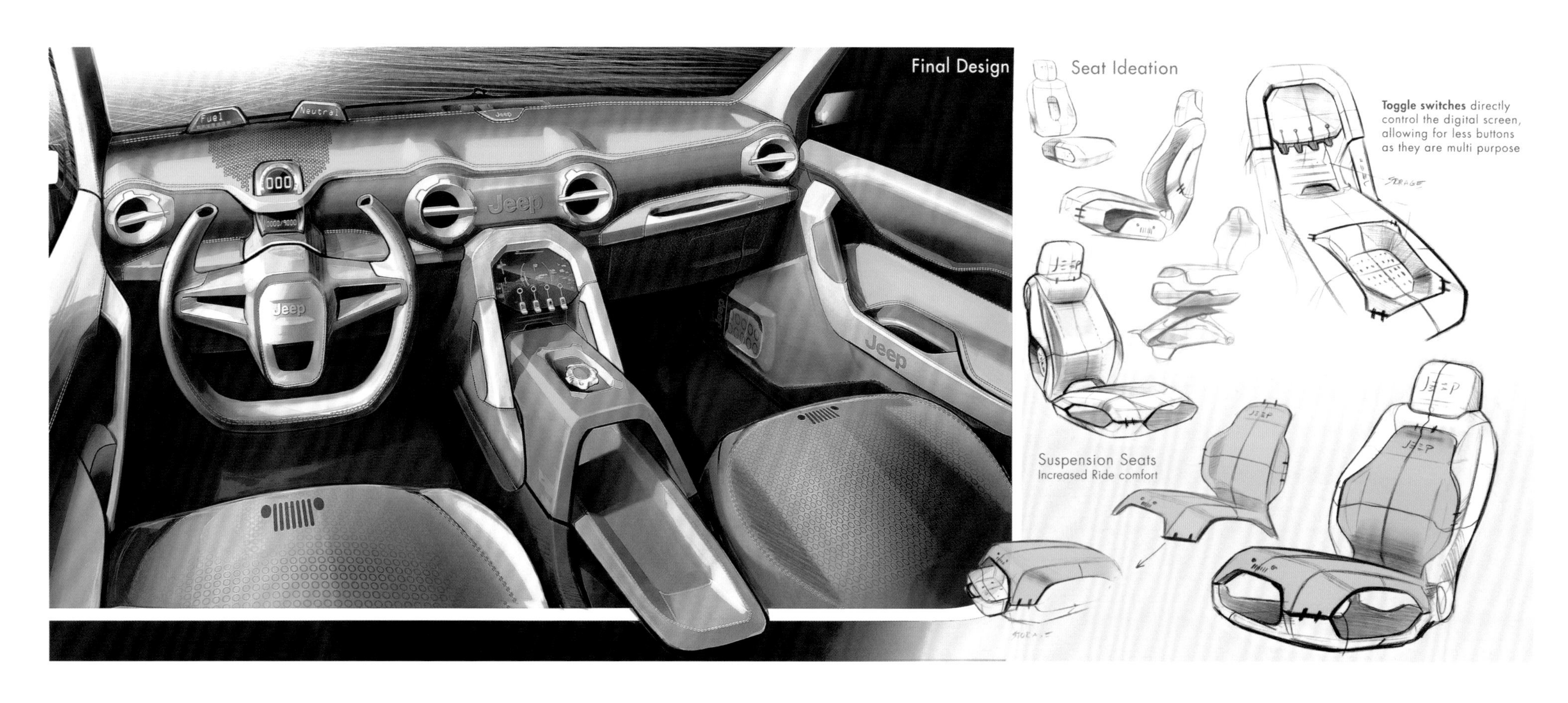
Final Design
Fuel
Neutral
Jeep
Seat Ideation
Toggle switches directly
control the digital screen,
allowing for less buttons
as they are multi purpose
Suspension Seats
Increased Ride comfort

Jeep
Jeep

Volkswagen Quantum Ambassador
Quantum Ambassador 概念车

Quantum Explorer
1 Science Lab
Big enough to carry a team of 5 fearless scientists and their instruments on the journey through a black hole
Quantum Explorer
3 Faraday Cage
Vehicle is encased in a faraday cage to protect the occupants from electro-magnetic signals and radiation
Periscope cover
Sensors
Faraday Cage
Roof Details 2
Sensors and cameras are mounted underneath the center cutout like a periscope positioned over the captain for direct use

设计师：西蒙 · 韦尔斯

客户：Volkswagen公司

这款车专为视频游戏设计，并获得Volkswagen汽车设计大赛提名。在游戏中，科学家可以乘坐这辆车进入黑洞进行探索。基于这个理念，设计师试图通过造型的创新从而打造未来交通工具的美感。

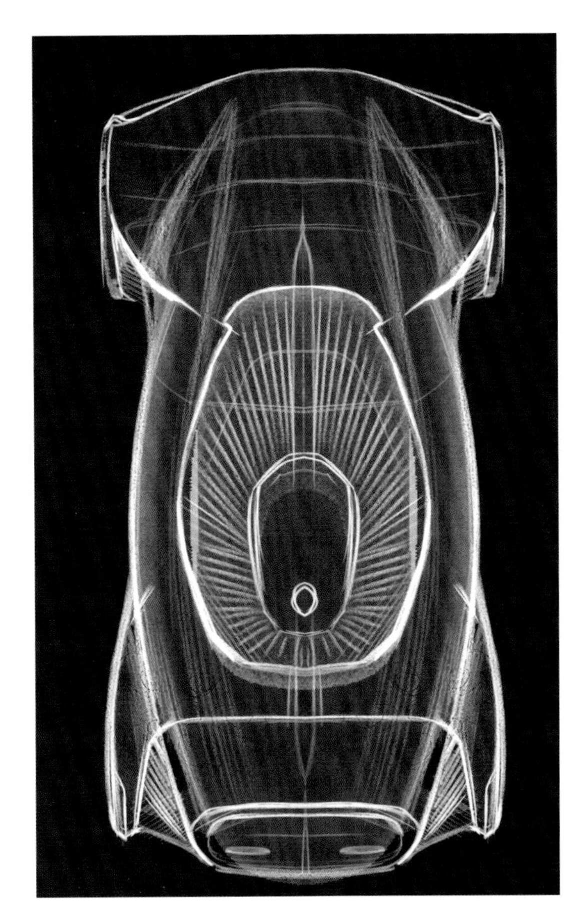

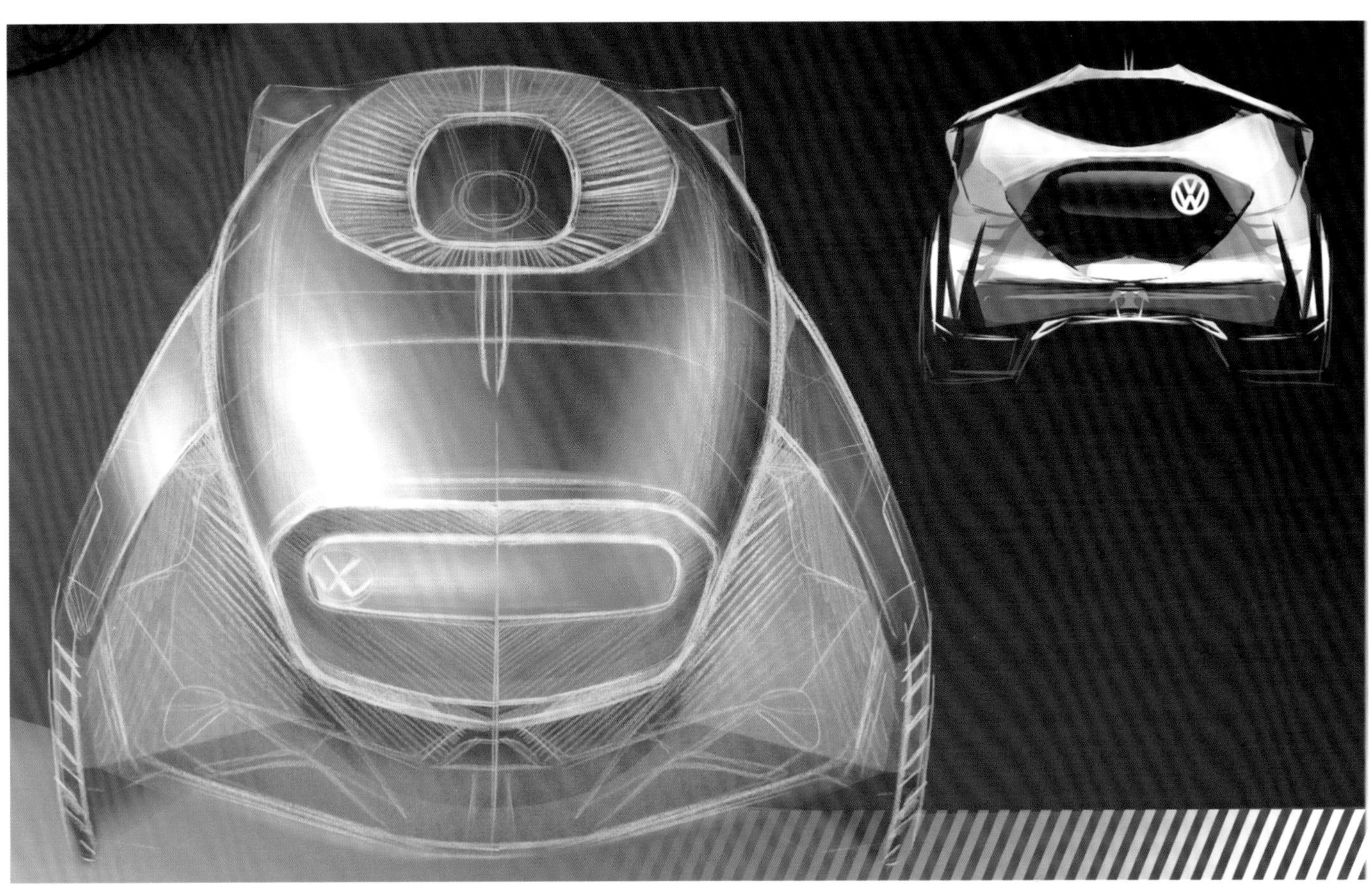

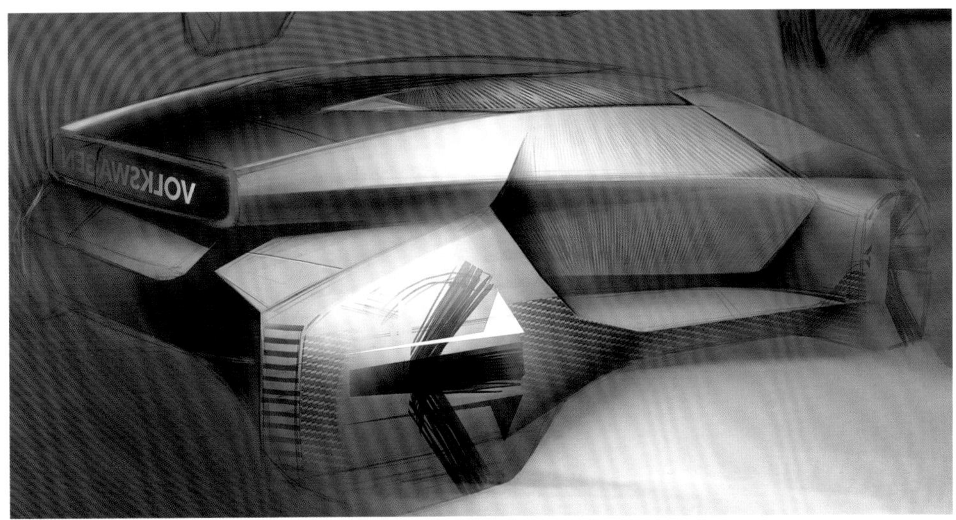

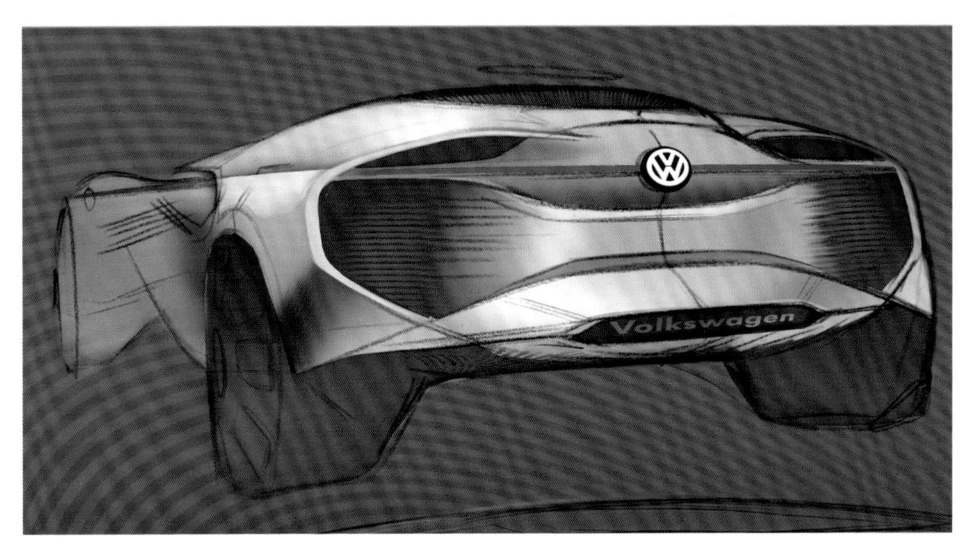
Volkswagen

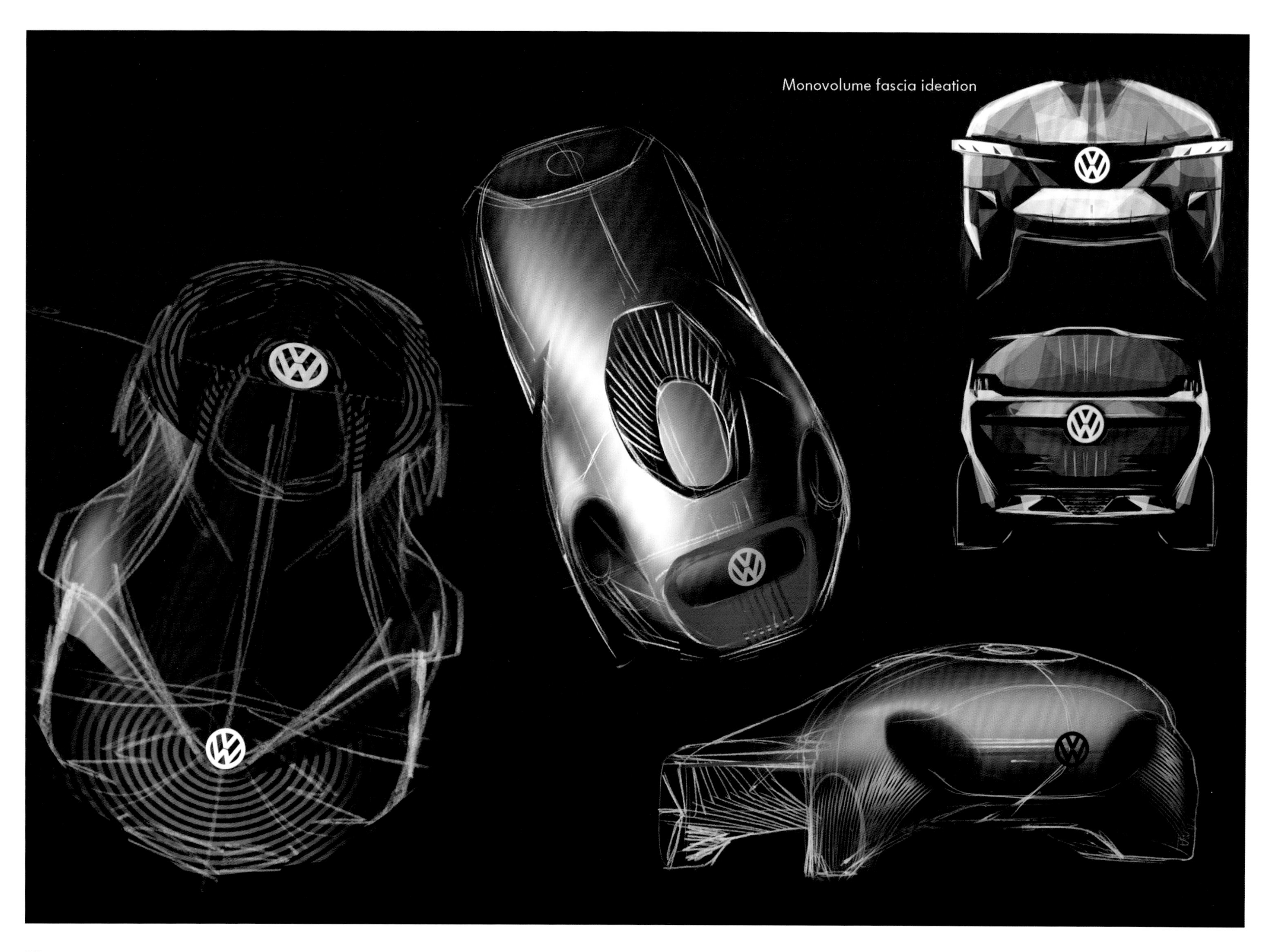
Monovolume fascia ideation

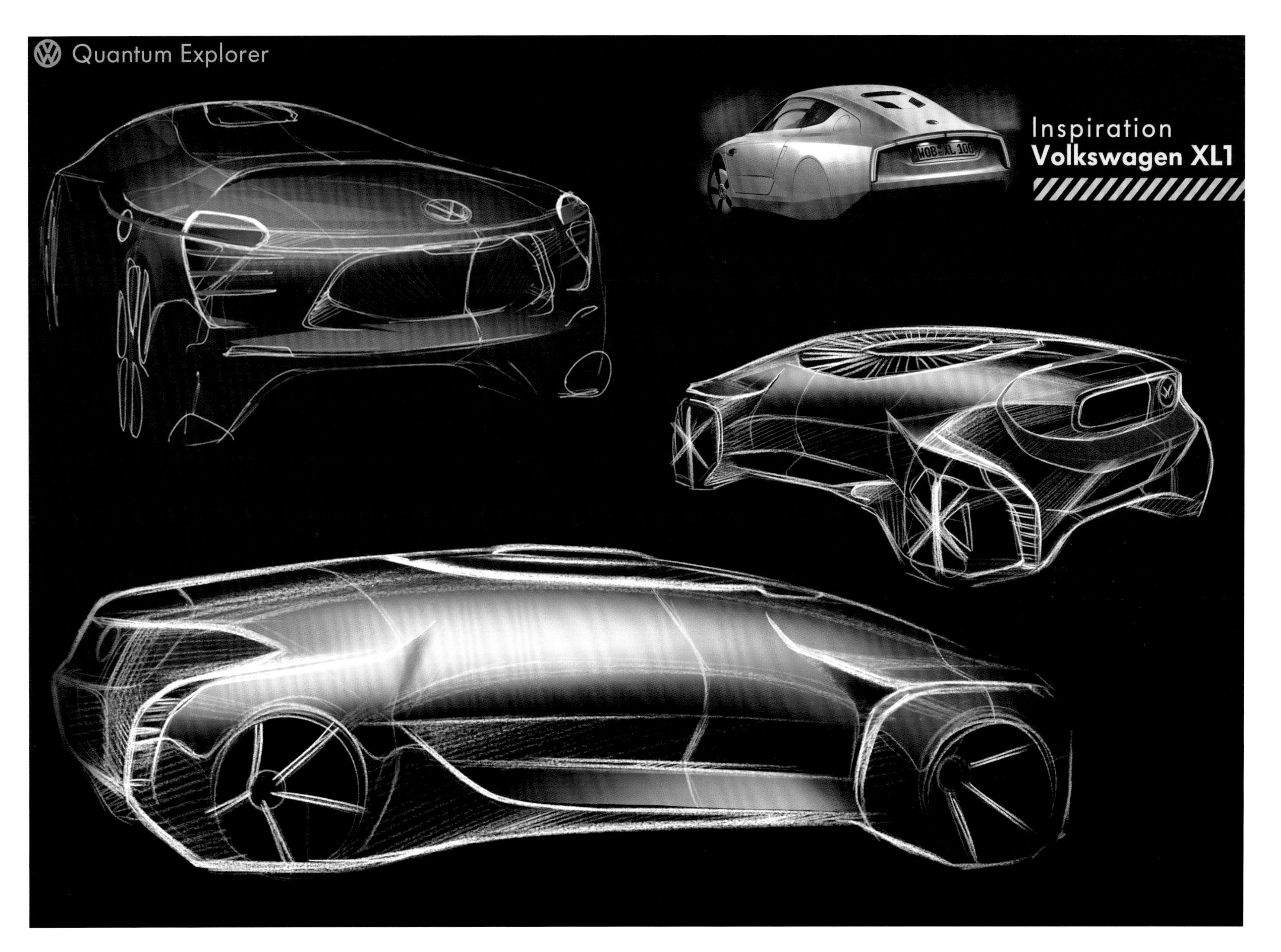
Quantum Explorer
Inspiration
Volkswagen XL1
WOB XL 100

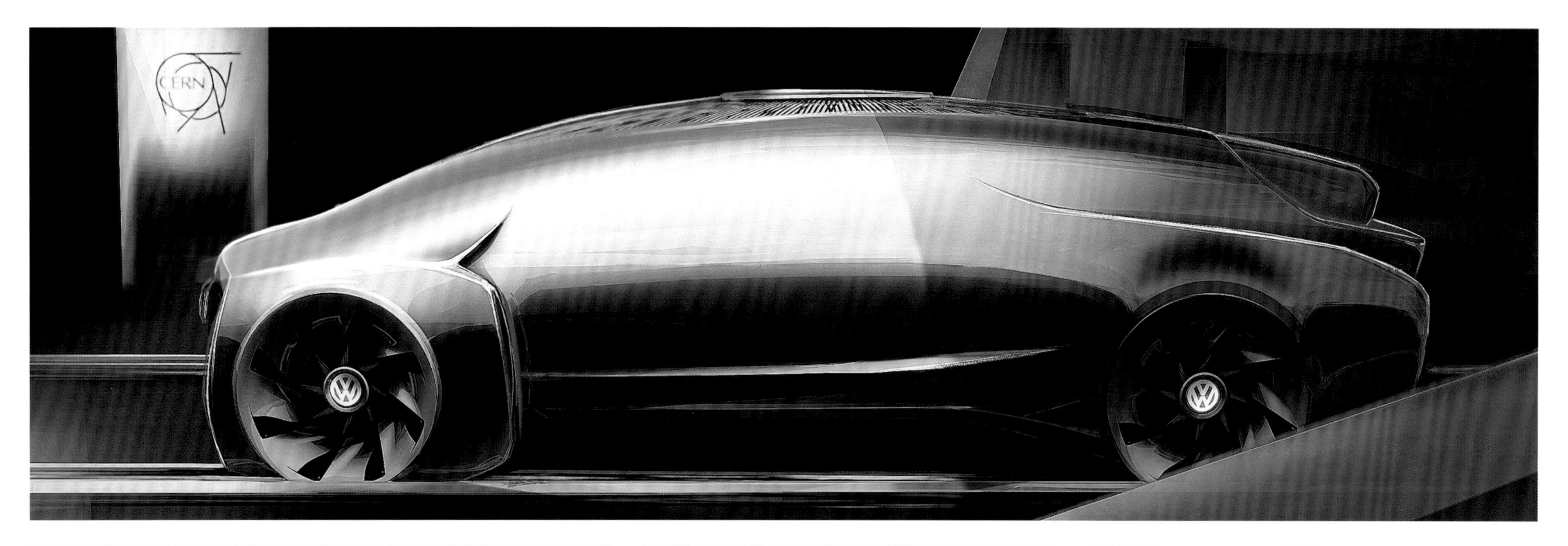
CERN

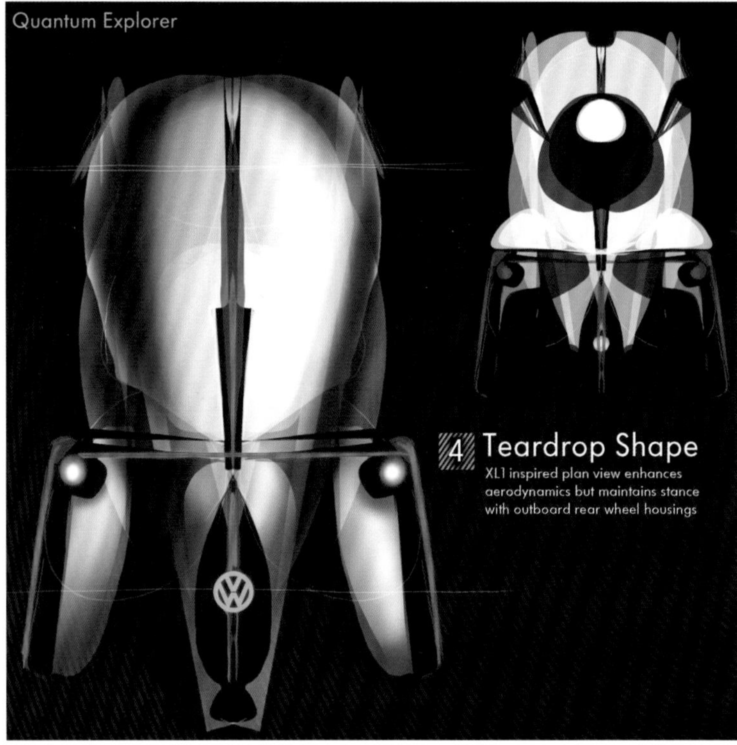
Quantum Explorer
4 Teardrop Shape
XL1 inspired plan view enhances aerodynamics but maintains stance with outboard rear wheel housings

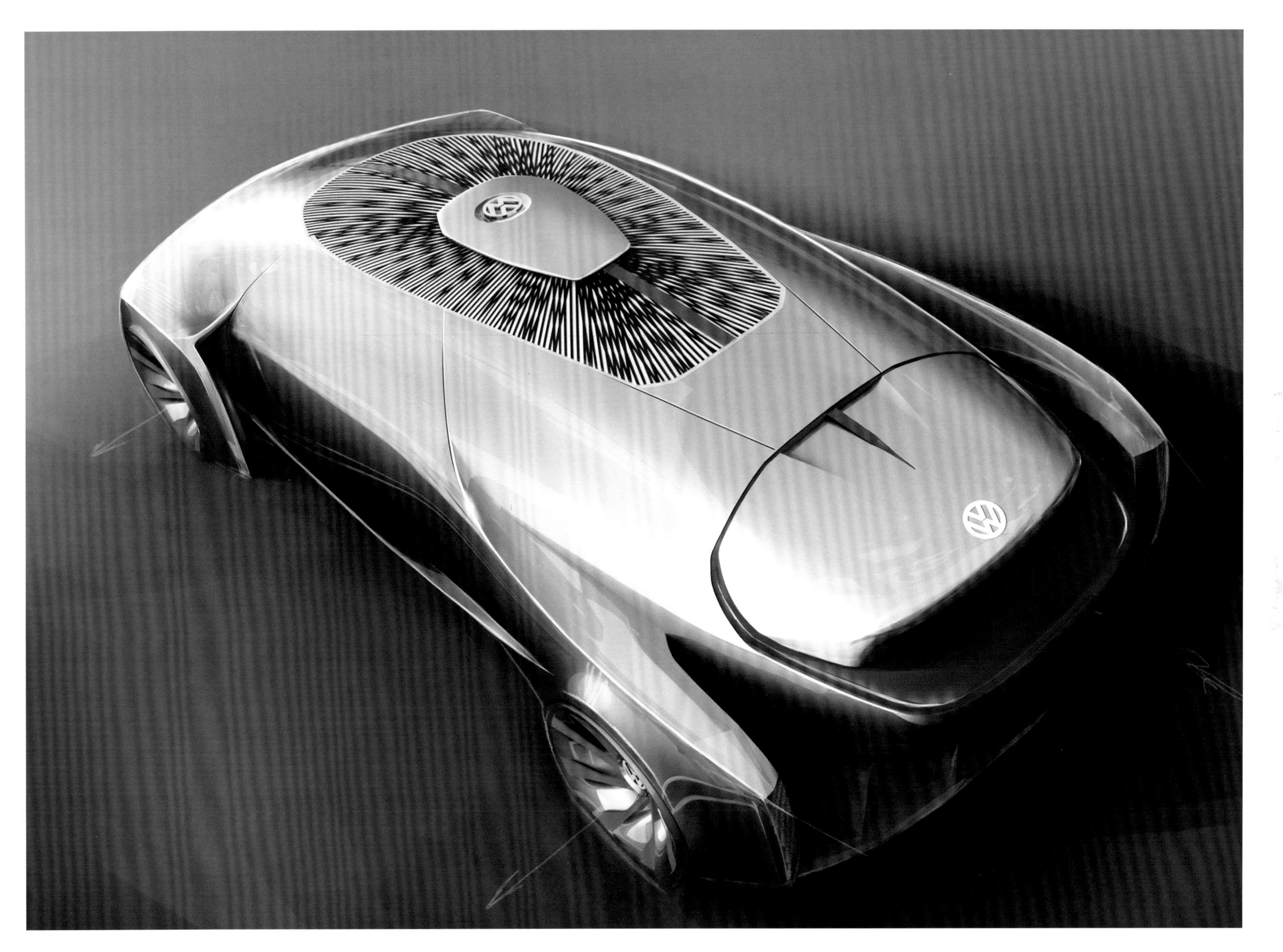

Concept Car
概念车

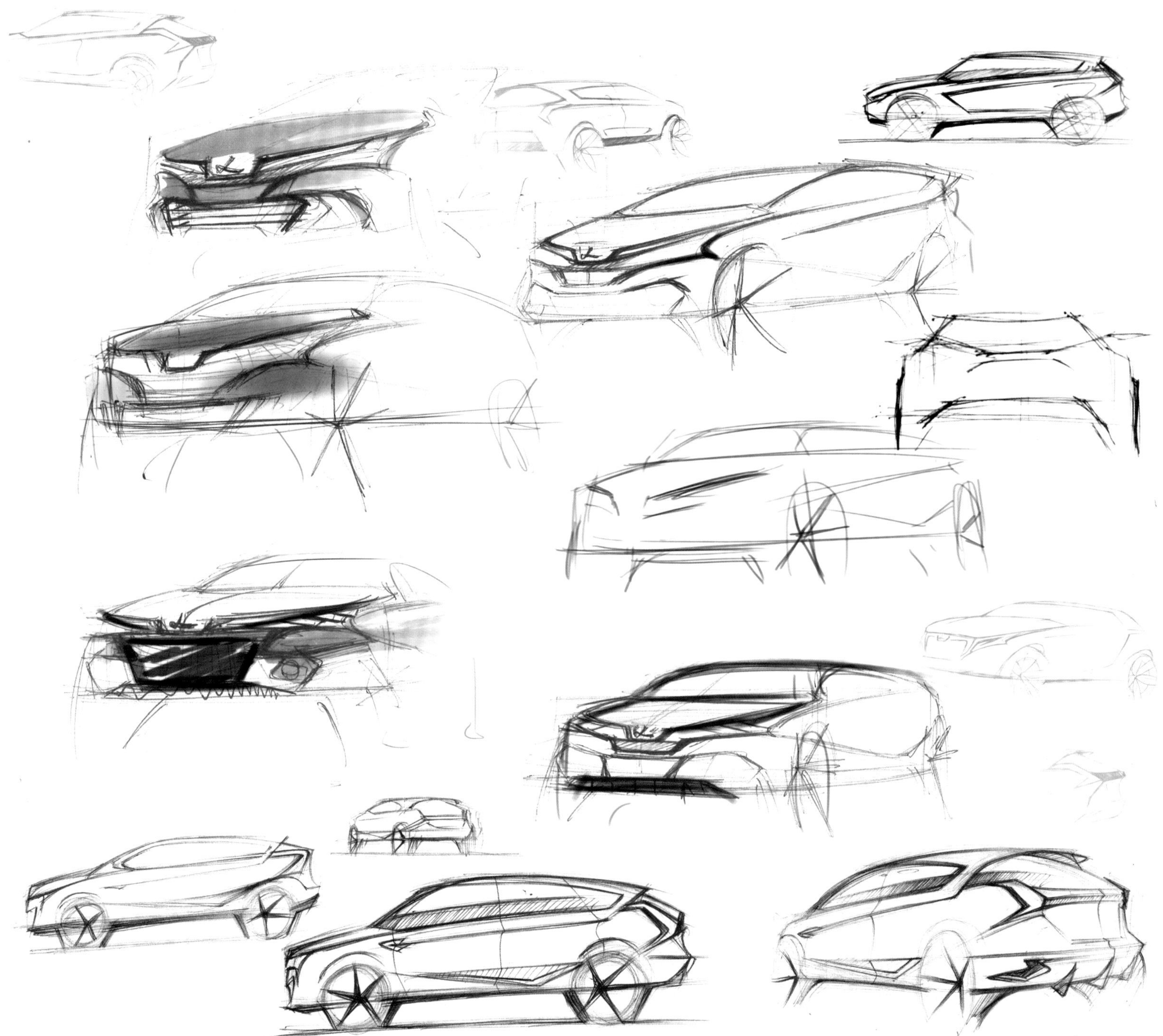

设计师：王子维
客户：个人作品

这是设计师在Luxgen公司实习期间完成的汽车外观设计作品，旨在研发一款新型的SUV（运动型多用途车）的模型。设计师首先将资料做成意向板，据此挖掘设计主题及设计理念。在确定了车身主体比例之后开始增添细节元素，最后再使用Photoshop软件进行完善。

Truck Exterior

卡车外形

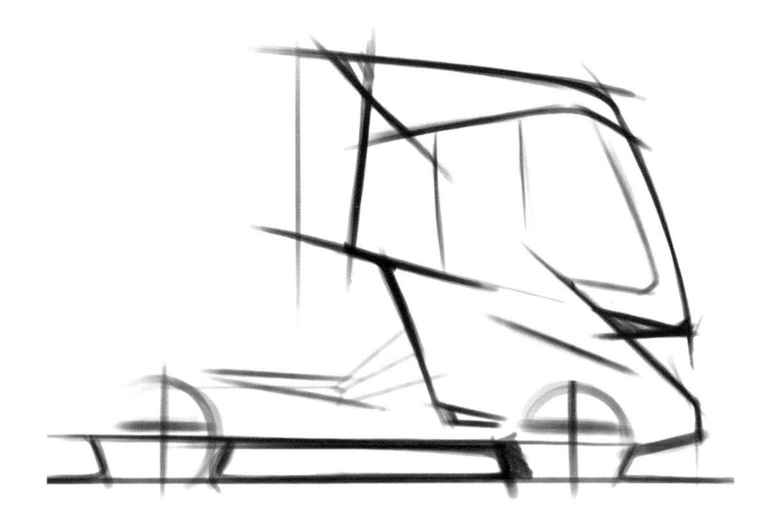

设计师：耶勒·特百思
客户：个人作品

这一设计的主要目标是探索一些全新的卡车外形，因此并没有最终确定的方向。多数草图都是使用圆珠笔勾画的，然后用马克笔快速表现出车身的立体感。在画卡车或者公交车草图的时候，应通过夸大透视效果来更好地展示尺度，使其看起来更加真实。在绘制最终效果图时，可以简单地画出周遭环境以烘托气氛。

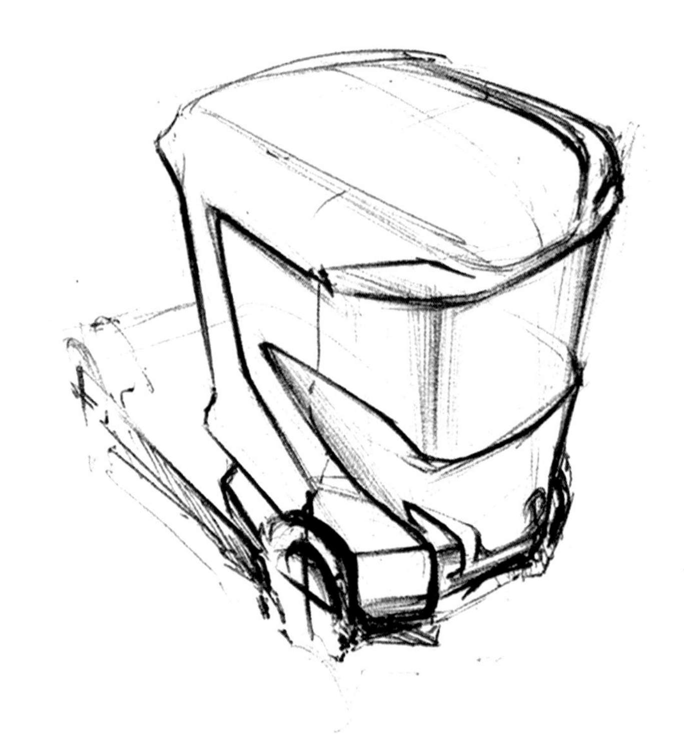

CAT

Multifunctional Waste Collection Vehicle

多功能垃圾运输车

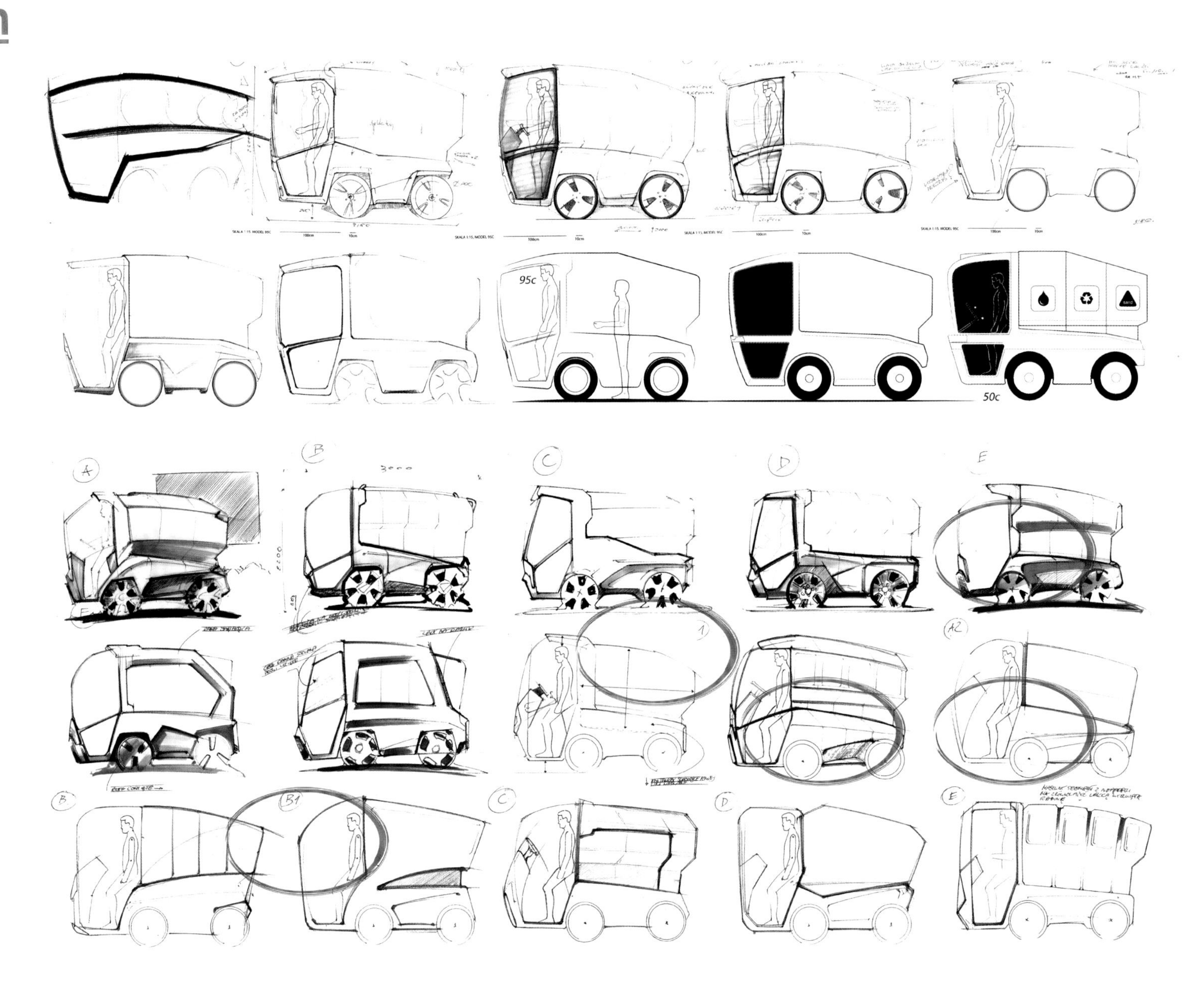

设计师：迈克尔·马基维茨

客户：学校项目

该毕业设计项目旨在将城市特种车辆的不同功能集合到一辆多功能车上，最重要的特色体现在驾驶舱的创新设计上。设计师通过一系列草图探讨了驾驶舱的尺度以及车辆的特殊功能。

HERCULES BEETLE

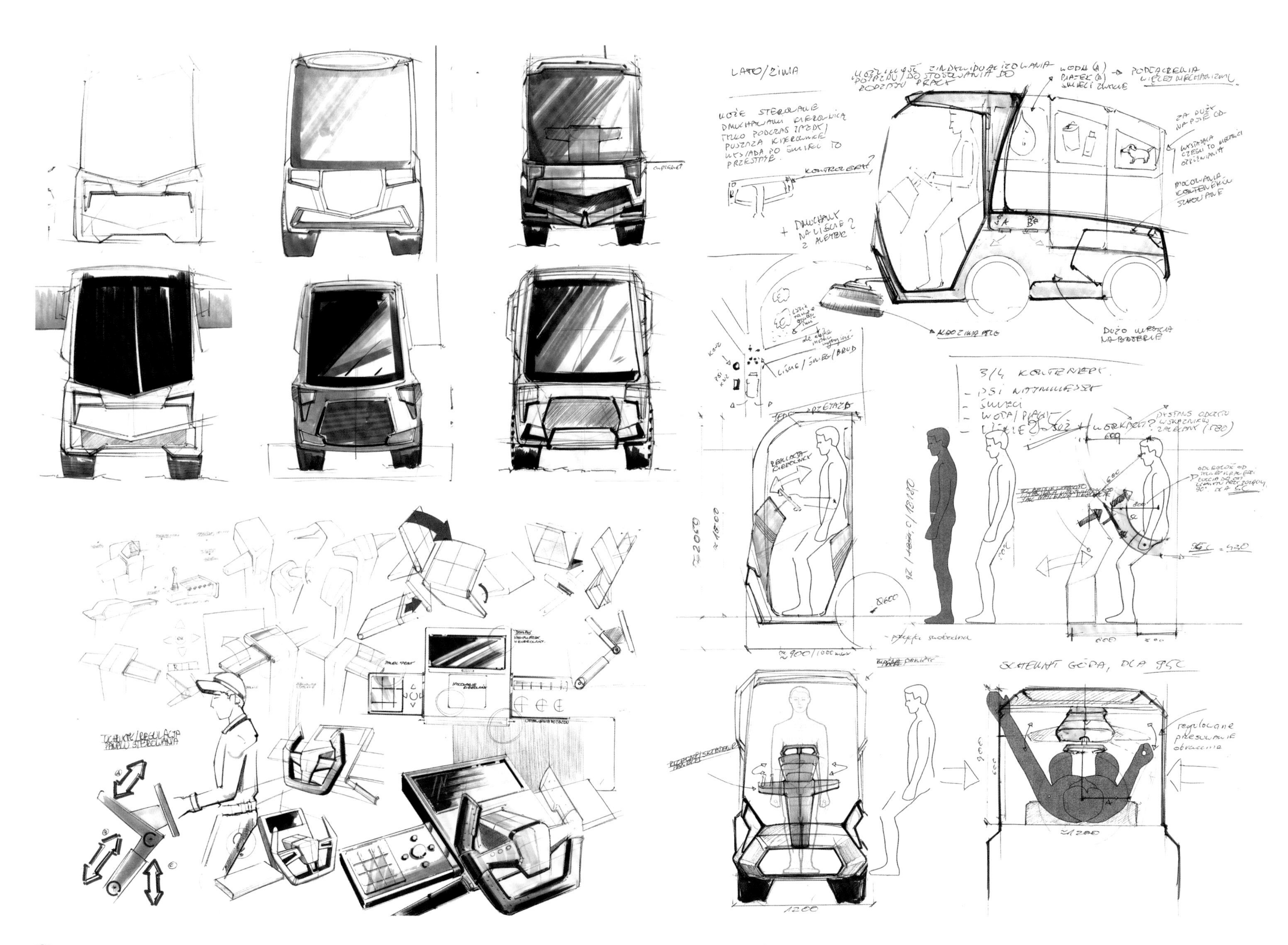
LATO/ZIMA
3/4 KONTENERY
ŚMIECI
SCHEMAT GÓRA, DLA 95C
regulowane
przesuwanie
obrotowe
≈1200
1200

The Next Iso Isetta

下一代Iso Isetta

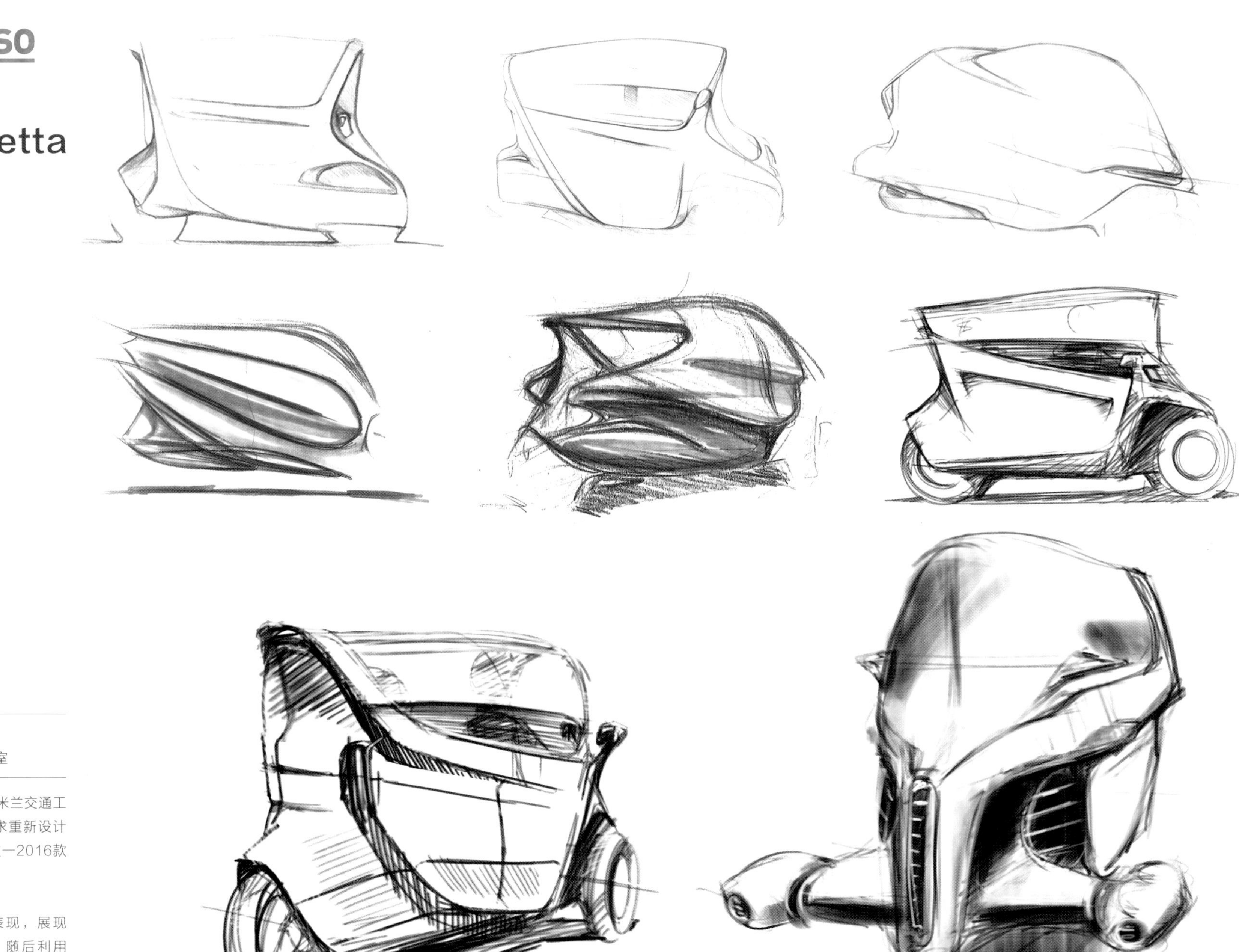

设计师：耶勒·特百思
客户：意大利Zagato设计工作室

设计师在Zagato设计工作室（米兰交通工具设计中心）毕业之际，被要求重新设计Isetta的标志性微型三轮车。这一2016款也是未来2025款的模型。

设计师通过几幅草图的快速表现，展现了汽车轻巧美观的造型语言。随后利用Photoshop软件对草图进行调整并深化。

HARD
AS SOFT AS
POSSIBLE
MAKE STEP IN HEIGHT
INCREASES
FEEL OF NOSE
SOFT
TRY TO ADD
A SPOILER TO
NARROW DOWN
LIGHT CATCHER!

Swift-Luggage Trailer
行李拖车

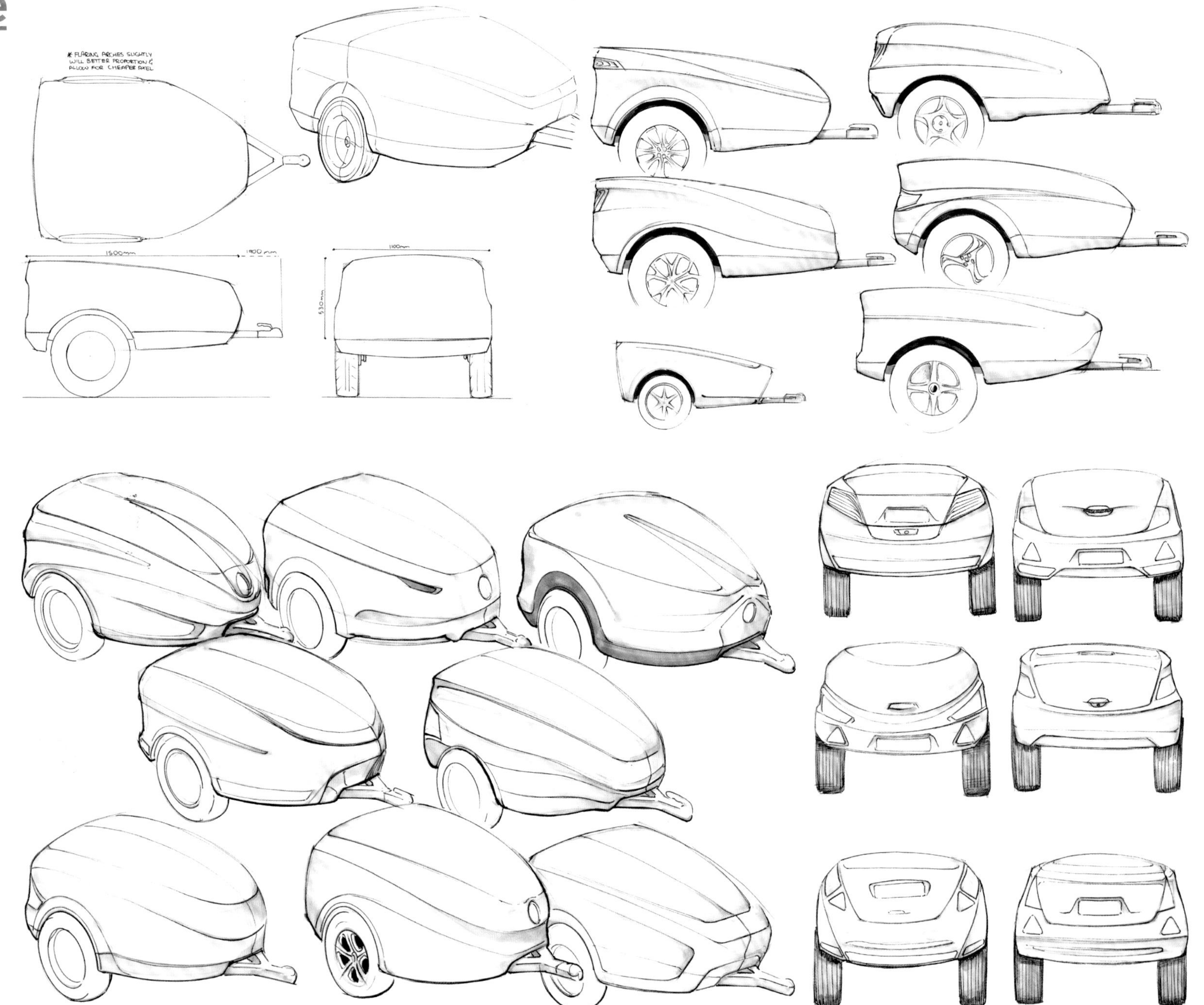

设计师：默里 · 夏普
客户：学校项目，约翰内斯堡大学

在南非，随着油价的迅速增长，很多消费者被迫寻找全新而环保的交通方式。随着市场上微型汽车的增多，购买一辆美观而节能的行李拖车成为了很多人的选择。

该行李拖车的设计不仅在视觉上和当代小汽车一样精美，还蕴含着丰富的附加市场价值，能够让使用者享受与其互动的完美体验。这一设计的成功意味着人们更乐于使用轻巧、便于携带的行李拖车出行，在节省成本的同时，更能提升旅行的舒适感。

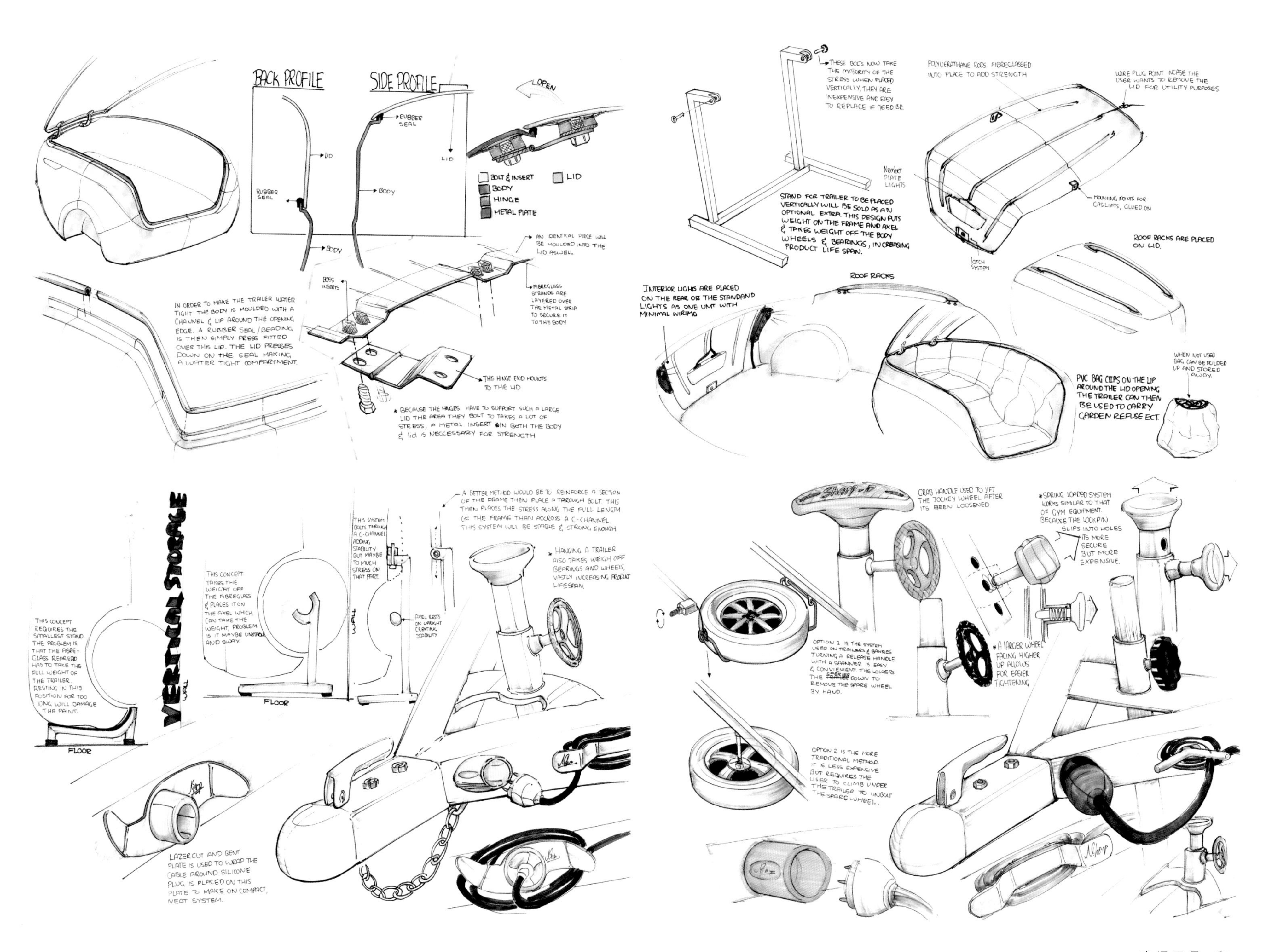

BACK PROFILE
SIDE PROFILE
OPEN
LID
RUBBER SEAL
BODY
BOLT & INSERT
HINGE
METAL PLATE
AN IDENTICAL PIECE WILL BE MOULDED INTO THE LID ASWELL
IN ORDER TO MAKE THE TRAILER WATER TIGHT THE BODY IS MOULDED WITH A CHANNEL & LIP AROUND THE OPENING EDGE. A RUBBER SEAL / BEADING IS THEN SIMPLY PRESS FITTED OVER THIS LIP. THE LID PRESSES DOWN ON THE SEAL MAKING A WATER TIGHT COMPARTMENT.
BOSS INSERTS
FIBREGLASS STRANDS ARE LAYERED OVER THE METAL STRIP TO SECURE IT TO THE BODY
THIS HINGE END MOUNTS TO THE LID
* BECAUSE THE HINGES HAVE TO SUPPORT SUCH A LARGE LID THE AREA THEY BOLT TO TAKES A LOT OF STRESS, A METAL INSERT IN BOTH THE BODY & lid IS NECCESSARY FOR STRENGTH
THESE BOB'S NOW TAKE THE MAJORITY OF THE STRESS WHEN PLACED VERTICALLY, THEY ARE INEXPENSIVE AND EASY TO REPLACE IF NEED BE
STAND FOR TRAILER TO BE PLACED VERTICALLY WILL BE SOLD AS AN OPTIONAL EXTRA. THIS DESIGN PUTS WEIGHT ON THE FRAME AND AXEL & TAKES WEIGHT OFF THE BODY WHEELS & BEARINGS, INCREASING PRODUCT LIFE SPAN.
POLYURETHANE RODS FIBREGLASSED INTO PLACE TO ADD STRENGTH
WIRE PLUG POINT INCASE THE USER WANTS TO REMOVE THE LID FOR UTILITY PURPOSES.
Number PLATE LIGHTS
MOUNTING POINTS FOR GASLIFTS, GLUED ON
latch SYSTEM
ROOF RACKS ARE PLACED ON LID.
INTERIOR LIGHTS ARE PLACED ON THE REAR OF THE STANDARD LIGHTS AS ONE UNIT WITH MINIMAL WIRING
ROOF RACKS
WHEN NOT USED BAG CAN BE FOLDED UP AND STORED AWAY.
PVC BAG CLIPS ON THE LIP AROUND THE LID OPENING THE TRAILER CAN THEN BE USED TO CARRY GARDEN REFUSE ECT.
VERTICAL STORAGE
THIS CONCEPT REQUIRES THE SMALLEST STAND. THE PROBLEM IS THAT THE FIBRE-GLASS REAR END HAS TO TAKE THE FULL WEIGHT OF THE TRAILER RESTING IN THIS POSITION FOR TOO LONG WILL DAMAGE THE PAINT.
WALL
FLOOR
THIS CONCEPT TAKES THE WEIGHT OFF THE FIBREGLASS & PLACES IT ON THE AXEL WHICH CAN TAKE THE WEIGHT, PROBLEM IS IT MAYBE UNSTABLE AND SWAY.
THIS SYSTEM BOLTS THROUGH A C-CHANNEL ADDING STABILITY BUT MAYBE TO MUCH STRESS ON THAT PART.
AXEL RESTS ON UPRIGHT CREATING STABILITY
A BETTER METHOD WOULD BE TO REINFORCE A SECTION OF THE FRAME THEN PLACE A THROUGH BOLT. THIS THEN PLACES THE STRESS ALONG THE FULL LENGTH OF THE FRAME THAN ACCROSS A C-CHANNEL. THIS SYSTEM WILL BE STABLE & STRONG ENOUGH.
HANGING A TRAILER ALSO TAKES WEIGH OFF BEARINGS AND WHEELS, VASTLY INCREASING PRODUCT LIFESPAN.
LAZERCUT AND BENT PLATE IS USED TO WRAP THE CABLE AROUND SILICONE PLUG IS PLACED ON THIS PLATE TO MAKE ON COMPACT, NEAT SYSTEM.
GRAB HANDLE USED TO LIFT THE JOCKEY WHEEL AFTER ITS BEEN LOOSENED
*SPRING LOADED SYSTEM WORKS SIMILAR TO THAT OF GYM EQUIPMENT. BECAUSE THE LOCKPIN SLIPS INTO HOLES ITS MORE SECURE BUT MORE EXPENSIVE.
OPTION 1 IS THE SYSTEM USED ON TRAILERS & BAKKIES TURNING A RELEASE HANDLE WITH A SPANNER IS EASY & CONVIENIENT. THIS LOWERS THE SPARE DOWN TO REMOVE THE SPARE WHEEL BY HAND.
* A LARGER WHEEL FACING HIGHER UP ALLOWS FOR EASIER TIGHTENING
OPTION 2 IS THE MORE TRADITIONAL METHOD. IT IS LESS EXPENSIVE BUT REQUIRES THE USER TO CLIMB UNDER THE TRAILER TO UNBOLT THE SPARE WHEEL.

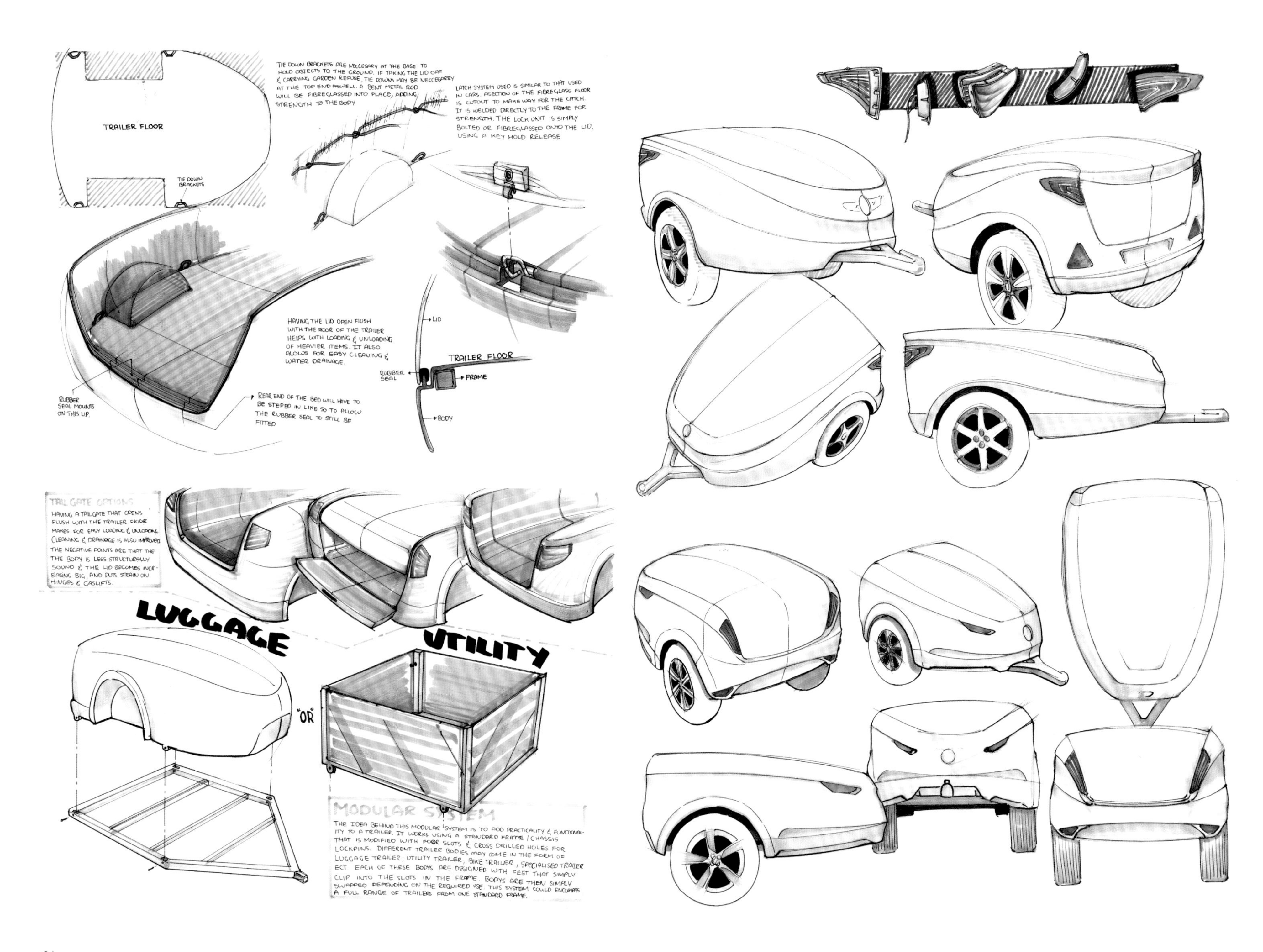
TRAILER FLOOR
TIE DOWN BRACKETS
TIE DOWN BRACKETS ARE NECCESARY AT THE BASE TO HOLD OBJECTS TO THE GROUND. IF TAKING THE LID OFF & CARRYING GARDEN REFUSE, TIE DOWNS MAY BE NECCESARRY AT THE TOP END ASWELL. A BENT METAL ROD WILL BE FIBREGLASSED INTO PLACE, ADDING STRENGTH TO THE BODY
LATCH SYSTEM USED IS SIMILAR TO THAT USED IN CARS. ASECTION OF THE FIBREGLASS FLOOR IS CUTOUT TO MAKE WAY FOR THE CATCH. IT IS WELDED DIRECTLY TO THE FRAME FOR STRENGTH. THE LOCK UNIT IS SIMPLY BOLTED OR FIBREGLASSED ONTO THE LID, USING A KEY HOLD RELEASE
HAVING THE LID OPEN FLUSH WITH THE FLOOR OF THE TRAILER HELPS WITH LOADING & UNLOADING OF HEAVIER ITEMS. IT ALSO ALLOWS FOR EASY CLEANING & WATER DRAINAGE.
LID
TRAILER FLOOR
RUBBER SEAL
FRAME
BODY
RUBBER SEAL MOUNTS ON THIS LIP.
REAR END OF THE BED WILL HAVE TO BE STEPED IN LIKE SO TO ALLOW THE RUBBER SEAL TO STILL BE FITTED
TAILGATE OPTIONS
HAVING A TAILGATE THAT OPENS FLUSH WITH THE TRAILER FLOOR MAKES FOR EASY LOADING & UNLOADING. CLEANING & DRAINAGE IS ALSO IMPROVED. THE NEGATIVE POINTS ARE THAT THE THE BODY IS LESS STRUCTURALLY SOUND & THE LID BECOMES INCR-EASING BIG, AND PUTS STRAIN ON HINGES & GASLIFTS.
LUGGAGE
"OR"
UTILITY
MODULAR SYSTEM
THE IDEA BEHIND THIS MODULAR SYSTEM IS TO ADD PRACTICALITY & FUNCTIONAL-ITY TO A TRAILER. IT WORKS USING A STANDARD FRAME / CHASSIS THAT IS MODIFIED WITH FOOR SLOTS & CROSS DRILLED HOLES FOR LOCKPINS. DIFFERENT TRAILER BODIES MAY COME IN THE FORM OF LUGGAGE TRAILER, UTILITY TRAILER, BIKE TRAILER, SPECIALISED TRAILER ECT. EACH OF THESE BODYS ARE DESIGNED WITH FEET THAT SIMPLY CLIP INTO THE SLOTS IN THE FRAME. BODYS ARE THEN SIMPLY SWAPPED DEPENDING ON THE REQUIRED USE. THIS SYSTEM COULD ENCOMPAS A FULL RANGE OF TRAILERS FROM ONE STANDARD FRAME.

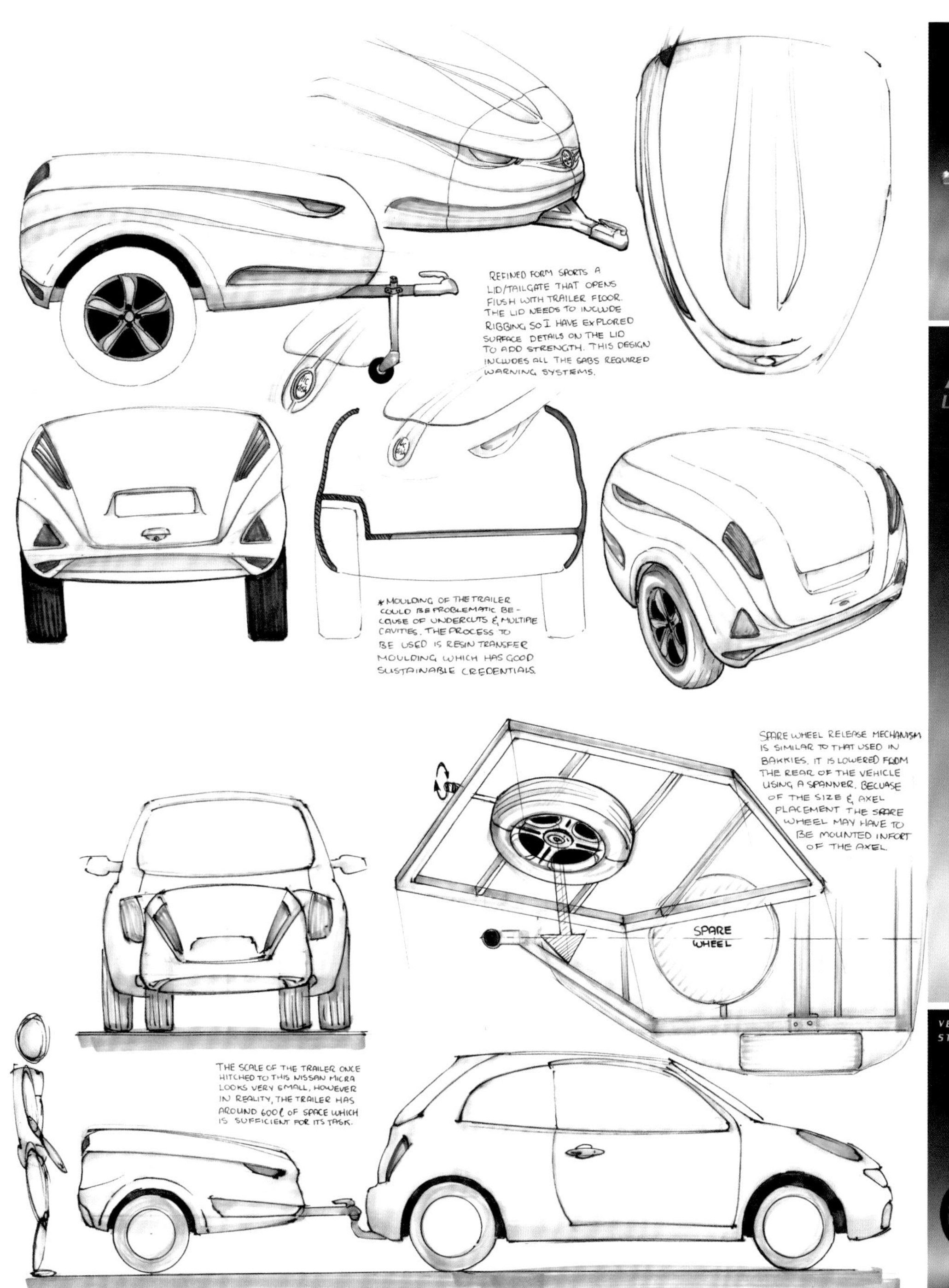
REFINED FORM SPORTS A LID/TAILGATE THAT OPENS FLUSH WITH TRAILER FLOOR. THE LID NEEDS TO INCLUDE RIBBING SO I HAVE EXPLORED SURFACE DETAILS ON THE LID TO ADD STRENGTH. THIS DESIGN INCLUDES ALL THE SABS REQUIRED WARNING SYSTEMS.
* MOULDING OF THE TRAILER COULD BE PROBLEMATIC BE-CAUSE OF UNDERCUTS & MULTIPLE CAVITIES. THE PROCESS TO BE USED IS RESIN TRANSFER MOULDING WHICH HAS GOOD SUSTAINABLE CREDENTIALS.
SPARE WHEEL RELEASE MECHANISM IS SIMILAR TO THAT USED IN BAKKIES. IT IS LOWERED FROM THE REAR OF THE VEHICLE USING A SPANNER. BECUASE OF THE SIZE & AXEL PLACEMENT THE SPARE WHEEL MAY HAVE TO BE MOUNTED INFORT OF THE AXEL.
SPARE WHEEL
THE SCALE OF THE TRAILER ONCE HITCHED TO THIS NISSAN MICRA LOOKS VERY SMALL, HOWEVER IN REALITY, THE TRAILER HAS AROUND 600ℓ OF SPACE WHICH IS SUFFICIENT FOR ITS TASK.

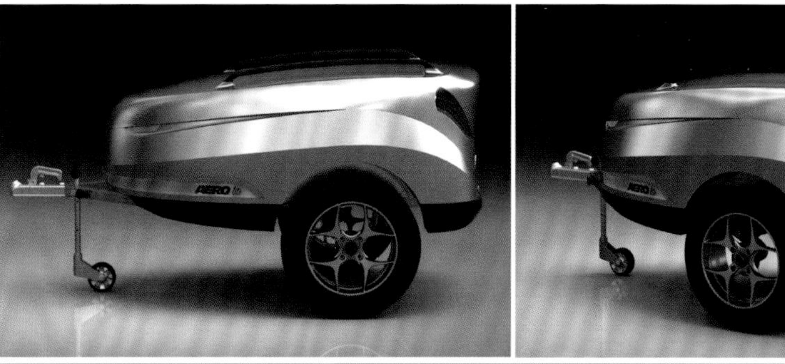

AERO lite
LIGHTEN YOUR LOAD
Introducing the all new Aero lite, a new breed of trailer that revolutionises the way we travel. Designed specifically for the small entry level car market, one can now travel efficiently and comfortably in their small car with all the necessary luggage. The aerodynamically tested design and advanced fibreglass moulded body means this trailer is more than just beautiful. A host of new innovations include: a spring loaded jockey adjustment system; a built-in, secure licence disk window (Bottom Right); a designated cable stay and silicone connector plug; concealed hinge and latch systems allowing a seamless form; a custom designed stand that not only allows for vertical storage within your garage, but also prolongs the life of the trailer by removing stress from key areas (Bottom Left).
AERO lite

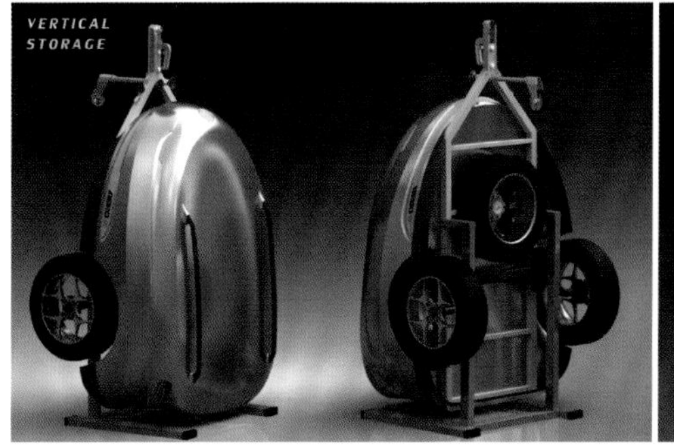
VERTICAL STORAGE

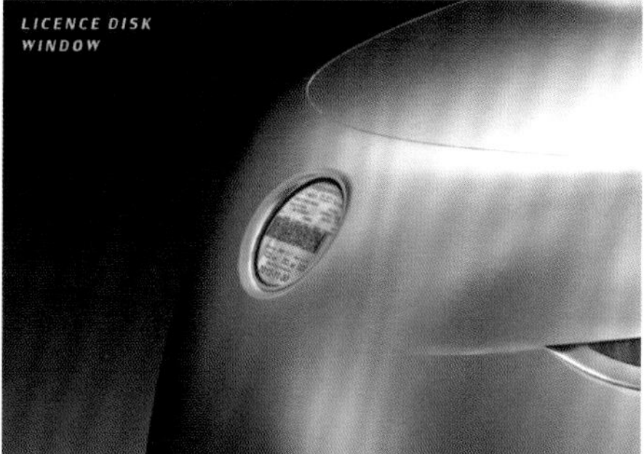
LICENCE DISK WINDOW

S3 Aircraft

S3飞机

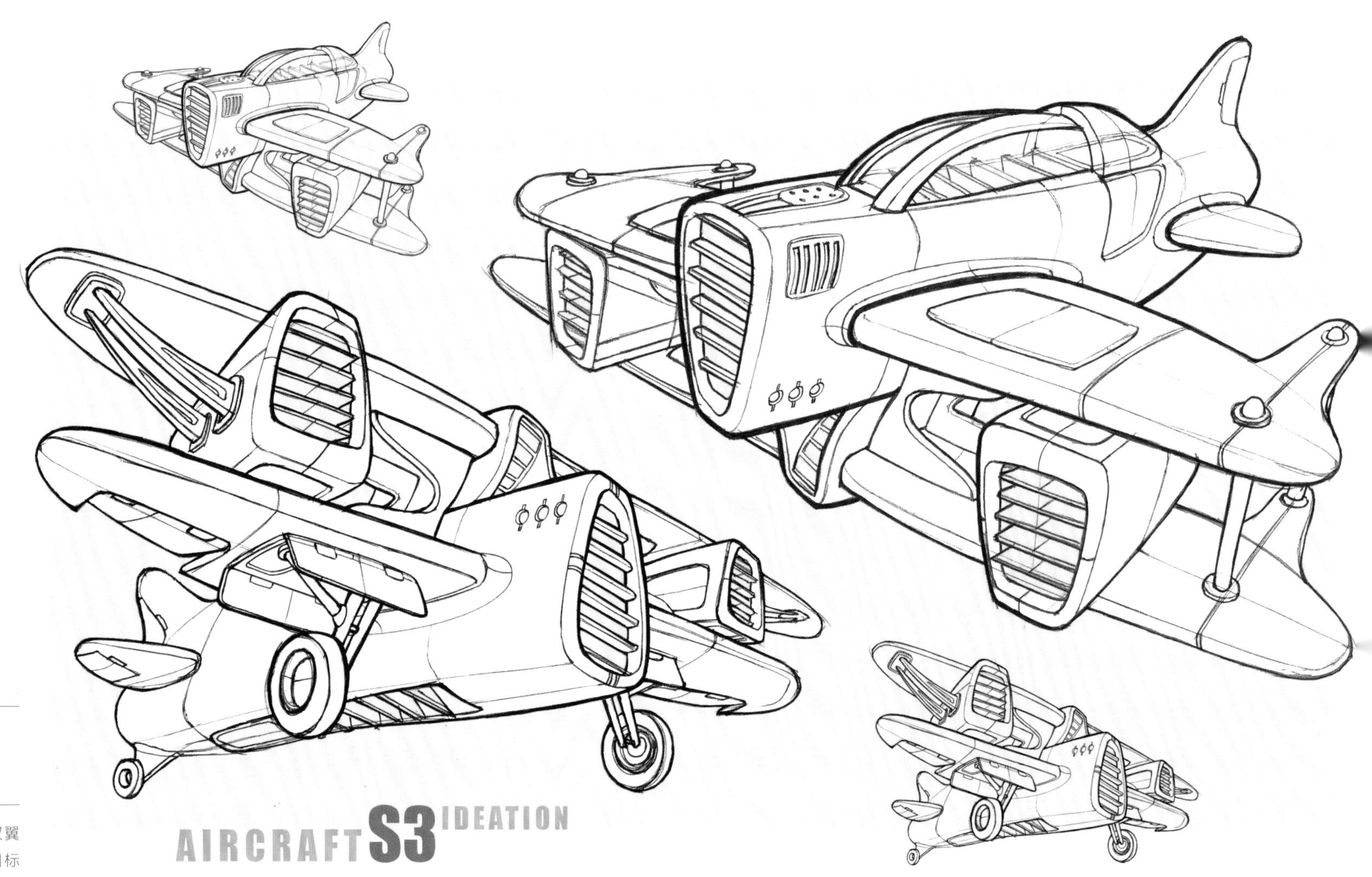

设计师：埃德加拉斯·瑟尼卡斯
（Edgaras Cernikas）
客户：Concept Art工作室

这是为一款视频游戏打造的一架特技双翼飞机的概念设计。在构思阶段，主要目标是将传统和现代的飞机特色相融合，即将经典的双翼飞机与喷气机的造型结合。

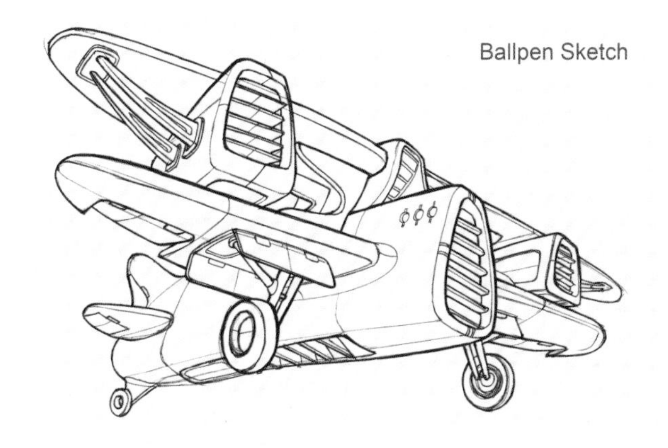
Ballpen Sketch

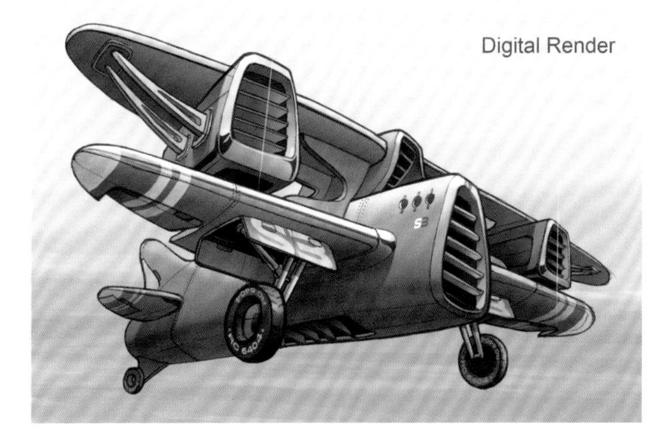
S3
Digital Render

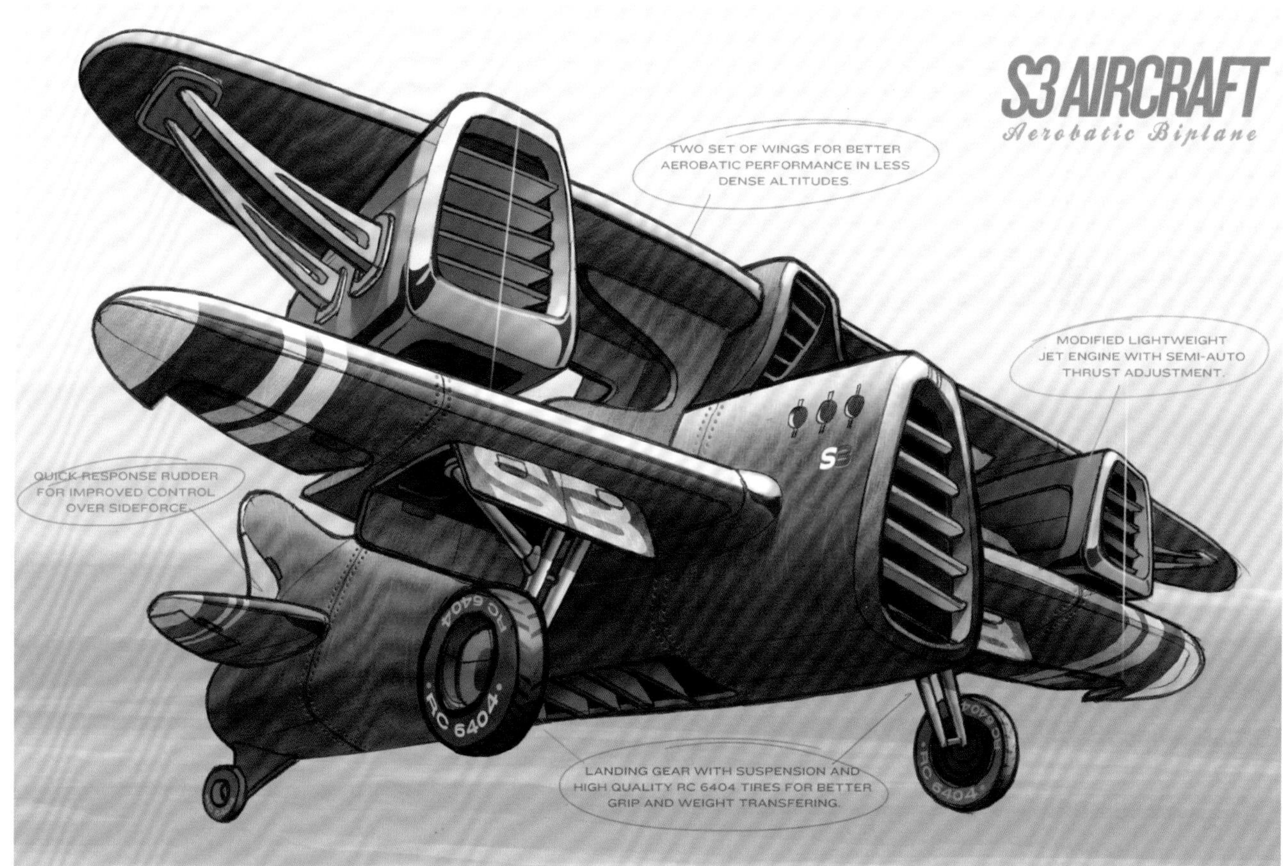
S3 AIRCRAFT
Aerobatic Biplane
TWO SET OF WINGS FOR BETTER AEROBATIC PERFORMANCE IN LESS DENSE ALTITUDES.
MODIFIED LIGHTWEIGHT JET ENGINE WITH SEMI-AUTO THRUST ADJUSTMENT.
QUICK RESPONSE RUDDER FOR IMPROVED CONTROL OVER SIDEFORCE.
LANDING GEAR WITH SUSPENSION AND HIGH QUALITY RC 6404 TIRES FOR BETTER GRIP AND WEIGHT TRANSFERING.
RC 6404

Plane Rendering
飞机效果图

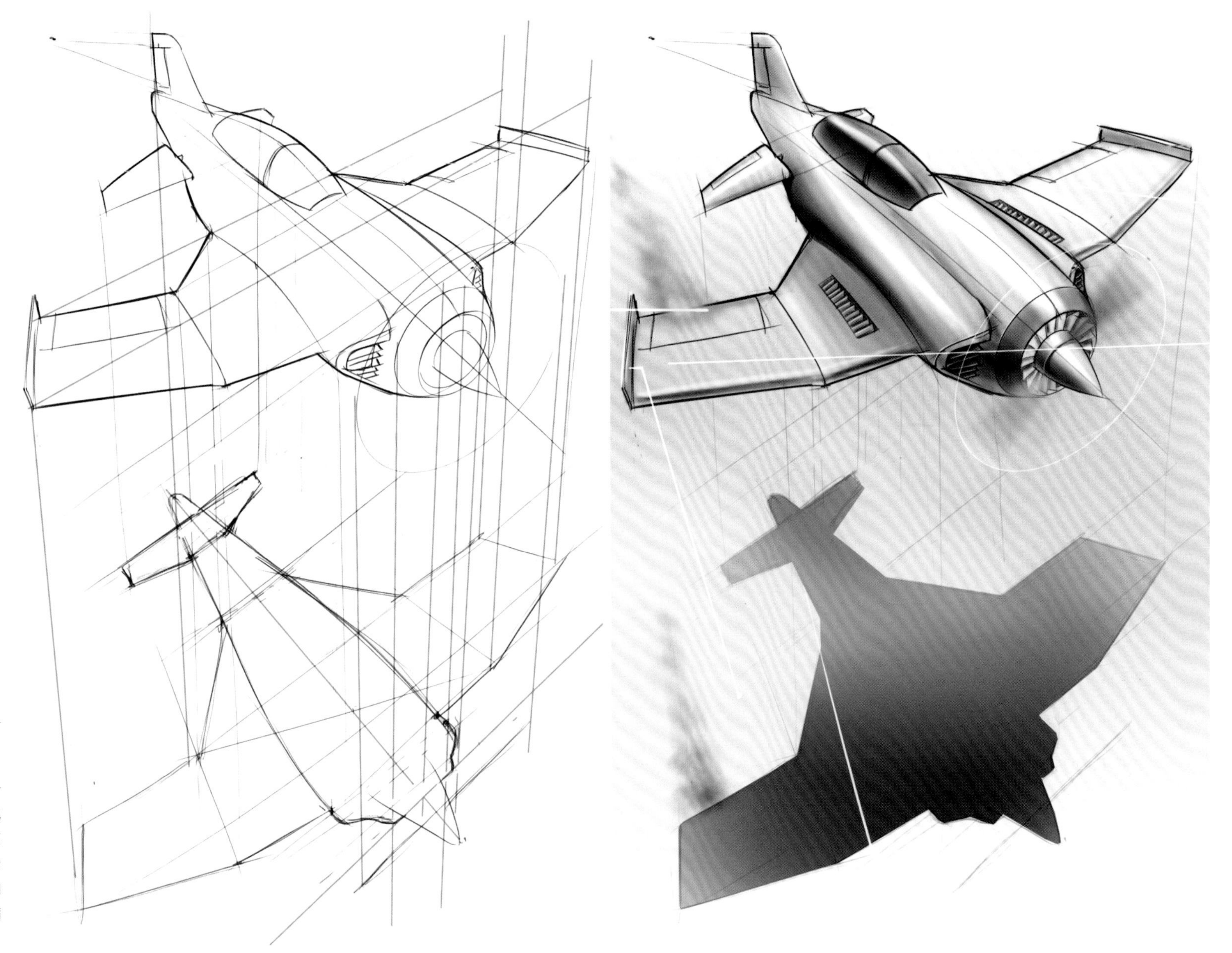

设计师：罗蒂米 · 索罗拉（Rotimi Solola）
客户：个人作品

设计师一直热爱汽车设计，执着于画汽车，但始终没有太大的突破。为了提高草图绘制水平，他转而开始对飞机进行绘制。在摆脱了长时间画汽车带来的枯燥和压力之后，罗蒂米重新获得了成就感与满足感。

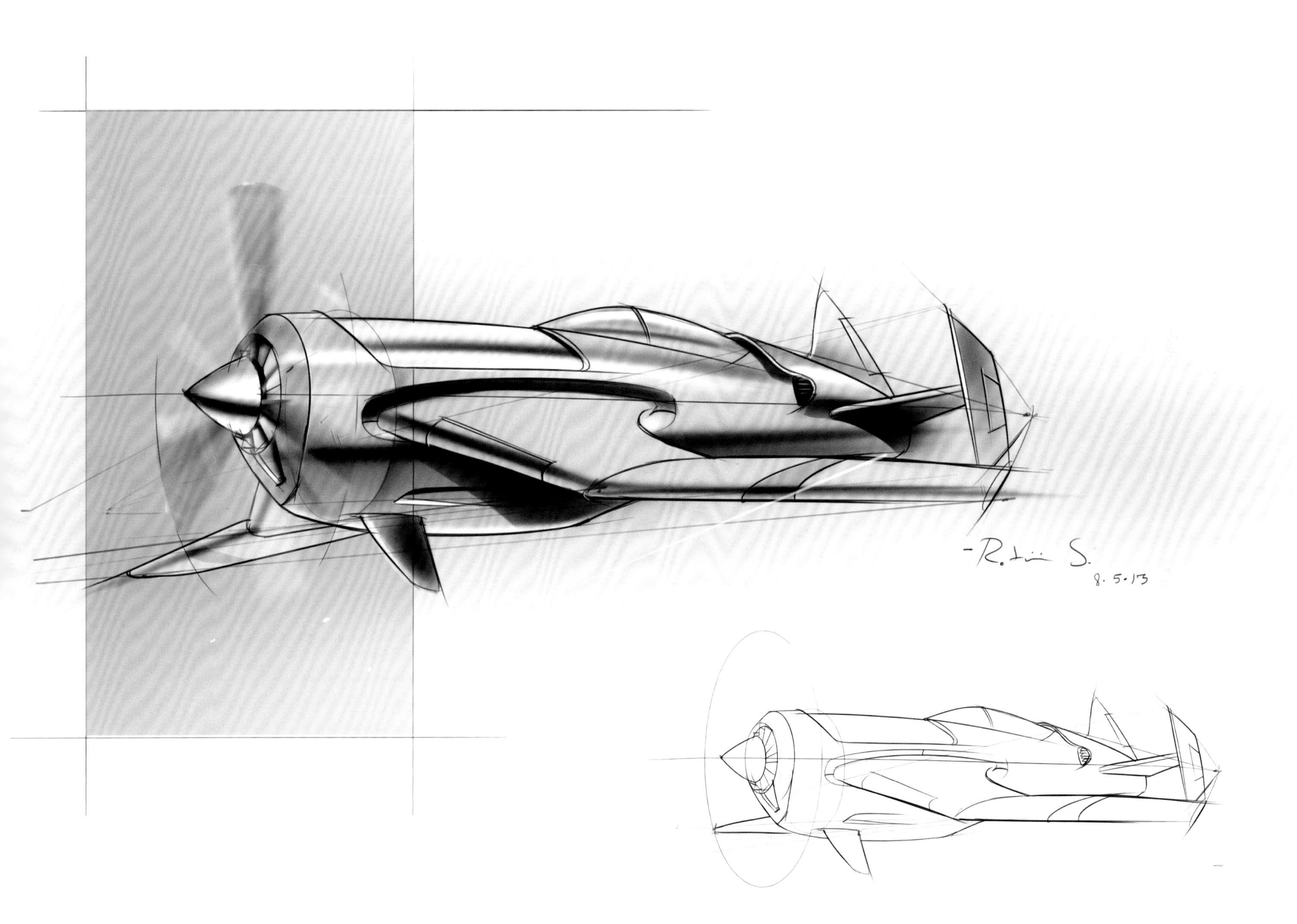
8.5.13

Tahr Quad

Tahr Quad
农用车

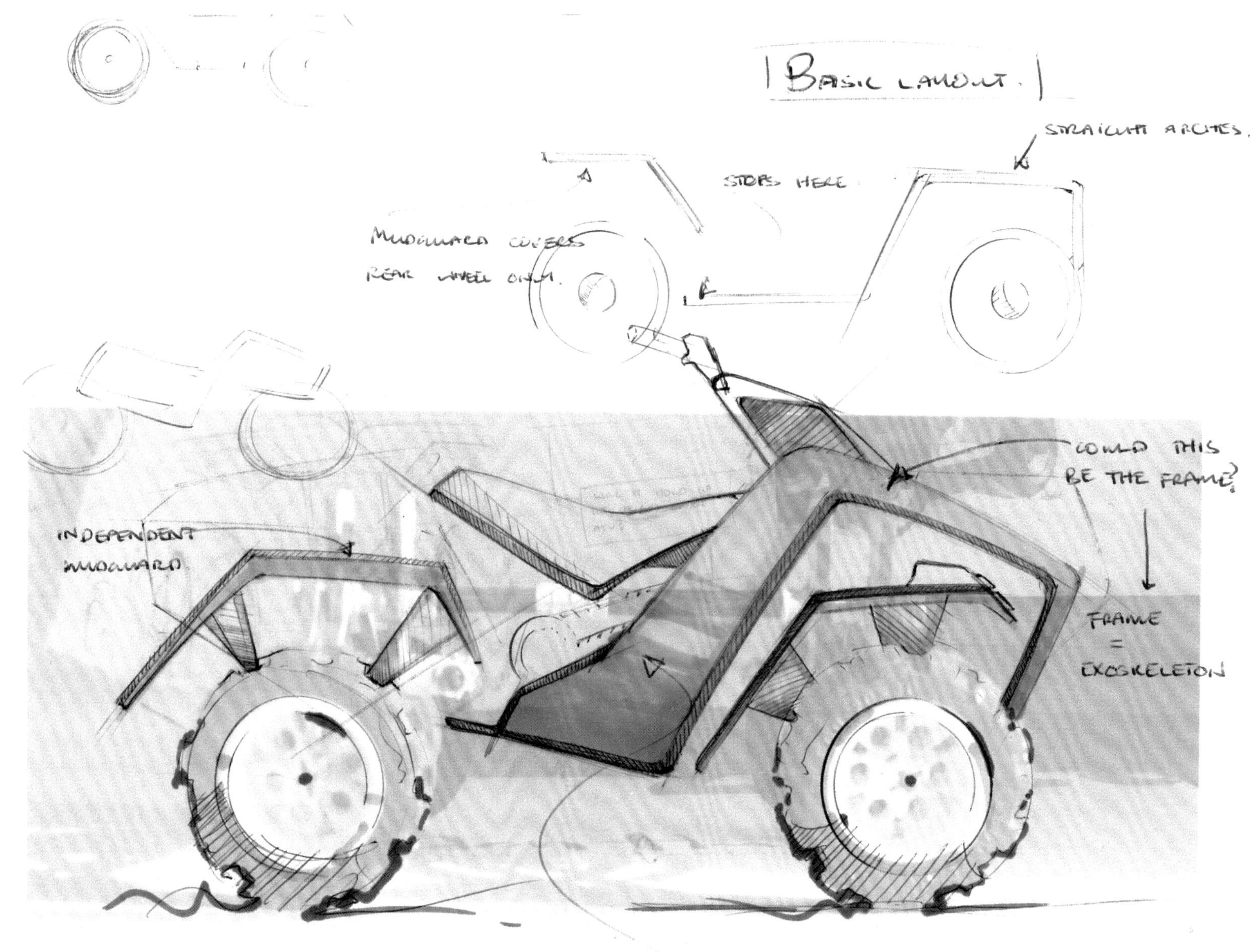

设计师：尼古拉斯·马克斯（Nicholas Marks）
客户：个人作品

这是一款农用多功能车，设计旨在对驾驶与行进方式的创新，减少对庄稼的破坏。

首先绘制多个正侧面车身的简略草图，有助于准确地把握车身比例。在创作这组草图的过程中，设计师运用了特别的绘画方式，即在照片或图片上直接勾画草图，这样做的好处就是可以落笔更加准确。

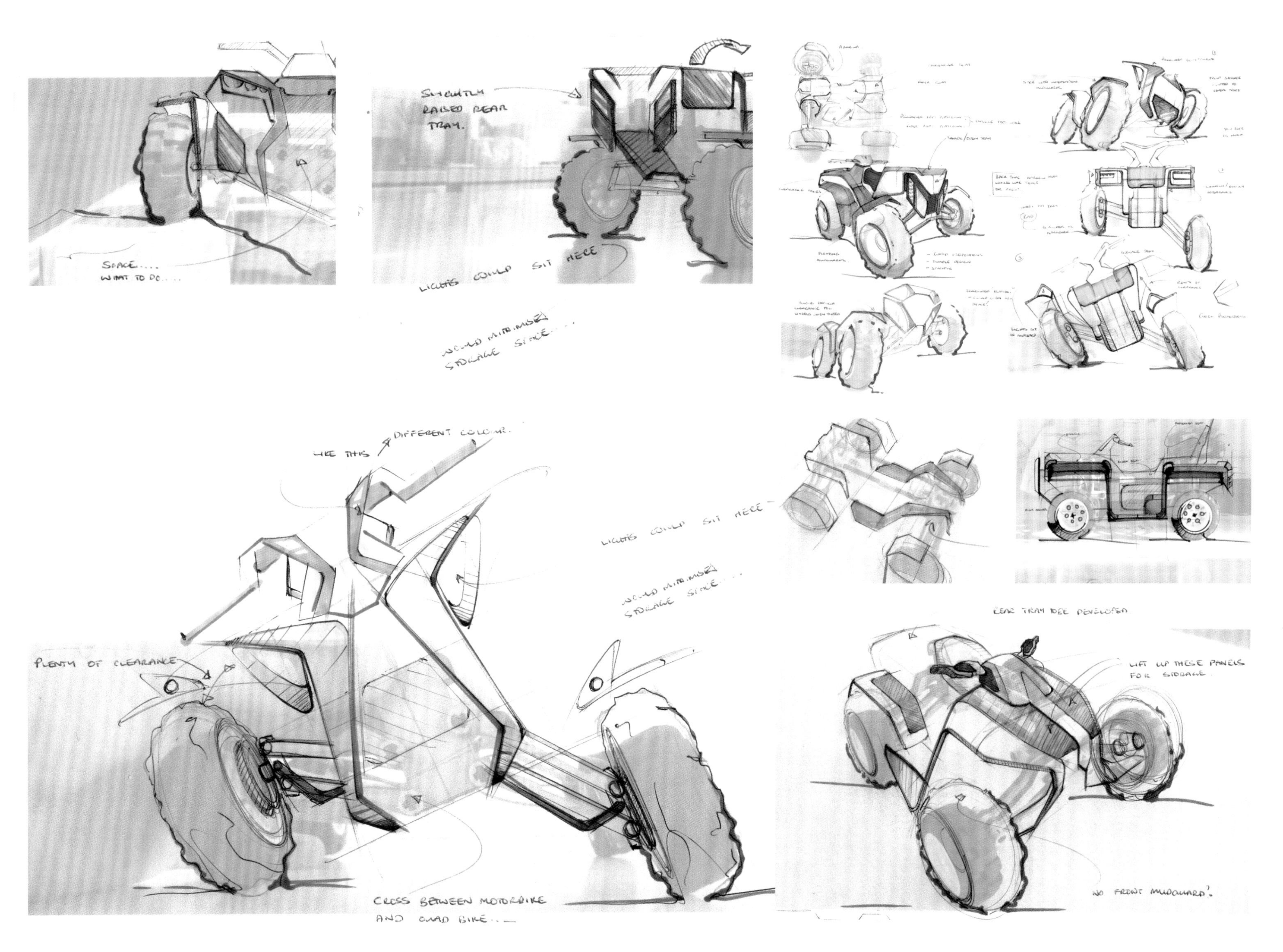
SPACE....
WHAT TO DO....
SLIGHTLY
RAISED REAR
TRAY.
LIGHTS COULD SIT HERE
WOULD MINIMISE
STORAGE SPACE....
LIKE THIS
DIFFERENT COLOUR.
LIGHTS COULD SIT HERE
WOULD MINIMISE
STORAGE SPACE....
PLENTY OF CLEARANCE
CROSS BETWEEN MOTORBIKE
AND QUAD BIKE...
REAR TRAY TO BE DEVELOPED
LIFT UP THESE PANELS
FOR STORAGE
NO FRONT MUDGUARD?

Volkswagen Go-2

Go-2折叠电动车

设计师：达尼洛·马乔·萨依托（Danilo Makio Saito）
客户：TalentoVolkswagen 2014汽车设计竞赛

这一项目是为Talento和Volkswagen联合举办的主题为Carry Me（带上我）的设计大赛，旨在设计一款便携的交通工具。该作品系本次大赛的第一名。

"Go-2"是一个可折叠电动自行车概念设计，也是一个可与汽车相集成的交通工具。其电池置于车架中间，整个车身可折叠并放置在汽车后部。

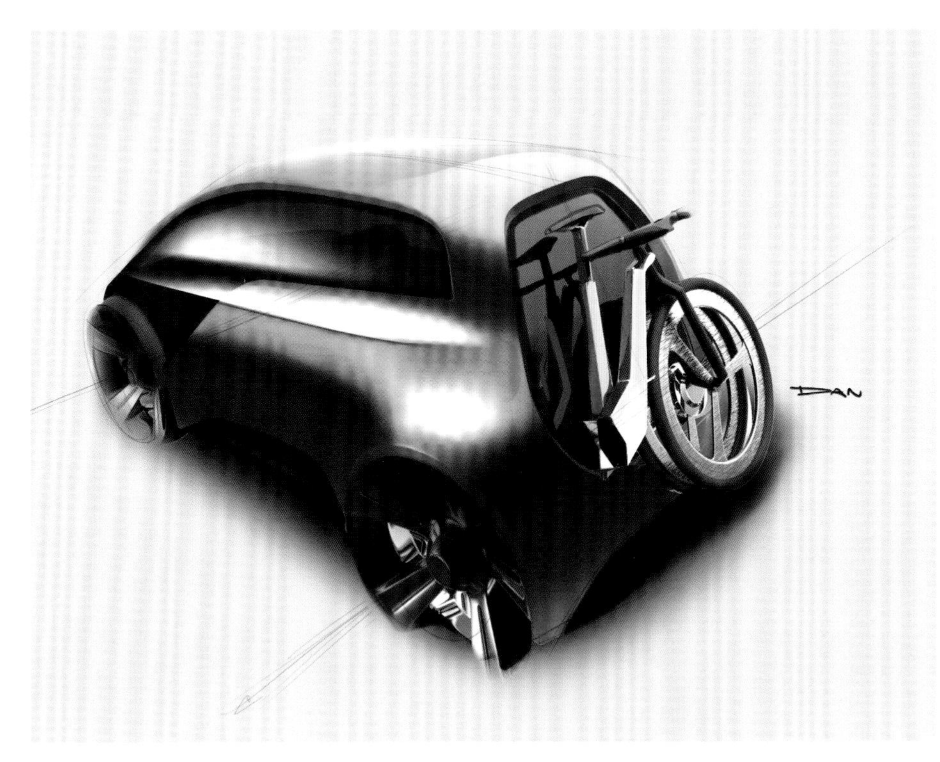
DAN

DAN
DAN

T-Wheel Folding Bike

T-Wheel 折叠自行车

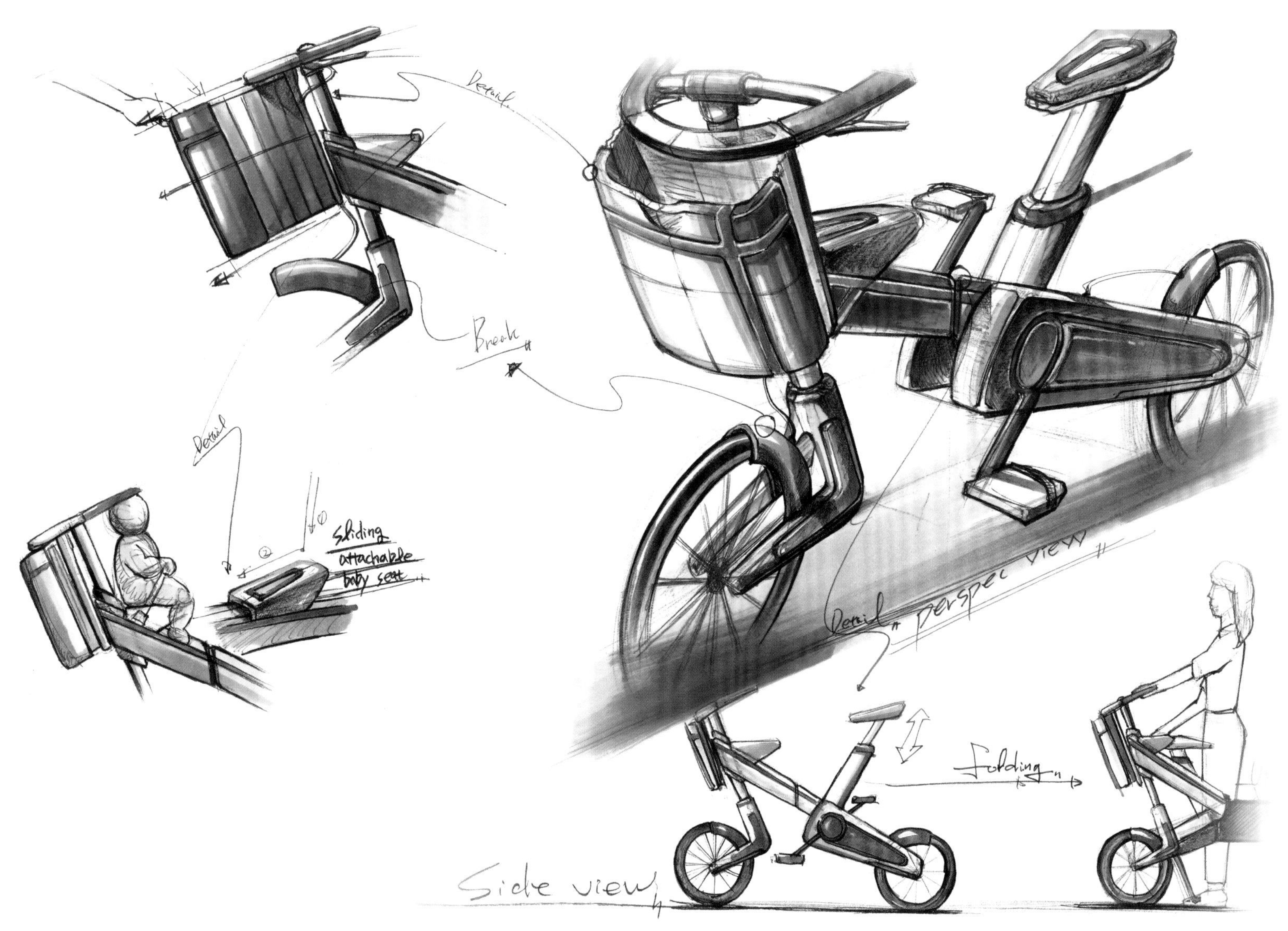

设计师：柳时彤

客户：某大学第三届产品设计竞赛暨自行车设计竞赛

这款自行车的纳米碳管轮胎耐磨性高，且便于更换。同时，可折叠的车身也更节省空间。

初期草图线条趋于模糊弱化；随着方案的深入，线条也更加清晰明确，以准确地表现出车身的结构及其折叠功能。

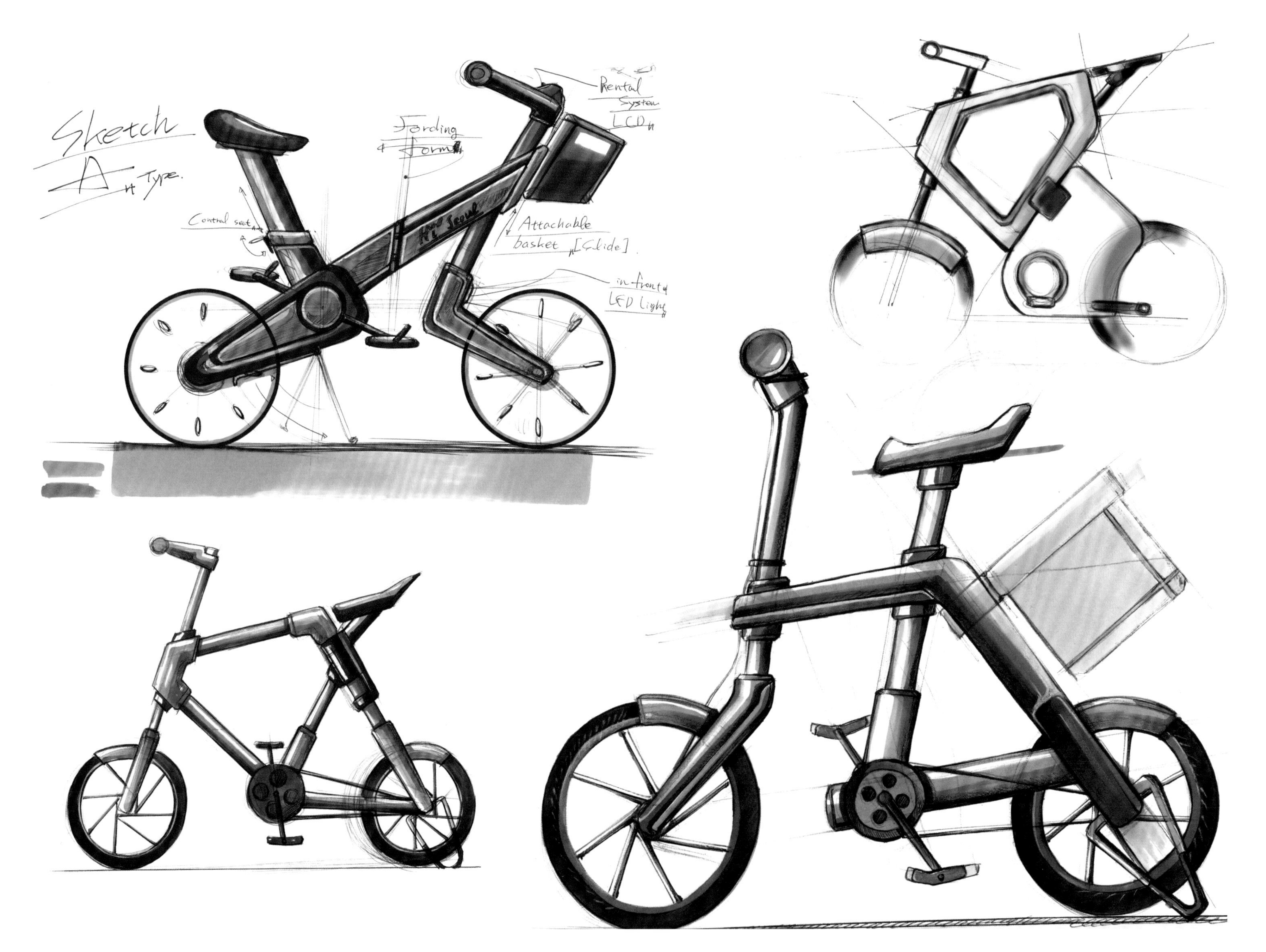
Sketch
A Type.
Fording Form
Rental System LCD
Contral seat
Attachable basket [Glide]
in-front LED Light
Hi Seoul

Toy Tractor RT

玩具拖拉机

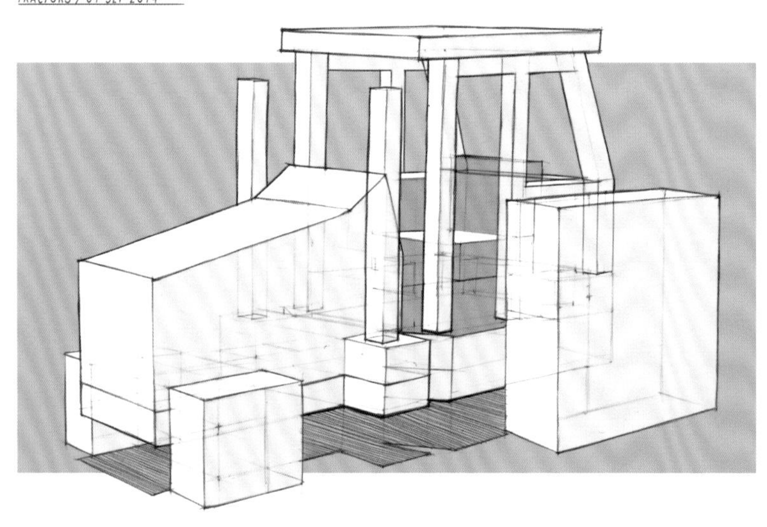

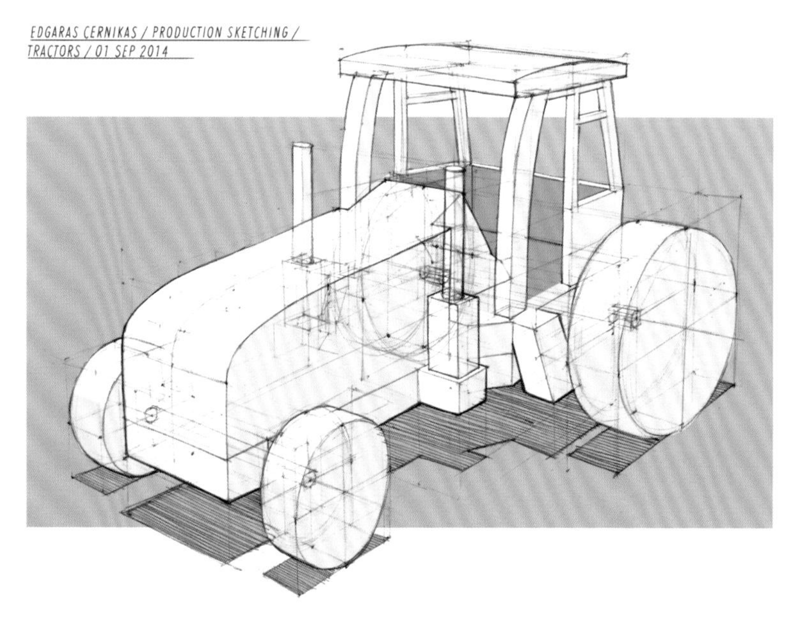

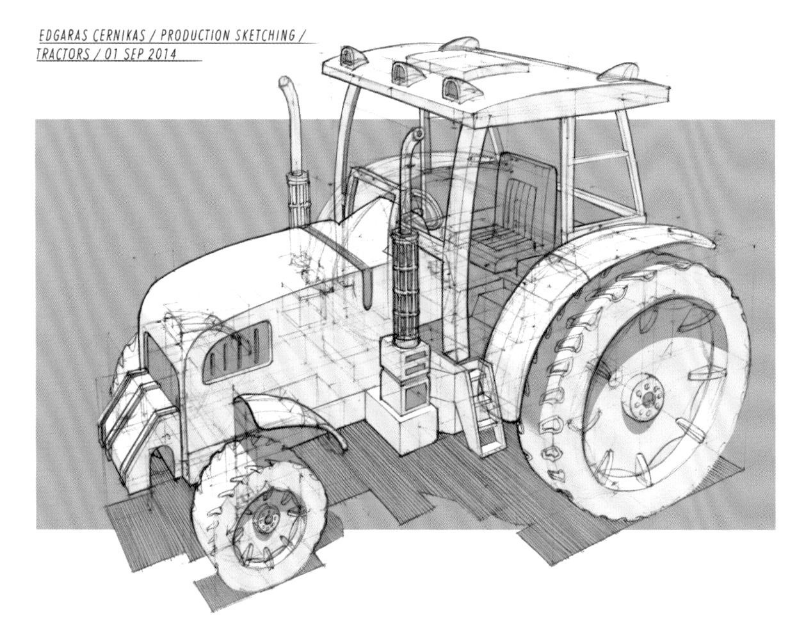

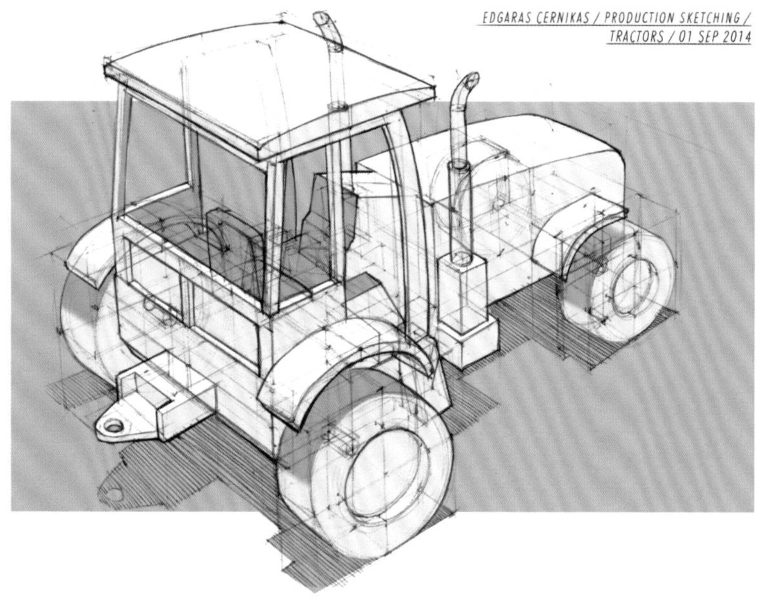

设计师：埃德加拉斯·瑟尼卡斯
客户：Concept Art工作室

在草图绘制阶段，从基本形态出发，逐步深化玩具拖拉机的细节。整个绘制过程均参考了真实的拖拉机形态，在此基础上进行了简化。

Sidewalk Snowplow

人行道扫雪车

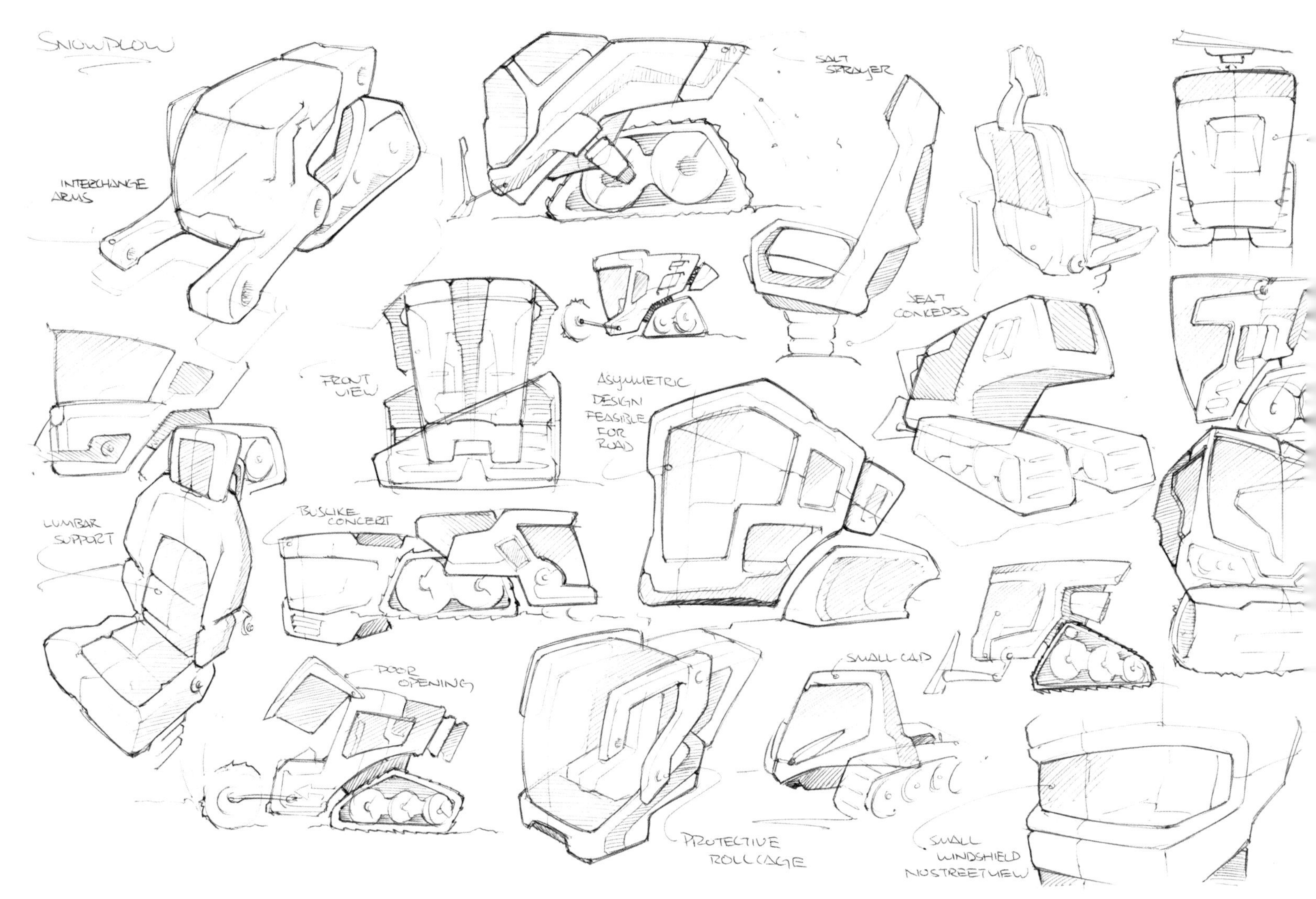

设计师：耿天亦
客户：个人作品

在除雪的过程中，人行道往往容易被忽略，这极易给行人带来危险。这款扫雪车就是专为清扫人行道上的积雪而设计的。设计师首先对人行道进行调研，从而确定了扫雪车的尺度及其行进方式，并在此基础上对其功能与外观进行了设计。

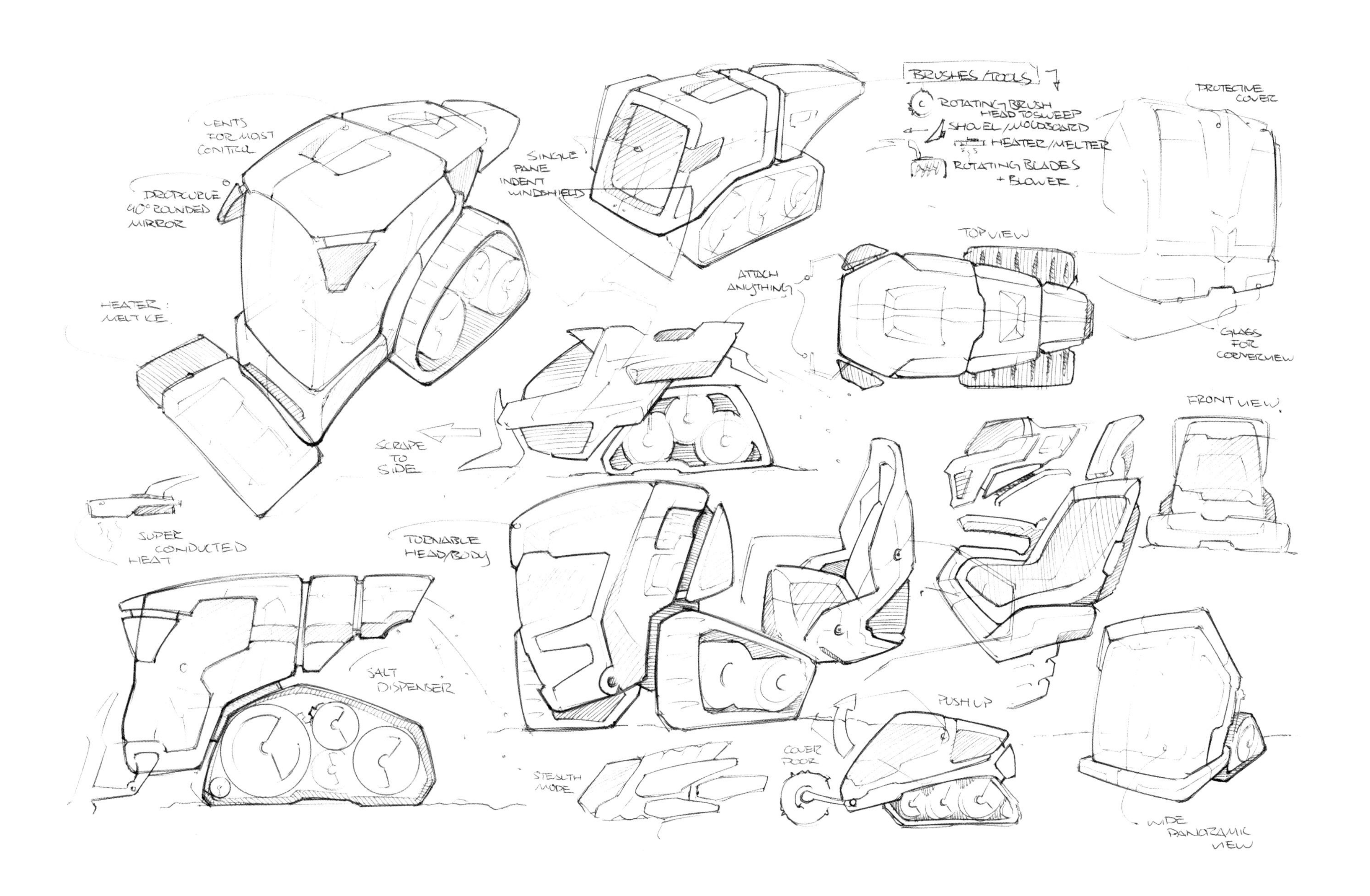
VENTS FOR MOIST CONTROL
HEATER: MELT ICE
SUPER CONDUCTED HEAT
SINGLE PANE INDENT WINDSHIELD
BRUSHES/TOOLS
ROTATING BRUSH HEAD TO SWEEP
SHOVEL/MOLDBOARD
HEATER/MELTER
ROTATING BLADES + BLOWER
PROTECTIVE COVER
TOP VIEW
ATTACH ANYTHING
GLASS FOR CORNER VIEW
SCRAPE TO SIDE
FRONT VIEW
TURNABLE HEAD/BODY
SALT DISPENSER
PUSH UP
COVER DOOR
STEALTH MODE
WIDE PANORAMIC VIEW

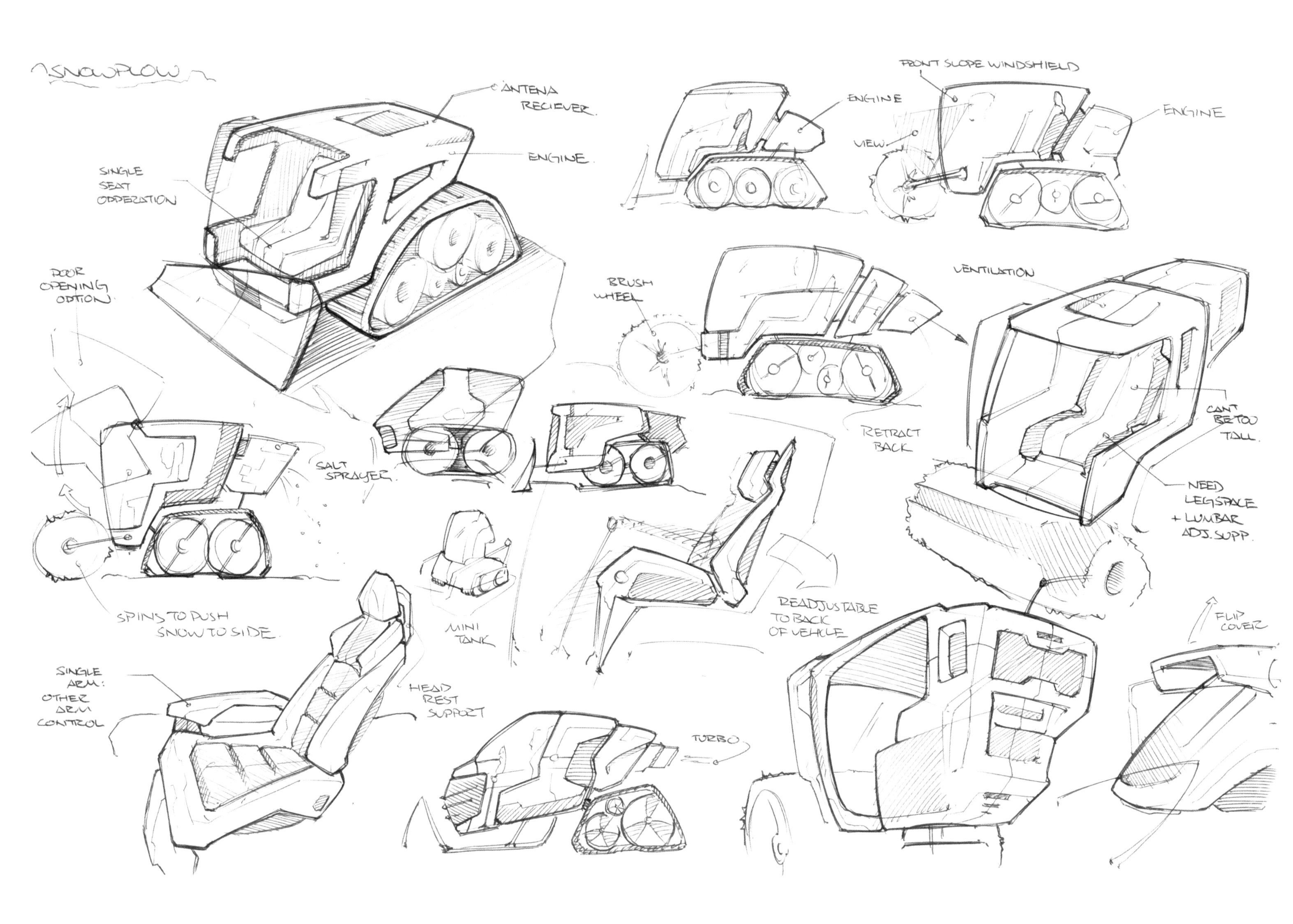
SNOWPLOW
ANTENA RECIEVER.
ENGINE.
SINGLE SEAT OPPERATION
DOOR OPENING OPTION
FRONT SLOPE WINDSHIELD
ENGINE
VIEW
ENGINE
BRUSH WHEEL
VENTILATION
SALT SPRAYER
RETRACT BACK
CAN'T BE TOO TALL
NEED LEGSPACE + LUMBAR ADJ. SUPP.
SPINS TO PUSH SNOW TO SIDE
MINI TANK
READJUSTABLE TO BACK OF VEHICLE
FLIP COVER
SINGLE ARM: OTHER ARM CONTROL
HEAD REST SUPPORT
TURBO

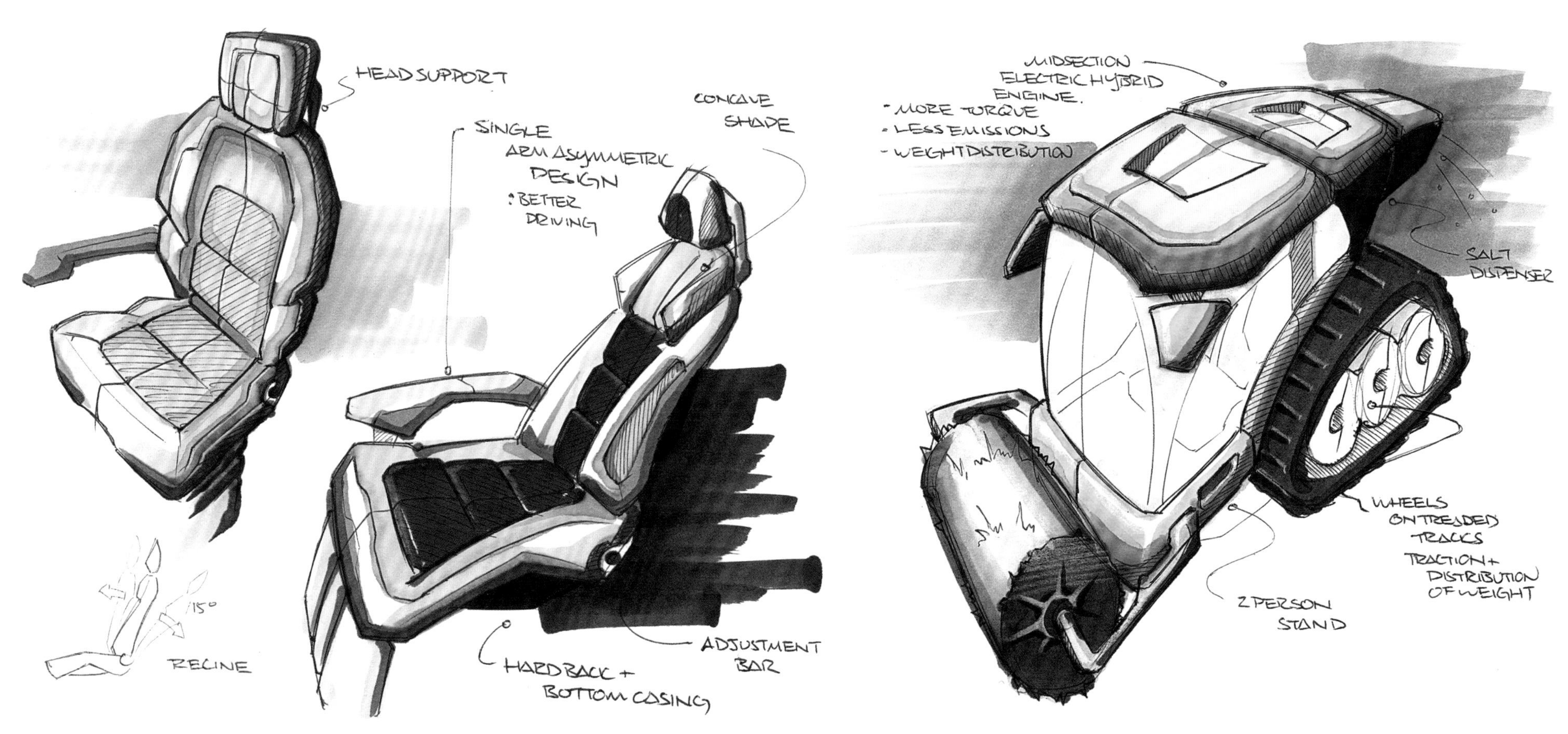
HEAD SUPPORT
SINGLE ARM ASYMMETRIC DESIGN
: BETTER DRIVING
CONCAVE SHADE
RECINE
15°
HARD BACK + BOTTOM CASING
ADJUSTMENT BAR
MIDSECTION ELECTRIC HYBRID ENGINE.
- MORE TORQUE
- LESS EMISSIONS
- WEIGHT DISTRIBUTION
SALT DISPENSER
WHEELS ON TREADED TRACKS
TRACTION + DISTRIBUTION OF WEIGHT
2 PERSON STAND

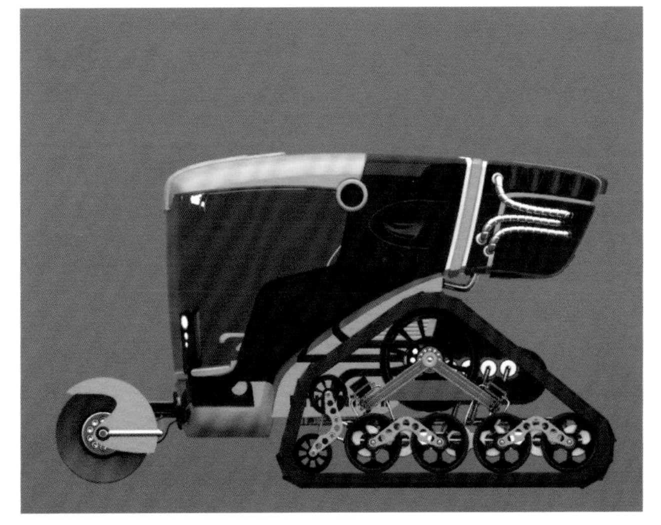

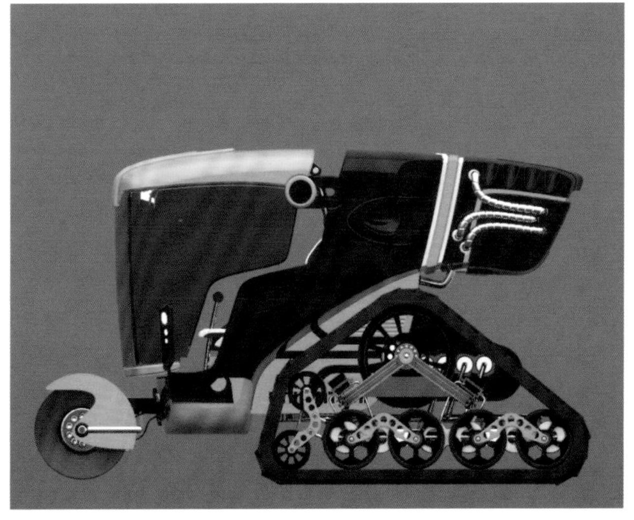

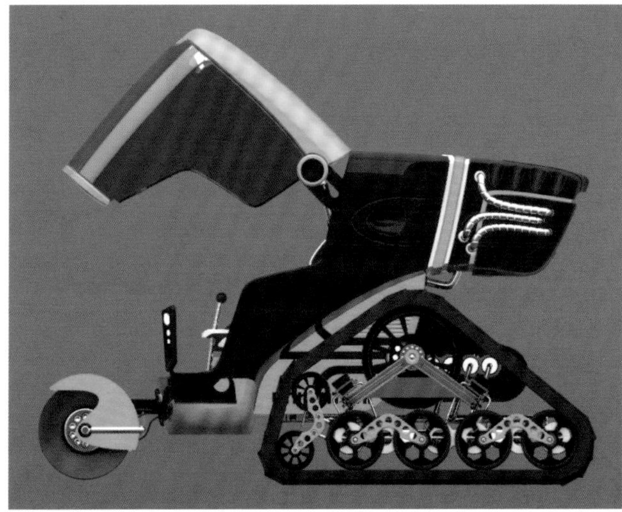

Belaz Cosmo

Belaz Cosmo
卡车

设计师：安娜斯塔西亚·埃尔夫罗瓦（Anastasiya Alforova）、亚历山大·巴比奇（Alexander Babich）
客户：Belaz公司

这一项目是为Belaz公司举办的“Dumpers 2020-2030”设计大赛创作的，获得了最佳创意设计奖。设计师首先设想了卡车在采石场中工作的场景，然后手绘了多个造型方案。

BELAZ COSMO
Product Architecture
Hybrid propulsion system: diesel engine charges the four geared wheels with traction motors.
The system of cab mounting. Entrance and exit of driver happens when the cab is leaned back.
2 Docking.
1 Entrance in a special mode.
3 Rise, turning.
4 Switching control
Unloading of the body.
50° max
COSMO
Airless bicomponent tires. Designed specifically for heavy equipment.
Retractable headlights. Closed at the day to shine brightly at night (dust and dirt protection).
Headlights are located on wheels. Turning together with wheels it gives visibility in the motion direction.
System of emergency shutdown.
Automatic stop-wheel chocks. The popularity of remote control technology was the main reason for creation of this element. It also serves as mudguards.
Maneuverability.
When the speed is high, front wheels are responsible for rotation.
Increased level of maneuvering is possible due to rotation of the central suspension axis.
Damping element.
Position during motion.
Position during parking.
The clearance changing.
max
min
Motor-wheel with electric motor and integrated accumulator. Recharging happens from dumper or domestic power grid.

The Future Combine Harvester
未来收割机

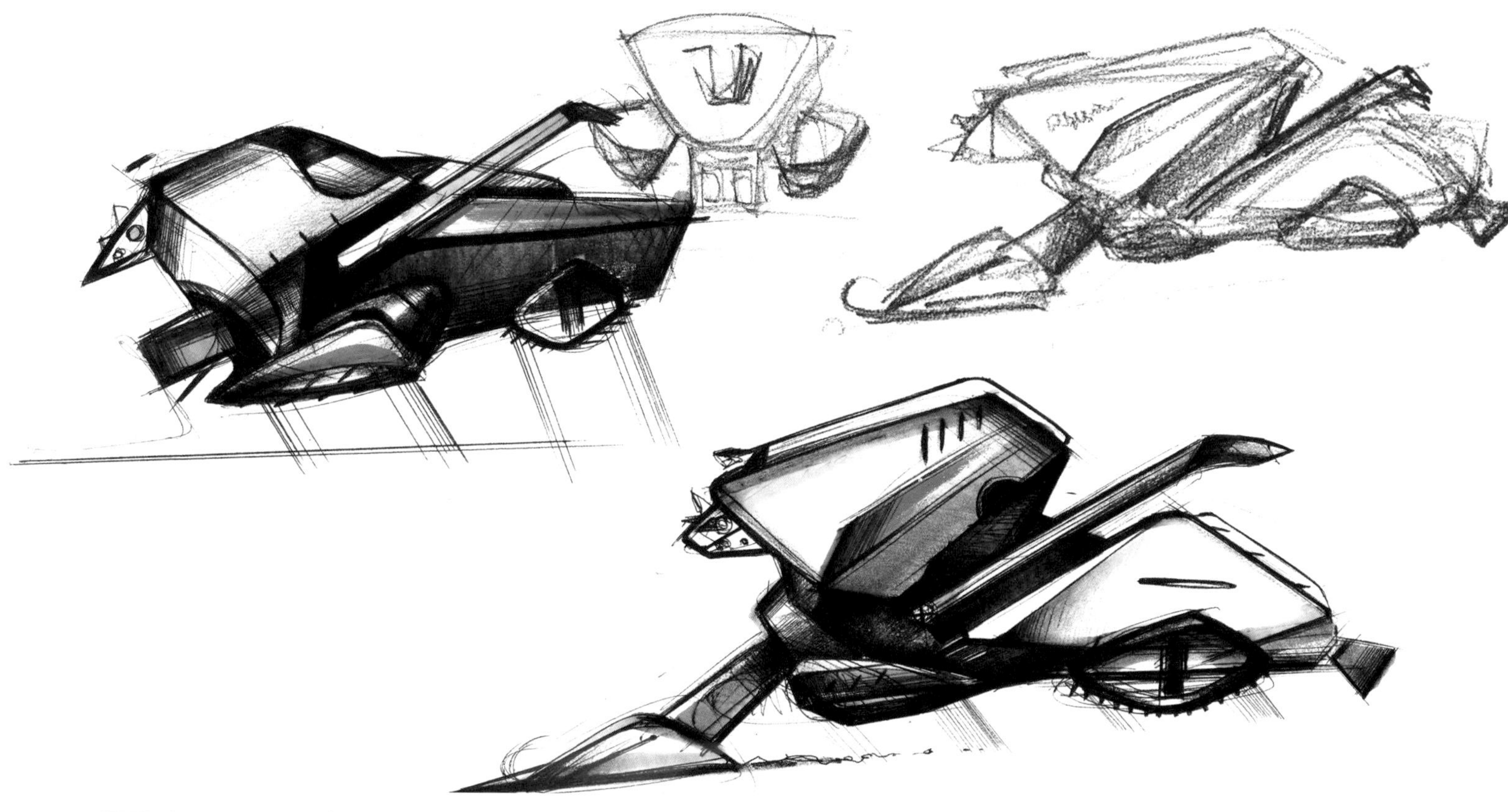

设计师：伊利亚·阿瓦科夫（Ilya Avakov）
客户：个人作品

随着收割机的不断发展，其重量对土地的压力也在不断增长。在此设计中，设计师想象使用“反重力”的引擎，从而减少收割机自身对耕地土壤的破坏。也许在未来的某一天，收割机可以变成“巨型剃须刀”的模样。

240

图书在版编目（CIP）数据

国际产品设计师手绘集：创意、深化、表达／度本图书编著；赵佚译．—北京：中国青年出版社，2015. 5
ISBN 978-7-5153-3367-0
I.①国… II.①度… ②赵… III. ①产品设计－作品集－世界－现代 IV. ①TB472
中国版本图书馆CIP数据核字（2015）第112324号

策划编辑：郭 光 莽 昱 赵媛媛 陈 皓
责任编辑：刘稚清 张 军
助理编辑：陈 皓
书籍设计：彭 涛

国际产品设计师手绘集：创意、深化、表达
度本图书／编著；赵佚／译

出版发行：中国青年出版社
地 址：北京市东四十二条21号
邮政编码：100708
电 话：（010）59521188／59521189
传 真：（010）59521111
企 划：北京中青雄狮数码传媒科技有限公司
印 刷：深圳市精彩印联合印务有限公司
开 本：889 x 1194 1/16
印 张：13
版 次：2015年7月北京第1版
印 次：2015年7月第1次印刷
书 号：ISBN 978-7-5153-3367-0
定 价：89.00元

本书如有印装质量等问题，请与本社联系
电话：（010）59521188 / 59521189
读者来信：reader@cypmedia.com
投稿邮箱：author@cypmedia.com
如有其他问题请访问我们的网站：http://www.cypmedia.com